## IMMANUEL VE



# WORLDS IN COLLISION

Immanuel Velikovsky was born in Vitebsk, Russia, in 1895 and has studied at the universities of Moscow, Berlin, Vienna and Edinburgh. From 1921 to 1924 he edited, with Albert Einstein, the Scripta Universitatis atque Bibliothecae Hierosolymitarum, from which the Hebrew University of Jerusalem was to grow. In 1939 he emigrated to America.

WORLDS IN COLLISION was first published in 1950. His other publications include AGES IN CHAOS (Vol. 1, 1952), EARTH IN UPHEAVAL (1955) and OEDIPUS AND AKHNATON (1960). He now lives in Princeton, New Jersey, U.S.A.

ABACUS

ABACUS edition published in 1972
by Sphere Books Ltd
30/32 Gray's Inn Road, London WC1X 8JL
Reprinted 1973 (twice), 1974, 1976

First published in Great Britain in 1950 by
Victor Gollancz Ltd
Reprinted eighteen times



To Elisheva



Set in Baskerville

*Printed in Great Britain by Hazell Watson & Viney Ltd
Aylesbury, Bucks.*

## PREFACE TO THE PAPERBACK EDITION

First published in 1950, this book was left unchanged in all subsequent printings[1]; nor have any textual changes been made in this paperbound edition. This was so by design: I wished to keep the text in its original form in order that, unaltered, it should face all subsequent discoveries in the fields it covers or touches upon. Should there have been changes, the reader of a new edition would be unable to judge to what extent a book, heretical in 1950, could measure up to later developments.

In 1950 it was generally assumed that the fundamentals of science were all known and that only details and decimals were left to fill in. In the same year, a cosmologist, certainly not of a conservative bent of mind, Fred Hoyle, wrote in the conclusion of his book, *The Nature of the Universe:* "Is it likely that any astonishing new developments are lying in wait for us? Is it possible that the cosmology of 500 years hence will extend as far beyond our present beliefs as our cosmology goes beyond that of Newton?" And he continued: "I doubt whether this will be so. I am prepared to believe that there will be many advances in the detailed understanding of matters that still baffle us. . . . But by and large I think that our present picture will turn out to bear an approximate resemblance to the cosmologies of the future," and he referred to the limitations of optical means in penetrating the depth of space.

The years that have passed since the publication of *Worlds in Collision* have seen the first great achievements in radio astronomy, the discoveries of the International Geophysical Year, and the dawn of the space age. The picture has changed completely. Signs of recent violence, disruption, and fragmentation have been observed on earth and elsewhere in the solar system: a submarine gigantic canyon that runs almost twice around the globe—a sign of a global twist; a layer of ash of extraterrestrial origin underlying all oceans; paleomagnetic evidence that the magnetic poles were suddenly and repeatedly reversed and, it is claimed, the terrestrial axis with them; gases escaping from some craters on the moon, thought to be cold to its center; an exceedingly high

[1]By the summer of 1964 15 hard-cover printings in the United States and 14 in Great Britain.

surface heat of Venus. Furthermore, with the discovery of radio signals arriving from Jupiter, of the existence of a magnetosphere surrounding the earth, of the solar plasma, of the net charge on the sun, and of the magnetic field permeating the interplanetary space, decisive evidence has come up that the solar system, and the universe in general, are not electromagnetically sterile—a basic change in the understanding of the universe, its nature, and the forces active in it.

The words found in the Preface to the 1950 edition, designating the work as heresy in realms where the names of Newton and Darwin reign supreme, should no longer evoke the same spontaneous rejection on the part of even the most conservative in science, unless it is a defense mechanism devised to protect an inner realization of incertitude.

"What, to the scientist, constitutes a really satisfactory sort of success for theory? The answer lies largely in the words generality, elegance, control, and prediction."[2] As to generality, hardly anyone raised an objection. Possibly there was some elegance in the timing: when these words were written in 1960, ten years after the publication of my book and the great opposition it provoked, some of the most compelling data were radioed by the space vehicle, *Pioneer V*. I would like to relate here a few details about the control and prediction of two crucial tests, decisive for this book.

Early in my work I came to the understanding that Venus is a newcomer to the planetary family, that it had a stormy if only short history, and that it must still be very hot and "giving off heat"; further, that it must be surrounded by a very extensive envelope of hydrocarbon (petroleum) gases and dust. Such claims were in total disagreement with what was known in 1946 when I completed the manuscript of the work or in 1950 when it was published. To stress the crucial nature of these claims, they were put under the headings "The Gases of Venus" and "The Thermal Balance of Venus" immediately preceding the section, "The End." Should I be right in these claims, the entire chain of deductions—of which the identification of the extraterrestrial agent of the paroxysms described is but the final ring—is strengthened. And since these crucial claims were in flagrant discord with accepted values, in case of confirmation they ought not to be denoted as lucky guesses.

As late as 1959, Venus' ground temperature was calculated to be only 17°C, three degrees above the mean annual temperature of the Earth. But by 1961, from the nature of the radio signals

[2]Warren Weaver, "The Imperfection of Science," *Proc. of the Amer. Philos. Soc.*, Oct. 17, 1960.

emitted by Venus, it was found that Venus' ground temperature is about 315°C, or 600°F. Dr. F. D. Drake of the National Radio Astronomy Observatory, responsible for this reading, wrote: "We would have expected a temperature only slightly greater than that of the earth," and the find was "a surprise . . . in a field in which the fewest surprises were expected."

There was admittedly no satisfactory explanation of such high temperature of Venus in the frame of the accepted notions. Greenhouse effect could not explain so high a temperature, nor could radioactivity, decaying for billions of years. The *Mariner II*, the space vehicle that passed Venus in December, 1962, was instrumented to detect whether the heat is real and as high as 600°. It found it real and a full 800°. It found, also, that the night side of Venus is, if anything, hotter than the day side and that light does not penetrate the cloud cover. It must be gloomy and bleak under this cover, it is stated in the *Mariner* report by the Jet Propulsion Laboratory; very little greenhouse effect could realize itself under such conditions.

The other crucial test concerned the gaseous envelope of the planet. In 1946, four years before the publication of this book, I directed a request and inquiry to Professor R. Wildt of Yale and the late Professor W. S. Adams of Mount Wilson and Palmonar observatories, foremost authorities on the subject of planetary atmosphere, indicating that the presence of hydrocarbon gases and dust in the cloud envelope of Venus would constitute a crucial test for the cosmological concepts evolved from the study of historical sources. Wildt wrote on September 13, 1946: "The absorption spectrum of Venus' atmosphere cannot be interpreted as resulting from gaseous hydrocarbons." Adams answered (September 9, 1946): "There is no evidence of the presence of hydrocarbon gas in the atmosphere of Venus."

I must have been completely firm in my belief of not having made a wrong deduction—from the first premise of global catastrophe to the last one, of identifying the agent—to have chosen to print, in disregard of the expert opinions: "On the basis of this research, I assume that Venus must be rich in petroleum gases."

On February 26, 1963, making known the results of the *Mariner* probe, Dr. Homer Newell of NASA announced that, in the judgment of those responsible for that part of the program, Venus is enshrouded in an envelope of hydrocarbon gases and dust, 15 miles thick, 45 miles above the ground of the planet.

It was acknowledged as very puzzling that Venus should have such a massive atmosphere a score of times heavier than the terrestrial atmosphere; that it should have taken the form of an envelope 45 miles above the surface of the planet; and that it should consist of heavy molecules of hydrocarbon gases and dust. It was also found that Venus rotates retrogradely, though very slowly, a sign of its having been disturbed in its motion in the past, or having been captured by the sun, or having originated in a way different from that of the other planets.

At the time of the *Mariner* probe, two prominent members of the American scientific community, V. Bargmann, professor of Physics, Princeton University, and Lloyd Motz, professor to Astronomy, Columbia University, wrote a letter to *Science* (December 21, 1962) claiming for me the correct prediction of the great heat of Venus, of the radionoises from Jupiter, of the existence of a magnetosphere around the Earth. A paper, "Some Additional Examples of Correct Prognosis," written by me, was printed in the September, 1963, issue of the *American Behavioral Scientist;* it contains a survey of various tests, confirmations, and supporting evidence. In that issue, sponsored by a group of eminent men in scholarship and public affairs, is also told the story of reception—or rejection—of this book, coupled with efforts toward its suppression: it was actually successfully suppressed while in the hands of its first publisher, who had to give it up, though a No. 1 national best-seller, under the exerted boycott of all this publisher's textbooks by certain groups organized for that purpose in some of the academic councils of the country.

Some attempts were made to minimize the value of the crucial tests claimed and confirmations obtained (a prominent astronomer wrote in the December, 1963, issue of *Harper's:* "As to the 'high temperature' of Venus, 'hot' is only a relative term; for example, liquid air is hot, relative to liquid helium," whereas I claimed an incandescent state of Venus [p.77] and a gaseous state of all hydrocarbons). Professor H. H. Hess, Chairman of the Space Board of the National Academy of Sciences, volunteered to write me a letter for publication: "Some of these predictions were said to be impossible when you made them; all of them were predicted long before proof that they were correct came to hand. Conversely I do not know of any specific prediction you made that has since proven to be false."

If my premises are wrong and only by sheer chance did I obtain such a score, then the theorists of probabilities ought to find out the odds involved; if, as some friendlier skeptics assume, the score is due to an unusual gift of intuition, then I should be accused of sorcery, not only of heresy. However, if the story is a reconstruction of events that took place and of logical implications of them, then the score is but a "natural fallout from a single central idea" (R. Juergens).

Nevertheless, more efforts were made to disqualify this work. But hardly any astronomical argument of 1950 could be brought profitably against my book in 1964 without denying all the important discoveries of the intervening years. Therefore, attempts were made to evade all these issues and to switch the debate, actually the campaign of depreciation, to questioning my proper use of sources. When a journal printed for physicists serves its readers with philological arguments in Egyptology and commits the task to a journalist "uniformed and rash," in the mild appraisal of Professor Moses Hadas, and prints a vulgar display of ignorance and distortion, then it is as good as an admission that none of the physical arguments employed earlier could carry weight and no new ones could be devised.

It is about such tactics that the students' paper, *The Daily Princetonian*, wrote editorially (February, 1964): "While it could have been assumed that anyone challenging the basic premises of Newton and Darwin might be laying himself open to a certain amount of argument, the personal vituperation, deliberate misrepresentation of facts, offhand misquotations, efforts at suppression of the books containing the theories, and the denial of the right to rebut opponents in professional journals that Dr. Velikovsky encountered indicate that far more was going on than 'mere' challenge to established ideas. What the Velikovsky affair made crystal clear . . . is that the theories of science may be held not only for the truth they embody, but because of the vested interests they represent for those who hold them."

The deplorable tactics of certain groups in the academia alienated the younger generation, and the historical and physical evidence accumulating with each passing year did not escape their sight, and conclusions were drawn. What was unbelievable and heretical in 1950 is making great inroads into the science that claimed dogmatic completeness and infallibility as recently as then.

On the eve of the publication of *Worlds in Collision*, the philosopher H. Butterfield wrote (*The Origin of Modern Science*, 1949): "But the supreme paradox of the scientific revolution is in the fact that things which we find it easy to instil into the boys at school . . . things which would strike us as the ordinary natural way of looking at the universe . . . defeated the greatest intellects for centuries."

IMMANUEL VELIKOVSKY

## PREFACE

*Worlds in Collision* is a book of wars in the celestial sphere that took place in historical times. In these wars the planet earth participated too. This book describes two acts of a great drama: one that occurred thirty-four to thirty-five centuries ago, in the middle of the second millennium before the present era; the other in the eighth and the beginning of the seventh century before the present era, twenty-six centuries ago. Accordingly, this volume consists of two parts, preceded by a prologue.

Harmony or stability in the celestial and terrestrial spheres is the point of departure of the present-day concept of the world as expressed in the celestial mechanics of Newton and the theory of evolution of Darwin. If these two men of science are sacrosanct, this book is a heresy. However, modern physics, of atoms and of the quantum theory, describes dramatic changes in the microcosm—the atom—the prototype of the solar system; a theory, then, that envisages not dissimilar events in the macrocosm—the solar system—brings the modern concepts of physics to the celestial sphere.

This book is written for the instructed and uninstructed alike. No formula and no hieroglyphic will stand in the way of those who set out to read it. If, occasionally, historical evidence does not square with formulated laws, it should be remembered that a law is but a deduction from experience and experiment, and therefore laws must conform with historical facts, not facts with laws.

The reader is not asked to accept a theory without question. Rather, he is invited to consider for himself whether he is reading a book of fiction or non-fiction, whether what he is reading is invention or historical fact. On one point alone, not necessarily decisive for the theory of cosmic catastrophism, I borrow credence: I use a synchronical scale of Egyptian and Hebrew histories which is not orthodox.

It was in the spring of 1940 that I came upon the idea that in the days of Exodus, as evident from many passages of the Scriptures, there occurred a great physical catastrophe, and that such an event could serve in determining the time of the Exodus in Egyptian history or in establishing a synchronical scale for the histories of the peoples concerned. Thus I started *Ages in Chaos*, a reconstruction of the history of the ancient world from the middle of the second millennium before the present era to the advent of Alexander the Great. Already in the fall of that same year, 1940, I felt that I had acquired an understanding of the real nature and extent of that catastrophe, and for nine years I worked on both projects, the political and the natural histories.

Although *Ages in Chaos* was finished first, in the order of publication it will follow this work.

*Worlds in Collision* comprises only the last two acts of the cosmic drama. A few earlier acts—one of them known as the Deluge—will be the subject of another volume of natural history.

The historical-cosmological story of this book is based on the evidence of historical texts of many peoples around the globe, on classical literature, on epics of northern races, on sacred books of the peoples of the Orient and Occident, on traditions and folklore of primitive peoples, on old astronomical inscriptions and charts, on archaeological finds, and also on geological and paleontological material.

If cosmic upheavals occurred in the historical past, why does not the human race remember them, and why was it necessary to carry on research to find out about them? I discuss this problem in the Section "The Collective Amnesia." The task I had to accomplish was not unlike that faced by a psychoanalyst who, out of disassociated memories and dreams, reconstructs a forgotten traumatic experience in the early life of an individual. In an analytical experiment on mankind, historical inscriptions and legendary motifs often play the same role as recollections (infantile memories) and dreams in the analysis of a personality.

Can we, out of this polymorphous material, establish actual facts? We shall check one people against another, one inscription against another, epics against charts, geology against legends, until we are able to extract the historical facts.

In a few cases it is impossible to say with certainty whether a record or a tradition refers to one or another catastrophe that took place through the ages; it is also probable that in some traditions various elements from different ages are fused together. In the final analysis, however, it is not so essential to segregate definitively the records of single world catastrophes. More important, it seems, is to establish (1) that there were physical upheavals of a global character in historical times; (2) that these catastrophes were caused by extraterrestrial agents; and (3) that these agents can be identified.

There are many implications that follow from these conclusions. I refer to them in the Epilogue, so that I can omit reference to them here.

A few readers went over this book in manuscript and made valuable suggestions and remarks. In chronological order of their reading they are:

Dr. Horace M. Kallen, formerly Dean of the Graduate Faculty of the New School for Social Research, New York; John J. O'Neill, Science Editor of the *New York Herald-Tribune*; James Putnam,

Associate Editor of the Macmillan Company; Clifton Fadiman, literary critic and commentator; Gordon A. Atwater, Chairman and Curator of the Hayden Planetarium of the American Museum of Natural History, New York. The last two read the work at their own request after Mr. O'Neill had discussed it in an article in the *Herald-Tribune* of August 11, 1946. I am indebted to all of them but I alone am responsible for content and form.

Miss Marion Kuhn cleared the manuscript of grammatical weeds and helped in reading the proofs.

Many an author has dedicated his book to his wife or mentioned her in the preface. I have always felt this was somewhat ostentatious, but now this work is being published, I feel I shall be most ungrateful if I fail to mention that my wife Elisheva spent almost as much time on it at our desk as I did. I dedicate this book to her.

The years when *Ages in Chaos* and *Worlds in Collision* were written were years of a world catastrophe created by man—of war that was fought on land, on sea, and in the air. During that time man learned how to take apart a few of the bricks of which the universe is built—the atoms of uranium. If one day he should solve the problem of the fission and fusion of the atoms of which the crust of the earth or its water and air are composed, he may perchance, by initiating a chain reaction, take this planet out of the struggle for survival among the members of the celestial sphere.

*New York, September* 1949. Immanuel Velikovsky.

# CONTENTS

PART II

MARS

## EPILOGUE

# PROLOGUE

CHAPTER I

# IN AN IMMENSE UNIVERSE

Quota pars operis tanti
nobis committitur?
—SENECA

IN AN IMMENSE universe a little globe revolves around a star; it is the third in the row—Mercury, Venus, Earth—of the planetary family. It is of a solid core covered over most of its surface with liquid, and it has a gaseous envelope. Living creatures fill the liquid; other living creatures fly in the gas; and still others creep and walk upon the ground on the bottom of the gaseous ocean. Man, a being of erect stature, thinks himself the prince of creation. He felt like this long before he, by his own efforts, came to know how to fly on wings of metal around the globe. He felt godlike long before he could talk to his fellow-man on the other side of the globe. Today he can see the microcosm in a drop and the elements in the stars. He knows the laws governing the living cell with its chromosomes, and the laws governing the macrocosm of the sun, moon, planets, and stars. He assumes that gravitation keeps the planetary system together, man and beast on their planet, the sea within its borders. For millions and millions of years, he maintains, the planets have rolled along on the same paths, and their moons around them, and man in these eons has arisen from a one-cell infusorium all the long way up the ladder to his status of Homo sapiens.

Is man's knowledge now nearly complete? Are only a few more steps necessary to conquer the universe: to extract the energy of the atom—since these pages were written this has already been done—to cure cancer, to control genetics, to communicate with other planets and learn if they have living creatures, too?

Here begins Homo ignoramus. He does not know what life is or how it came to be and whether it originated from inorganic matter. He does not know whether other planets of this sun or of other suns have life on them, and if they have, whether the forms of life there are like those around us, ourselves included.

He does not know how this solar system came into being, although he has built up a few hypotheses about it. He knows only that the solar system was constructed billions of years ago. He does not know what this mysterious force of gravitation is that holds him and his fellow man on the other side of the planet with their feet on the ground, although he regards the phenomenon itself as "the law of laws." He does not know what the earth looks like five miles under his feet. He does not know how mountains came into existence or what caused the emergence of the continents, although he builds hypotheses about these, nor does he know from where oil came—again hypotheses. He does not know why, only a short time ago, a thick glacial sheet pressed upon most of Europe and North America, as he believes it did; nor how palms could grow above the polar circle, nor how it came about that the same fauna fill the inner lakes of the Old and the New World. He does not know where the salt in the sea came from.

Although man knows that he has lived on this planet for millions of years, he finds a recorded history of only a few thousand years. And even these few thousand years are not sufficiently well known.

Why did the Bronze Age precede the Iron Age even though iron is more widely distributed over the world and its manufacture is simpler than that of the alloy of copper and tin? By what mechanical means were structures of immense blocks built on the high mountains of the Andes?

What caused the legend of the Flood to originate in all the countries of the world? Is there any adequate meaning to the term "antediluvian"? From what experiences grew the eschatological pictures of the end of the world?

In this work, of which the present book is the first part, some of these questions will be answered, but only at the cost of giving up certain notions now regarded as sacred laws in science —the millions of years of the present constitution of the solar system and the harmonious revolution of the earth—with all their implications as regards the theory of evolution.

## THE CELESTIAL HARMONY

The sun rises in the east and sets in the west. The day consists of twenty-four hours. The year consists of 365 days, 5 hours, and

49 minutes. The moon circles around the earth, changing its phases—crescent, full, decrescent. The terrestrial axis points in the direction of the polar star. After winter comes spring, then summer and fall. These are common facts. Are they invariable laws? Must it be so forever? Was it so always?

The sun has nine planets. Mercury has no satellites; Venus has no satellites; the earth has a moon; Mars has two small trabants, mere pieces of rock, and one of them completes its month before Mars ends its day; Jupiter has eleven moons and eleven different kinds of months to count; Saturn has nine moons, Uranus has five moons,[1] Neptune one, Pluto none.[2] Was it always so? Will it be so forever?

The sun rotates in an easterly direction. All planets revolve in their orbits in the same direction (counterclockwise if seen from the north) around the sun. Most of their moons revolve counterclockwise (in direct motion), but there are a few that revolve in the opposite direction (in retrograde motion).

No orbit is an exact circle; there is no regularity in the eccentrical shapes of the planetary orbits; each elliptical curve verges in a different direction.

It is not known for certain, but it is assumed that Mercury permanently shows the same face to the sun, as our moon does with respect to the earth. Information obtained by different methods of observation of Venus is contradictory; it is not known whether Venus rotates so slowly that its day equals its year, or so rapidly that the night side is never sufficiently cooled. Mars rotates in 24 hours, 37 minutes, 22.6 seconds (mean period), a period comparable to the terrestrial day. Jupiter, which in volume is thirteen hundred times larger than the earth, completes a rotation in the short space of 9 hours and 50 minutes. What causes this variability? It is not a law that a planet must rotate or have days and nights; still less that its day and night must return every twenty-four hours.

If Pluto rotates from east to west,[3] it has the sun rising in the west. Uranus has the sun rising and setting neither in the east nor in the west. So it is not a law that a planet of the solar

[1] The fifth satellite of Uranus was discovered in 1948.

[2] Due to the great distance of Neptune and Pluto from the earth, smaller satellites around these planets may have remained undiscovered.

Note: While this book was on the press another satellite of Neptune was discovered by G. P. Kuiper.

[3] G. Gamow, *Biography of the Earth* (1941), p. 24.

system must rotate from west to east and that the sun must rise in the east.

The equator of the earth is inclined to the plane of its ecliptic at an angle of 23½°; this causes the change of seasons during the annual revolution around the sun. The axes of other planets point in the directions of seemingly deliberate choice. It is not a general law for all planets that winter must follow fall and summer the spring.

The axis of Uranus is placed almost in the plane of its orbit;[1] for about twenty years one of its polar regions is the hottest place on the planet. Then night gradually descends and twenty years later the other pole enters the tropics for an equal length of time.

The moon has no atmosphere. It is not known whether Mercury has any atmosphere. Venus is covered with dense clouds, but not of water vapour. Mars has a transparent atmosphere, but almost without oxygen or water vapour, and its composition is unknown. Jupiter and Saturn have gaseous envelopes; it is not known whether they have solid cores. It is not a general law that a planet must have atmosphere or water.

Mars is 0.15 of the volume of the earth; the next planet, Jupiter, is about 8,750 times as large as Mars. There is no regularity of, or relation between, the size of the planets and their position in the system.

On Mars are seen "canals" and polar caps; on the moon, craters; the earth has reflecting oceans; Venus has brilliant clouds; Jupiter has belts and a red spot; Saturn has rings.

The celestial harmony is composed of bodies different in size, different in form, different in the velocity of rotation, with differently directed axes of rotation, with different directions of rotation, with differently composed atmospheres or without atmospheres, with a varying number of moons or without moons, and with satellites revolving in either direction.

It appears then to be by chance that the earth has a moon, that we have day and night and that their combined length is equal to twenty-four hours, that we have a sequence of seasons, that we have oceans and water, atmosphere and oxygen, and probably also that our planet is placed between Venus at our left and Mars at our right.

[1] The equator of Uranus is inclined at an angle of 82° to the plane of its orbit.

# THE ORIGIN OF THE PLANETARY SYSTEM

All theories of the origin of the planetary system and the motive forces that sustain the motion of its members go back to the gravitational theory and the celestial mechanics of Newton. The sun attracts the planets, and if it were not for a second urge, they would fall into the sun; but each planet is impelled by its momentum to proceed in a direction away from the sun, and as a result, an orbit is formed. Similarly, a satellite or a moon is subject to an urge that drives it away from its primary, but the attraction of the primary bends the path on which the satellite would have proceeded if there had been no attraction between the bodies, and out of these urges a satellite orbit is traced. The inertia or persistence of motion implanted in planets and satellites was postulated by Newton, but he did not explain how or when the initial pull or push occurred.[1]

The theory of the origin of the planetary system which dominated the entire nineteenth century was proposed by Swedenborg, the theologian, and Kant, the philosopher. It was put into scientific terms by Laplace,[2] although not explored by him quantitatively, and in brief is as follows :

Hundreds of millions of years ago the sun was nebulous and very large and had a form approaching that of a disc. This disc was as wide as the whole orbit of the farthest of the planets. It rotated around its centre. Owing to the process of compression caused by gravitation, a globular sun shaped itself in the centre of the disc. Because of the rotating motion of the whole nebula, a centrifugal force was in action; parts of matter more on the periphery resisted the retracting action directed toward the centre and broke up into rings which balled into globes—these were the planets in the process of shaping. In other words, as a result of the shrinkage of the rotating sun, matter broke away and portions of this solar material developed into planets. The plane in which the planets revolve is the equatorial plane of the sun.

This theory is now regarded as unsatisfactory. Three objections stand out above others. First, the velocity of the axial rotation of the sun at the time the planetary system was built could not have been sufficient to enable bands of matter to

[1] Isaac Newton, *Principia* (Mathematical Principles) (1686), Bk. III.
[2] P. S. Laplace, *Exposition du système du monde* (1796).

break away; but even if they had broken away, they would not have balled into globes. Second, the Laplace theory does not explain why the planets have larger angular velocity of daily rotation and yearly revolution than the sun could have imparted to them. Third, what made some of the satellites revolve retrogradely, or in a direction opposite to that of most of the members of the solar system?

"It appears to be clearly established that, whatever structure we assign to a primitive sun, a planetary system cannot come into being merely as the result of the sun's rotation. If a sun, rotating alone in space, is not able of itself to produce its family of planets and satellites, it becomes necessary to invoke the presence and assistance of some second body. This brings us at once to the tidal theory."[1]

The tidal theory, which, in its earlier stage, was called the planetesimal theory,[2] assumes that a star passed close to the sun. An immense tide of matter arose from the sun in the direction of the passing star and was torn from the body of the sun but remained in its domain, being the material out of which the planets were built. In the planetesimal theory the mass that was torn out broke into small parts which solidified in space; some were driven out of the solar system, and some fell back into the sun, but the rest moved around it because of its gravitational pull. Sweeping in elongated orbits around the sun, they conglomerated, rounded out their orbits as a result of mutual collisions, and grew to form planets and satellites around the planets.

The tidal theory[3] does not allow the matter torn from the sun to disperse first and reunite later; the tide broke into a few portions that rather quickly changed from gaseous to fluid, and then to the solid state. In support of this theory it was indicated that such a tide, when broken into a number of "drops," would probably build the largest "drops" out of its middle portion, and small "drops" from its beginning (near the sun) and its end (most remote from the sun). Actually, Mercury, nearest to the sun, is a small planet. Venus is larger; earth is a little larger than Venus; Jupiter is three hundred and twenty times as large as the earth (in mass); Saturn is somewhat smaller than Jupiter; Uranus and Neptune, though large planets, are

[1] Sir James H. Jeans, *Astronomy and Cosmogony* (1929), p. 409.
[2] The planetesimal hypothesis was developed by T. C. Chamberlin and F. R. Moulton.
[3] The tidal theory was developed by J. H. Jeans and H. Jeffreys.

not as large as Jupiter and Saturn. Pluto is quite as small as Mercury.

The first difficulty of the tidal hypothesis lies in the very point adduced in its support, the mass of the planets. Between the earth and Jupiter there revolves a small planet, Mars, a tenth part of the earth in mass, where, according to the scheme, a planet ten to fifty times as large as the earth should be expected. Again, Neptune is larger and not smaller than Uranus.

Another difficulty is the allegedly rare chance of an encounter between two stars. One of the authors of the tidal theory gave this estimate of its probability:[1]

"At a rough estimate we may suppose that a given star's chance of forming a planetary system is one in 5,000,000,000,000,000,000 years." But since the life span of a star is much shorter than this figure, "only about one star in 100,000 can have formed a planetary system in the whole of its life." In the galactic system of one hundred million stars, planetary systems "form at the rate of about one per five billion years. . . . our own system, with an age of the order of two billion years, is probably the youngest system in the whole galactic system of stars."

The nebular and tidal theories alike regard the planets as derivatives of the sun, and the satellites as derivatives of the planets.

The problem of the origin of the moon can be regarded as disturbing to the tidal theory. Being smaller than the earth, the moon completed earlier the process of cooling and shrinking, and the lunar volcanoes had already ceased to be active. It is calculated that the moon possesses a lighter specific weight than the earth. It is assumed that the moon was produced from the superficial layers of the earth's body, which are rich in light silicon, whereas the core of the earth, the main portion of its body, is made of heavy metals, particularly iron. But this assumption postulates the origin of the moon as not simultaneous with the origin of the earth; the earth, being formed out of a mass ejected from the sun, had to undergo a process of levelling, which placed the heavy metals in the core and silicon at the periphery, before the moon parted from the earth by a new tidal distortion. This would mean two consecutive tidal distortions in a system where the chance of even one is held extremely rare. If the passing of one star near another happens

[1] Jeans, *Astronomy and Cosmogony*, p. 409.

among one hundred million stars once in five billion years, two occurrences like this for one and the same star seem quite incredible. Therefore, as no better explanation is available, the satellites are supposed to have been torn from the planets by the sun's attraction on their first perihelion passage, when, sweeping along on stretched orbits, the planets came close to the sun.

The circling of the satellites around the planets also confronts existing cosmological theories with difficulties. Laplace built his theory of the origin of the solar system on the assumption that all planets and satellites revolve in the same direction. He wrote that the axial rotation of the sun and the orbital revolutions and axial rotations of the six planets, the moon, the satellites, and the rings of Saturn present forty-three movements, all in the same direction. "One finds by the analysis of the probabilities that there are more than four thousand billion chances to one that this arrangement is not the result of chance; this probability is considerably higher than that of the reality of historical events with regard to which no one would venture a doubt."[1] He deduced that a common and primal cause directed the movements of the planets and satellites.

Since the time of Laplace, new members of the solar system have been discovered. Now we know that though the majority of the satellites revolve in the same direction as the planets revolve and the sun rotates, the moons of Uranus revolve in a plane almost perpendicular to the orbital plane of their planet, and three of the eleven moons of Jupiter, one of the nine moons of Saturn, and the one moon of Neptune revolve retrogradely. These facts contradict the main argument of the Laplace theory: a rotating nebula could not produce satellites revolving in two directions.

In the tidal theory the direction of the planets' movements depended on the star that passed: it passed in the plane in which the planets now revolve and in a direction which determined their circling from west to east. But why should the satellites of Uranus revolve perpendicularly to that plane and some moons of Jupiter and Saturn in reverse directions? This the tidal theory fails to explain.

According to all existing theories, the angular velocity of the revolution of a satellite must be slower than the velocity of

[1] Laplace, *Théorie analytique des probabilités* (3rd ed., 1820), p. lxi; cf. H. Faye *Sur l'Origine du monde* (1884), pp. 131-132.

rotation of its parent. But the inner satellite of Mars revolves more rapidly than Mars rotates.

Some of the difficulties that confront the nebular and tidal theories also confront another theory that has been proposed in recent years.[1] According to it, the sun is supposed to have been a member of a double star system. A passing star crushed the companion of the sun, and out of its debris planets were formed. In further development of this hypothesis, it is maintained that the larger planets were built out of the debris, and the smaller ones, the so-called "terrestrial" planets, were formed from the larger ones by a process of cleavage.

The birth of smaller, solid planets out of the larger, gaseous ones is conjectured in order to explain the difference in the relation of weight to volume in the larger and smaller planets; but this theory is unable to explain the difference in the specific weights of the smaller planets and their satellites. By a process of cleavage, the moon was born of the earth; but since the specific weight of the moon is greater than that of the larger planets and smaller than that of the earth, it would seem to be more in accord with the theory that the earth was born of the moon, despite its smallness. This confuses the argument.

The origin of the planets and their satellites remains unsolved. The theories not only contradict one another, but each of them bears within itself its own contradictions. "If the sun had been unattended by planets, its origin and evolution would have presented no difficulty."[2]

## THE ORIGIN OF THE COMETS

The nebular and tidal theories endeavour to explain the origin of the solar system but do not include the comets in their schemes. Comets are more numerous than planets. More than sixty comets are known to belong definitely to the solar system. These are the comets of short periods (less than eighty years); they revolve in stretched ellipses and all but one do not go beyond the line marked by the orbit of Neptune. It is estimated that, besides the comets of short periods, several hundred thousand comets visit the solar system; however it is not known for

[1] By Lyttleton and, independently, by Russell.
[2] Jeans, *Astronomy and Cosmogony*, p. 395.

certain that they return periodically. They are seen presently at an approximate rate of five hundred in a century, and are said to have an average period of tens of thousands of years.

A few theories of the origin of comets have been proposed, but aside from one attempt to see in them planetesimals that did not receive a side pull sufficiently strong to bring them into circular orbits,[1] no scheme has been developed that explains the origin of the solar system in its entirety, with its planets and comets; yet no cosmic theory can persist which limits itself to the problem of either planets or comets exclusively.

One theory sees in the comets errant cosmic bodies arriving from interstellar space. After approaching the sun, they turn away on an open (parabolic) curve. But if they happen to pass close to one of the larger planets, they may be compelled to change their open curves to ellipses and become comets of short period.[2] This is the theory of capture: comets of long periods or of no period are dislodged from their paths to become short-period comets. What the origin of long-period comets is remains an unanswered question.

The short-period comets apparently have some relation to the larger planets. About fifty comets move between the sun and the orbit of Jupiter; their periods are under nine years. Four comets reach the orbit of Saturn; two comets revolve inside the circle described by Uranus; and nine comets, with an average period of seventy-one years, move within the orbit of Neptune. These comprise the system of the short-period comets as it is known at present. To the last group belongs the Halley comet, which, among the comets of short periods, has the longest period of revolution—about seventy-six years. Then there is a great gap, after which there are comets that require thousands of years before they return to the sun, if they return at all.

The distribution of the short-period comets suggested the

[1] An attempt to explain the comets, in the frame of the planetesimal theory, as scattered debris of a great wreck, was made by T. C. Chamberlin, *The Two Solar Families* (1928).

[2] That planets are able to change the path of a comet is not only known from observation but has even been calculated in advance. In 1758 Clairaut predicted the retardation of Halley's comet, on its first return foretold by Halley, for a period of 618 days, because it had to pass near Jupiter and Saturn. It was retarded for almost the computed length of time. Similarly, the orbits of other comets were occasionally distorted. Lexell's comet was disturbed by Jupiter in 1767 and in 1770 by the earth, D'Arest's comet was disturbed in 1860, Wolf's comet in 1875 and 1922. By an encounter with Jupiter in 1886, Brook's comet changed its period from 29 years to 7 years; the period of Jupiter was not altered by more than two or three minutes, and probably less.

idea that they were "captured" by the large planets. This theory has for its support the direct observation that comets are disturbed on their path by the planets.

Another theory of the comets supposes their origin to have been in the sun, but in a manner unlike that conceived of in the tidal theory of the origin of planets. Mighty whirls on the surface of the sun sweep ignited gases into great protuberances; these are observed daily. Matter is driven off from the sun and returns to the sun. It is calculated that if the velocity of the ejection were to exceed 384 miles per second, the speed of motion in a parabola, the matter would not return to the sun but would become a long-range comet. Then the path of the ejected mass might become perturbed as a result of its passage near one of the larger planets, and the comet would become one of a short period.

Birth of a comet in this manner has never been observed, and the probability that matter in explosion may reach a speed of 384 miles per second is highly questionable. It was therefore supposed alternatively that millions of years ago, when the activity of their gaseous masses was more dynamic, the large planets expelled comets from their bodies. The speed required for the ejected mass to overcome the gravitational pull of the ejecting body is less in the case of the planets than in the case of the sun, owing to their smaller gravitational pull. It is calculated that a mass hurled from Jupiter at a speed of about 38 miles per second, or at only a little more than a third of this velocity in the case of Neptune, would become expelled.

This variant of the theory neglects the question of the origin of the long-period comets. However, an explanation was offered, according to which the large planets throw the comets that pass close to them from their short orbits into elongated ones, or even expel them entirely from the solar system.

When passing close to the sun, comets emit tails. It is assumed that the material of the tail does not return to the comet's head but is dispersed in space; consequently, the comets as luminous bodies must have a limited life. If Halley's comet has pursued its present orbit since late pre-Cambrian times, it must "have grown and lost eight million tails, which seems improbable."[1] If comets are wasted, their number in the solar system must permanently diminish, and no comet of short period could have preserved its tail since geological times.

[1] H. N. Russell, *The Solar System and Its Origin* (1935), p. 40.

But as there are many luminous comets of short period, they must have been produced or acquired at some time when other members of the system, the planets and the satellites, were already in their places. A theory has been offered that once the solar system moved through a nebula and obtained its comets there.

Did the sun emit planets by shrinkage or by tide, and comets by explosion? Did the comets come from interstellar space and were they captured into the solar system by larger planets? Did the larger planets produce the smaller planets by cleavage, or did they expel the short-period comets from their bodies?

It is admitted that we cannot know the truth about the origin of the planetary and cometary systems billions of years ago. "The problem of the origin and development of the solar system suffers from the label 'speculative.' It is frequently said that as we were not there when the system was formed, we cannot legitimately arrive at any idea of how it was formed." [1] The most we can do, it is believed, is to investigate one planet, the one under our feet, in order to learn its past; and then, by the deductive method, to apply the results to other members of the solar system.

[1] Harold Jeffreys, "The Origin of the Solar System" in *Internal Constitution of the Earth*, B. Gutenberg, ed. (1939).

## CHAPTER II

# THE PLANET EARTH

THE PLANET EARTH has a stony shell—the lithosphere; it consists of igneous rock, like granite and basalt, with sedimentary rock on top. The igneous rock is the original crust of the earth; sedimentary rock is deposited by water.

The inner composition of the earth is not known. The propagation of seismic waves gives support to the assumption that the shell of the earth is over 2,000 miles thick; on the basis of the gravitational effect of mountain masses (the theory of isostasy), the shell is estimated to be only sixty miles thick.

The presence of iron in the shell or the migration of heavy metals from the core to the shell has not been sufficiently explained. For these metals to have left the core, they must have been ejected by explosions, and in order to remain spread through the crust, the explosions must have been followed immediately by cooling.

If, in the beginning, the planet was a hot conglomerate of elements, as the nebular as well as the tidal theories assume, then the iron of the globe should have become oxidized and combined with all available oxygen. But for some unknown reason this did not take place; thus the presence of oxygen in the terrestrial atmosphere is unexplained.

The water of the oceans contains a large amount of soluble sodium chloride, common salt. Sodium might have come from rocks eroded by rain; but rocks are poor in chlorine and the proportion of sodium and chlorine in sea water calls for fifty times more chlorine in the igneous rock than it actually contains.

The deep strata of igneous rock contain no signs of fossil life. Incased in sedimentary rock are skeletons of marine and land animals, often in many layers one upon the other. Not infrequently igneous rock is found protruding into sedimentary rock or even covering it over large areas, pointing to successive eruptions of igneous rock that became heated and molten after there was life on the earth.

Upon strata which show no signs of fossil life are strata containing shells, and sometimes the shells are so numerous as to constitute the entire mass of rock. They are often found in the hardest rock. Higher strata contain skeletons of land animals, often of extinct species, and not infrequently, above the strata with the remains of land animals are other strata with marine fauna. The species of the animals, and even their genera, change with the strata. The strata often assume an oblique position, sometimes being almost vertical; frequently they are faulted and overturned in many ways.

Cuvier (1769—1832), the founder of vertebrate paleontology, or the science of petrified skeletons of animals possessing vertebrae, from fish to man, was much impressed by the picture presented by the sequence of the layers of earth.

"When the traveller passes over these fertile plains where gently flowing streams nourish in their course an abundant vegetation, and where the soil, inhabited by a numerous population, adorned with flourishing villages, opulent cities, and superb monuments, is never disturbed, except by the ravages of war, or by the oppression of the powerful, he is not led to suspect that Nature also has had her intestine wars, and that the surface of the globe has been broken up by revolutions and catastrophes. But his ideas change as soon as he digs into that soil which now presents so peaceful an aspect." [1]

Cuvier thought that great catastrophes had taken place on this earth, repeatedly changing sea beds into continents and continents into sea beds. He held that genera and species were unchangeable since Creation; but, observing different animal remains in various levels of earth, he concluded that catastrophes must have annihilated life in vast areas, leaving the ground for other forms of life. Where did these other genera come from? Either they were newly created or, more likely they migrated from other parts of the world, which were not at that time also visited by cataclysms.

He could not find the cause of these cataclysms. He saw in their traces "the problem in geology it is of most importance to solve," but he realized that "in order to resolve it satisfactorily, it would be necessary to discover the cause of these

[1] G. Cuvier, *Essay on the Theory of the Earth* (5th ed. 1827) (English transl. of *Discours sur les révolutions de la surface du globe, et sur les changements qu'elles ont produits dans le règne animal*).

events—an undertaking which presents a difficulty of quite a different kind." He knew only of "many fruitless attempts" already made and he did not find himself able to offer a solution. "These ideas have haunted, I may almost say have tormented me during my researches among fossil bones."[1]

Cuvier's theory of stabilized forms of life and of annihilating catastrophes was supplanted by a theory of evolution in geology (Lyell) and biology (Darwin). The mountains are what is left of plateaus eroded by wind and water in a very slow process. Sedimentary rock is detritus of igneous rock eroded by rain, then carried to sea, and there slowly deposited. Skeletons of birds and of land animals in these rocks are presumed to have belonged to animals that waded close to the shore of the sea in shallow water, died while wading, and were covered by sediment before fish destroyed the cadavers or the water separated the bones of their skeletons. No widespread catastrophes disrupted the slow and steady process. The theory of evolution, which can be traced to Aristotle, and which was the teaching of Lamarck in the days of Cuvier and of Darwin after him, has been generally accepted as truth by natural sciences for almost a hundred years.

Sedimentary rock covers high mountains and the highest of all, the Himalayas. Shells and skeletons of sea animals are found there. This means that at some early time fish swam over these mountains. What caused the mountains to rise?

A force pushing from within or pulling from without or twisting on the sides must have elevated the mountains and lifted continents from the bottom of the sea and submerged other land masses.

If we do not know what these forces are, we cannot answer the problem of the origin of the mountains and of continents, wherever on the globe we are faced with it. Here is how the question is put concerning the eastern coast of North America.

"Not long ago in a geological sense, the flat plain from New Jersey to Florida was under the sea. At that time the ocean surf broke directly on the Old Appalachian Mountains. Previously the south-eastern part of the mountain structure had sunk below the sea and become covered with a layer of sand and mud, thickening seaward. The wedgelike mass of marine

[1] G. Cuvier, *Essay on the Theory of the Earth*, pp. 240-242.

sediments was then uplifted and cut into by rivers, giving the Atlantic coastal plain of the United States. Why was it uplifted? To the westward are the Appalachians. The geologist tells us of the stressful times when a belt of rocks extending from Alabama to Newfoundland was jammed, thrust together, to make this mountain system. Why? How was it done? In former times the sea flooded the region of the great plains from Mexico to Alaska, and then withdrew. Why this change?" [1]

The birth of the Cordilleras—"again the mystery of mountain-making clamours for solution."

And so on all over the world. The Himalayas were under the sea. Now Eurasia is three miles or more above the bottom of the Pacific. Why?

"The problem of mountain-making is a vexing one: many of them [mountains] are composed of tangentially compressed and over-thrust rocks that indicate scores of miles of circumferential shortening in the Earth's crust. Radial shrinkage is woefully inadequate to cause the observed amount of horizontal compression. Therein lies the real perplexity of the problem of mountain-making. Geologists have not yet found a satisfactory escape from this dilemma." [2]

Even authors of textbooks confess their ignorance. "Why have sea floors of remote periods become the lofty highlands of today? What generates the enormous forces that bend, break, and mash the rocks in mountain zones? These questions still await satisfactory answers." [3]

The process of raising the mountains is supposed to have been very slow and gradual. On the other hand, it is clear that igneous rock, already hard, had to become fluid in order to penetrate sedimentary rock or cover it. It is not known what initiated this process, but it is asserted that it must have happened long before man appeared on the earth. So when skulls of early man are found in late deposits, or skulls of modern man are found together with bones of extinct animals in early deposits, difficult problems are presented. Occasionally, also, during mining operations, a human skull is found in the middle of a mountain, under a thick cover of basalt or granite, like the Calaveras skull of California.

Human remains and human artifacts of bone, polished stone,

[1] R. A. Daly, *Our Mobile Earth* (1926), p. 90.

[2] F. K. Mather, Review of *Biography of the Earth* by G. Gamow, *Science*, Jan. 16, 1942.

[3] C. R. Longwell, A. Knopf, and R. F. Flint, *A Textbook of Geology* (1939), p. 405.

or pottery are found under great deposits of till and gravel, sometimes under as much as a hundred feet.

The origin of clay, sand, and gravel on igneous and sedimentary rock, offers a problem. The theory of Ice Ages was put forth (1840) to explain this and other enigmatic phenomena. As far north as Spitzbergen, in the polar circle, at some time in the past, coral reefs were formed, which do not occur except in tropical regions; palms also grew on Spitzbergen. The continent of Antarctica, which today has not a single tree on it, must have been covered at one time by forests, since it has coal deposits.

As we see, the planet earth is full of secrets. We have not come closer to solving the problem of the origin of the solar system by investigating the planet under our feet; on the contrary, we have found many other unsolved problems concerning the lithosphere, hydrosphere, and atmosphere of the earth. Shall we be more fortunate if we try to understand the process that caused the changes on the globe in the most recent geological epoch, the time of the last glacial period, a period close to the time which is regarded as historical?

## ICE AGES

Not many thousands of years ago, we are taught, great areas of Europe and of North America were covered with glaciers. Perpetual ice lay not only on the slopes of high mountains, but loaded itself in heavy masses upon continents even in moderate latitudes. Where today the Hudson, the Elbe, and the upper Dnieper flow, there were then frozen deserts. They were like the immense glacier of Greenland that covers that island. There are signs that a retreat of the glaciers was interrupted by a new massing of ice, and that their borders differed at various times. Geologists are able to find the boundaries of the glaciers. Ice moves slowly, pushing stones before it, and accumulations of stones or moraines remain when the glacier retreats melting away.

Traces have been found of five or six consecutive displacements of the ice sheet during the Ice Age, or of five or six glacial periods. Some force repeatedly pushed the ice sheet toward moderate latitudes. Neither the cause of the ice ages nor the cause of the retreat of the icy desert is known; the time of these retreats is also a matter of speculation.

Many ideas were offered and guesses made to explain how the glacial epochs originated and why they terminated. Some supposed that the sun at different times emits more or less heat, which causes periods of heat and cold on the earth; but no evidence that the sun is such a "variable star" was adduced to support this hypothesis.

Others conjectured that cosmic space has warmer and cooler areas, and that when our solar system travels through the cooler areas, ice descends upon latitudes closer to the tropics. But no physical agents were found responsible for such hypothetical cold and warm areas in space.

A few wondered whether the precession of the equinoxes or the slow change in the direction of the terrestrial axis might cause periodic variations in the climate. But it was shown that the difference in insolation could not have been great enough to have been responsible for the glacial ages.

Still others thought to find the answer in the periodic variations in the eccentricity of the ecliptic (terrestrial orbit), with glaciation at the maximal eccentricity. Some of them supposed that winter in aphelion, the remotest part of the ecliptic, would cause glaciation; and some thought that summer in aphelion would produce that effect.

Some scholars thought about the changes in the position of the terrestrial axis. If the planet earth is rigid, as it is regarded to be (L. Kelvin), the axis could not have shifted in geological times by more than three degrees (George Darwin); if it were elastic, it could have shifted up to ten or fifteen degrees in a very slow process.

The cause of the ice ages was seen by a few scholars in the decrease of the original heat of the planet; the warm periods between the ice ages were attributed to the heat set free by a hypothetical decomposition of organisms in the strata close to the surface of the ground. The increase and decrease in the action of warm springs were also considered.

Others supposed that dust of volcanic origin filled the terrestrial atmosphere and hindered insolation, or, contrariwise, that an increased content of carbon dioxide in the atmosphere obstructed the reflection of heat rays from the surface of the planet. A decrease in the amount of carbon dioxide in the atmosphere would cause a fall of temperature (Arrhenius), but calculations were made to show that this could not be the real cause of the glacial ages (Angström).

Changes in the direction of warm currents in the Atlantic Ocean were brought into the discussion, and the Isthmus of Panama was theoretically removed to allow the Gulf Stream to pass into the Pacific at the time of the glacial periods. But it was proved that the two oceans were already divided in the Ice Age; besides, a part of the Gulf Stream would have remained in the Atlantic anyway. The periodic retreats of ice between the glacial periods would have required periodic removal and replacement of the Isthmus of Panama.

Other theories of equally hypothetical nature were proposed; but the phenomena held responsible for the changes have not been proved to have existed, or to have been able to produce the effect.

All the above-mentioned theories and hypotheses fail if they cannot meet a most important condition: In order for ice masses to have been formed, increased precipitation must have taken place. This requires an increased amount of water vapour in the atmosphere, which is the result of increased evaporation from the surface of oceans; but this could be caused by heat only. A number of scientists pointed out this fact, and even calculated that in order to produce a sheet of ice as large as that of the Ice Age, the surface of all the oceans must have evaporated to a depth of many feet. Such an evaporation of oceans followed by a quick process of freezing, even in moderate latitudes, would have produced the ice ages. The problem is: What could have caused the evaporation and immediately subsequent freezing? As the cause of such quick alteration of heating and freezing of large parts of the globe is not apparent, it is conceded that "at present the cause of excessive ice-making on the lands remains a baffling mystery, a major question for the future reader of earth's riddles." [1]

Not only are the causes of the appearance and later disappearance of the glacial sheet unknown, but the geographical shape of the area covered by ice is also a problem. Why did the glacial sheet, in the southern hemisphere, move from the tropical regions of Africa toward the south polar region and not in the opposite direction, and, similarly, why, in the northern hemisphere, did the ice move in India from the equator toward the Himalaya mountains and the higher latitudes? Why did the glaciers of the Ice Age cover the greater part of North America and Europe, while the north of Asia remained free?

[1] R. A. Daly, *The Changing World of the Ice Age* (1934), p. 16.

In America the plateau of ice stretched up to latitude 40° and even passed across this line; in Europe it reached latitude 50°; while north-eastern Siberia, above the polar circle, even above latitude 75°, was not covered with this perennial ice. All hypotheses regarding increased and diminished insolation due to solar alterations or the changing temperature of the cosmic space, and other similar hypotheses, cannot avoid being confronted with this problem.

Glaciers are formed in the regions of eternal snow; for this reason they remain on the slopes of the high mountains. The north of Siberia is the coldest place in the world. Why did not the Ice Age touch this region, whereas it visited the basin of the Mississippi and all Africa south of the equator? No satisfactory solution to this question has been proposed.

## THE MAMMOTHS

Northeast Siberia, which was not covered by ice in the Ice Age, conceals another enigma. The climate there has apparently changed drastically since the end of the Ice Age, and the yearly temperature has dropped many degrees below its previous level. Animals once lived in this region that do not live there now, and plants grew there that are unable to grow there now. The change must have occurred quite suddenly The cause of this *Klimasturz* has not been explained. In this catastrophic change of climate and under mysterious circumstances, all the mammoths of Siberia perished.

The mammoth belonged to the family of elephants. Its tusks were sometimes as much as ten feet long. Its teeth were highly developed and their "density" was greater than in any other stage in the evolution of the elephants; apparently they did not succumb in the struggle for survival as an unfit product of evolution. The extinction of the mammoth is thought to have coincided with the end of the last glacial period.

Tusks of mammoths have been found in large numbers in north-east Siberia; this well-preserved ivory has been an object of export to China and Europe ever since the Russian conquest of Siberia and was exploited in even earlier times. In modern times the ivory market of the world still found its main source of supply in the tundras of north-east Siberia.

In 1799 the frozen bodies of mammoths were found in these

tundras. The corpses were well preserved, and the sledge dogs ate the flesh unharmed. "The flesh is fibrous and marbled with fat" and "looks as fresh as well frozen beef." [1]

What was the cause of their death and the extinction of their race?

Cuvier wrote of the extinction of the mammoths: "Repeated irruptions and retreats of the sea have neither all been slow nor gradual; on the contrary, most of the catastrophes which have occasioned them have been sudden; and this is especially easy to be proved with regard to the last of these catastrophes, that which, by a twofold motion, has inundated, and afterwards laid dry, our present continents, or at least a part of the land which forms them at the present day. In the northern regions it has left the carcasses of large quadrupeds which became enveloped in the ice, and have thus been preserved even to our own times, with their skin, their hair, and their flesh. If they had not been frozen as soon as killed, they would have been decomposed by putrefaction. And, on the other hand, this eternal frost could not previously have occupied the places in which they had been seized by it, for they could not have lived in such a temperature. It was, therefore, at one and the same moment that these animals were destroyed and the country which they inhabited became covered with ice. This event has been sudden, instantaneous, without any gradation, and what is so clearly demonstrated with respect to this last catastrophe, is not less so with reference to those which have preceded it." [2]

The theory of repeated catastrophes annihilating life on this planet and repeated creations or restorations of life, offered by Deluc [3] and expanded by Cuvier, did not convince the scientific world. Like Lamarck before Cuvier, Darwin after him thought that an exceedingly slow evolutional process governs genetics, and that there were no catastrophes interrupting this process of infinitesimal changes. According to the theory of evolution, these minute changes came as a result of adaptation to living conditions in the struggle of the species for survival.

Like the theories of Lamarck and Darwin, which postulate slow changes in animals, with tens of thousands of years required for a minute step in evolution, the geological theories

[1] Observation of D. F. Hertz in B. Digby, *The Mammoth* (1926), p. 9.
[2] Cuvier, *Essay on the Theory of the Earth,* pp. 14-15.
[3] J. A. Deluc (1727-1817), *Letters on the Physical History of the Earth* (1831).

of the nineteenth century, and of the twentieth as well, regard the geological processes as exceedingly slow and dependent on erosion by rain, wind, and tides.

Darwin admitted that he was unable to find an explanation for the extermination of the mammoth, an animal better developed than the elephant which survived.[1] But in conformity with the theory of evolution, his followers supposed that a gradual sinking of the land forced the mammoths to the hills, where they found themselves isolated by marshes. However, if geological processes are slow, the mammoths would not have been trapped on the isolated hills. Besides, this theory cannot be true because the animals did not die of starvation. In their stomachs and between their teeth undigested grass and leaves were found. This, too, proves that they died from a sudden cause. Further investigations showed that the leaves and twigs found in their stomachs do not now grow in the regions where the animals died, but far to the south, a thousand or more miles away. It is apparent that the climate has changed radically since the death of the mammoths; and as the bodies of the animals were found not decomposed but well preserved in blocks of ice, the change in temperature must have followed their death very closely or even caused it.

There remains to be added that after storms in the Arctic, tusks of mammoths are washed up on the shores of arctic islands; this proves that a part of the land where the mammoths lived and were drowned is covered by the Arctic Ocean.

## THE ICE AGE AND THE ANTIQUITY OF MAN

The mammoth lived in the age of man. Man pictured it on the walls of caves; remains of men have repeatedly been found in Central Europe together with remains of mammoths; occasionally the settlements of the neolithic man of Europe are found strewn with the bones of mammoths. [2] Man moved southward when Europe was covered with ice and returned when the ice retreated. Historical man witnessed great variation in climate.

[1] See G. F. Kunz, *Ivory and the Elephant in Art, in Archaeology, and in Science* (1916), p. 236.

[2] In Předmost in Moravia a settlement has been excavated in which remnants of a human culture and remains of men were found together with skeletons of eight hundred to one thousand mammoths. Shoulder blades of mammoths were used in the construction of human graves.

The mammoth of Siberia, the meat of which is still fresh, is supposed to have been destroyed at the end of the last glacial period, simultaneously with the mammoths of Europe and Alaska. If this is so, the Siberian mammoth was also the contemporary of a rather modern man. At a time when in Europe, close to the ice sheet, man was still in the later stages of neolithic culture, in the Near and Middle East—the region of the great cultures of antiquity—he may already have progressed well into the metal age. There exists no chronological table of neolithic culture because the art of writing was invented approximately at the advent of the copper—the early—period of the Bronze Age. It is presumed that the neolithic man of Europe left pictures but no inscriptions, and consequently there are no means of determining the end of the Ice Age in terms of chronology.

Geologists have tried to find the time of the end of the last glacial period by measuring the detritus carried by rivers from the glaciers and the deposits of detritus in lakes. The quantity carried by the Rhone from the glaciers of the Alps and the amount on the bottom of the Lake of Geneva, through which the Rhone flows, were calculated, and from the figures obtained the time and velocity of the retreat of the glacial sheet of the last glacial period were estimated. According to the Swiss scholar François Forel, twelve thousand years have passed since the time the ice sheet of the last glacial period began to melt, an unexpectedly low figure, as it was thought that the ice age ended thirty to fifty thousand years ago.

Such calculations suffer from being only indirect evaluations; and since the velocity at which the glacial mud had been deposited in the lakes was not constant and the amount varied, the mud must have assembled on the bottom of a lake at a faster rate in the beginning when the glaciers were larger; and if the Ice Age terminated suddenly, the deposition of detritus would have been much heavier at first, and there would be little analogy to the accumulation of detritus from the seasonal melting of snow in the Alps. Therefore, the time that has elapsed since the end of the last glacial period must have been even shorter than reckoned.

Geologists regard the Great Lakes of America as having been formed at the end of the Ice Age when the continental glacier retreated and the depressions freed from the glacier became lakes. In the last two hundred years Niagara Falls has retreated from Lake Ontario toward Lake Erie at the rate of five feet annually, washing down the rocks of the bed of the falls. [1] If this process has been going on at the same rate since the end of the last glacial period, about seven thousand years were needed to move Niagara Falls from the mouth of the gorge at Queenston to its present position. The assumption that the quantity of water moving through the gorge has been uniform since the end of the Ice Age is the basis of this calculation, and therefore, it was concluded, seven thousand years may constitute "the maximum length of time since the birth of the falls." [2] In the beginning, when immense masses of water were released by the retreat of the continental glacier, the rate of movement of Niagara Falls must have been much more rapid; the time estimate "may need significant reduction," and is sometimes lowered to five thousand years. [3] The erosion and sedimentation on the shores and the bottom of Lake Michigan also suggest a lapse of time counted in thousands, but not in tens of thousands, of years. Also the result of paleontological research in America carries evidence which constitutes "a guarantee that before the last period of glaciation, modern man, in the form of that highly developed race, the American Indian, was living on the eastern seaboard of North America" (A. Keith). [4] It is assumed that with the advent of the last glacial period the Indians retreated southward, returning to the north when the ice uncovered the ground and when the Great Lakes emerged, the basin of the St. Lawrence was formed, and Niagara Falls began its retreat toward Lake Erie.

If the end of the last glacial period occurred only a few thousand years ago, in historical times or at a time when the art of writing may have been already employed in the centres of ancient civilization, the records written in rocks by nature and the records written by man must give a co-ordinated picture. Let us, therefore, investigate the traditions and the

[1] The recession has been 5 feet per year since 1764; at present it is 2.3 feet on the sides of the horseshoe cataract, but substantially more in the centre.
[2] G. F. Wright, "The Date of the Glacial Period," *The Ice Age in North America and Its Bearing upon the Antiquity of Man* (5th ed., 1911).
[3] *Ibid.*, p. 539. Cf. also W. Upham in *American Geologist*, XXVIII, 243, and XXXVI, 288. He dates the uprise of the St. Lawrence basin 6,000 to 7,000 years ago; the St. Lawrence must have been freed from ice before Niagara Falls could come into full action. Not dissimilar figures were obtained from the retreat of the Falls of St. Anthony on the Mississippi at Minneapolis.
[4] Keith thinks that the development of the human skull went through a process of advance and retrogression during exceedingly long ages.

literary records of ancient man, and compare them with the records of nature.

## THE WORLD AGES

A conception of ages that were brought to their end by violent changes in nature is common all over the world. The number of ages differs from people to people and from tradition to tradition. The difference depends on the number of catastrophes that the particular people retained in its memory, or on the way it reckoned the end of an age.

In the annals of ancient Etruria, according to Varro, were records of seven elapsed ages. Censorinus, an author of the third Christian century and compiler of Varro, wrote that "men thought that different prodigies appeared by means of which the gods notified mortals at the end of each age. The Etruscans were versed in the science of the stars, and after having observed the prodigies with attention, they recorded these observations in their books."[1]

The Greeks had similar traditions. "There is a period," wrote Censorinus, "called 'the supreme year' by Aristotle, at the end of which the sun, moon, and all the planets return to their original position. This 'supreme year' has a great winter, called by the Greeks *kataklysmos*, which means deluge, and a great summer, called by the Greeks *ekpyrosis*, or combustion of the world. The world, actually, seems to be inundated and burned alternately in each of these epochs."

Anaximenes and Anaximander in the sixth pre-Christian century, and Diogenes of Apollonia in the fifth century, assumed the destruction of the world with subsequent re-creation. Heraclitus (−540 to −475) taught that the world is destroyed in conflagration after every period of 10,800 years. Aristarchus of Samos in the third century before the present era taught that in a period of 2,484 years the earth undergoes two destructions—of combustion and deluge. The Stoics generally believed in periodic conflagrations by which the world was consumed, to be shaped anew. "This is due to the forces of ever-active fire which exists in things and in the course of long cycles of time resolves everything into itself and out of it is constructed a reborn world"—so Philo presented the

[1] Censorinus *Liber de die natali* xviii.

notion of the Stoics that our world is refashioned in periodic conflagrations.[1] In one such catastrophe the world will meet its ultimate destruction; colliding with another world, it will fall apart into atoms out of which, in a long process, a new earth will be created somewhere in the universe. "Democritus and Epicurus," explained Philo, "postulate many worlds, the origin of which they ascribe to the mutual impacts and interlacing of atoms, and their destruction to the counterblows and collisions by the bodies so formed." As this earth goes to its ultimate destruction, it passes through recurring cosmic catastrophes and is re-formed with all that lives on it.

Hesiod, one of the earliest Greek authors, wrote about four ages and four generations of men that were destroyed by the wrath of the planetary gods. The third age was the age of bronze; when it was destroyed by Zeus, a new generation repeopled the earth, and using bronze for arms and tools, they began to use iron, too. The heroes of the Trojan War were of this fourth generation. Then a new destruction was decreed, and after that came "yet another generation, the fifth, of men who are upon the bounteous earth"—the generation of iron.[2] In another work of his, Hesiod described the end of one of the ages. "The life-giving earth crashed around in burning . . . all the land seethed, and the Ocean's streams . . . it seemed even as if Earth and wide Heaven above came together; for such a mighty crash would have arisen if Earth were being hurled to ruin, and Heaven from on high were hurling her down." [3]

Analogous traditions of four expired ages persist on the shores of the Bengal Sea and in the highland of Tibet—the present age is the fifth.[4]

The sacred Hindu book *Bhagavata Purana* tells of four ages and of pralayas or cataclysms in which, in various epochs, mankind was nearly destroyed; the fifth age is that of the present. The world ages are called Kalpas or Yugas. Each world age met its destruction in catastrophes of conflagration, flood, and hurricane. *Ezour Vedam* and *Bhaga Vedam*, sacred Hindu books, keeping to the scheme of four expired ages, differ

[1] Philo, *On the Eternity of the World* (transl. F. H. Colson, 1941), Sec. 8.
[2] Hesiod, *Works and Days* (transl. H. G. Evelyn-White, 1914), l. 169.
[3] Hesiod, *Theogony* (transl. Evelyn-White, 1914), ll. 693 ff.
[4] E. Moor, *The Hindu Pantheon* (1810), p. 102; A. von Humboldt, *Vues des Cordillères* (1816), English transl.: *Researches Concerning the Institutions and Monuments of the Ancient Inhabitants of America* (1814), Vol. II, pp. 15 ff.

only in the number of years ascribed to each.[1] In the chapter, "World Cycles," in *Visuddhi-Magga*, it is said that "there are three destructions: the destruction by water, the destruction by fire, the destruction by wind," but that there are seven ages, each of which is separated from the previous one by a world catastrophe.[2]

Reference to ages and catastrophes is found in *Avesta* (*Zend-Avesta*), the sacred scriptures of Mazdaism, the ancient religion of the Persians.[3] "Bahman Yast," one of the books of *Avesta*, counts seven world ages or millennia.[4] Zarathustra (Zoroaster), the prophet of Mazdaism, speaks of "the signs, wonders, and perplexity which are manifested in the world at the end of each millennium."[5]

The Chinese call the perished ages *kis* and number ten *kis* from the beginning of the world until Confucius.[6] In the ancient Chinese encyclopedia, *Sing-li-ta-tsiuen-chou*, the general convulsions of nature are discussed. Because of the periodicity of these convulsions, the span of time between two catastrophes is regarded as a "great year." As during a year, so during a world age, the cosmic mechanism winds itself up and "in a general convulsion of nature, the sea is carried out of its bed, mountains spring out of the ground, rivers change their course, human beings and everything are ruined, and the ancient traces effaced."[7]

An old tradition, and a very persistent one, of world ages that went down in cosmic catastrophes was found in the Americas among the Incas,[8] the Aztecs, and the Mayas.[9] A major part of stone inscriptions found in Yucatan refer to world catastrophes. "The most ancient of these fragments [*katuns* or calendar stones of Yucatan] refer, in general, to great catastrophes which, at intervals and repeatedly, convulsed the

[1] See C. F. Volney, *New Researches on Ancient History* (1856), p. 157.
[2] H. C. Warren, *Buddhism in Translations* (1896), pp. 320 ff.
[3] F. Cumont, "La Fin du monde selon les mages occidentaux," *Revue de l'histoire des religions* (1931), p. 50; H. S. Nyberg, *Die Religionen des alten Iran* (1938), pp. 28 ff.
[4] "Bahman Yast" (transl. E. W. West), in *Pahlavi Texts* (*The Sacred Books of the East*, ed. F. M. Müller, V [1880]), 191. See W. Bousset, "Die Himmelsreise der Seele," *Archiv für Religionswissenschaft*, IV (1901).
[5] "Dinkard," Bk. VIII, Chap. XIV (transl. West), in *Pahlavi Texts* (*The Sacred Books of the East*, XXXVII [1892]), 33.
[6] H. Murray, J. Crawfurd, and others, *An Historical and Descriptive Account of China* (2nd ed., 1836), I, 40.
[7] G. Schlegel, *Uranographie chinoise* (1875), p. 740, with reference to Wou-foung.
[8] H. B. Alexander, *Latin American Mythology* (1920), p. 240.
[9] Humboldt, *Researches*, II, 15.

American continent, and of which all nations of this continent have preserved a more or less distinct memory."[1] Codices of Mexico and Indian authors who composed the annals of their past give a prominent place to the tradition of world catastrophes that decimated humankind and changed the face of the earth.

In the chronicles of the Mexican kingdom it is said: "The ancients knew that before the present sky and earth were formed, man was already created and life had manifested itself four times."[2]

A tradition of successive creations and catastrophes is found in the Pacific—on Hawaii[3] and on the islands of Polynesia: there were nine ages and in each age a different sky was above the earth.[4] Icelanders, too, believed that nine worlds went down in a succession of ages, a tradition that is contained in the *Edda*.[5]

The rabbinical conception of ages crystallized in the post-Exilic period. Already before the birth of our earth, worlds had been shaped and brought into existence, only to be destroyed in time. "He made several worlds before ours, but He destroyed them all." This earth, too, was not created at the beginning to satisfy the Divine Plan. It underwent reshaping, six consecutive remouldings. New conditions were created after each of the catastrophes. On the fourth earth lived the generation of the Tower of Babel; we belong to the seventh age. Each of the ages or "earths" has a name.

Seven heavens were created and seven earths were created: the most removed, the seventh, Eretz; the sixth, Adamah; the fifth, Arka; the fourth, Harabah; the third, Yabbashah; the second, Tevel; and "our own land called Heled, and like the others, it is separated from the foregoing by abyss, chaos, and water."[6] Great catastrophes changed the face of the earth. "Some perished by deluge, others were consumed by conflagration," wrote the Jewish philosopher Philo.[7]

According to the rabbinical authority Rashi, ancient tradition knows of periodic collapses of the firmament, one of which

[1] C. E. Brasseur de Bourbourg, *S'il existe des Sources de l'histoire primitive du Mexique dans les monuments égyptiens, etc.* (1864), p. 19.
[2] Brasseur, *Histoire des nations civilisées du Mexique* (1857-1859), I, 53.
[3] R. B. Dixon, *Oceanic Mythology* (1916), p. 15.
[4] R. W. Williamson, *Religious and Cosmic Beliefs of Central Polynesia* (1933), I, 89.
[5] *The Poetic Edda: Vōluspa* (transl. from the Icelandic by H. A. Bellows, 1923), 2nd stanza.
[6] Louis Ginzberg, *Legends of the Jews* (1925), I, 4, 9-10, 72; V, 1, 10.
[7] Philo, *Moses*, II, x, 53.

occurred in the days of the Deluge, and which repeated themselves at intervals of 1,656 years.[1] The duration of the world ages varies in Armenian and Arabian traditions.[2]

## THE SUN AGES

An oft-repeated occurrence in the traditions of the world ages is the advent of a new sun in the sky at the beginning of every age. The word "sun" is substituted for the word "age" in the cosmogonical traditions of many peoples all over the world.

The Mayas counted their ages by the names of their consecutive suns. These were called Water Sun, Earthquake Sun, Hurricane Sun, Fire Sun. "These suns mark the epochs to which are attributed the various catastrophes the world has suffered."[3]

Ixtlilxochitl (circa 1568—1648), the native Indian scholar, in his annals of the kings of Tezcuco, described the world ages by the names of "suns."[4] The Water Sun (or Sun of Waters) was the first age, terminated by a deluge in which almost all creatures perished; the Earthquake Sun or age perished in a terrific earthquake when the earth broke in many places and mountains fell. The world age of the Hurricane Sun came to its destruction in a cosmic hurricane. The Fire Sun was the world age that went down in a rain of fire.[5]

"The nations of Culhua or Mexico," Humboldt quoted Gómara, the Spanish writer of the sixteenth century, "believe according to their hieroglyphic paintings, that, previous to the sun which now enlightens them, four had already been successively extinguished. These four suns are as many ages, in which our species has been annihilated by inundations, by earthquakes, by a general conflagration, and by the effect of destroying tempests."[6] Every one of the four elements participated in each of the catastrophes; deluge, hurricane, earthquake, and fire gave their names to the catastrophes because of the predominance of one of them in the upheavals. Symbols of the

[1] Commentary to Genesis 11 : 1.
[2] See R. Eisler, *Weltmantel und Himmelszelt* (1910), II, 451.
[3] Brasseur, *Sources de l'histoire primitive du Mexique*, p. 25.
[4] Fernando de Alva Ixtlilxochitl, *Obras Históricas* (1891-1892), Vol. II, *Historia Chichimeca*.
[5] Alexander, *Latin American Mythology*, p. 91.
[6] Humboldt, *Researches*, II, 16.

successive suns are painted on the pre-Columbian literary documents of Mexico.[1]

"*Cinco soles que son edades*," or "five suns that are epochs," wrote Gómara in his description of the conquest of Mexico.[2] An analogy to this sentence of Gómara may be found in Lucius Ampelius, a Roman author, who, in his book *Liber memorialis*, wrote:[3] "*Soles fuere quinque*" (There were five suns): It is the same belief that Gómara found in the New World.

The Mexican *Annals of Cuauhtitlan*, written in Nahua-Indian (circa 1570) and based on ancient sources, contains the tradition of seven sun epochs. Chicon-Tonatiuh or "the Seven Suns" is the designation for the world cycles or acts in the cosmic drama.[4]

The Buddhist sacred book of *Visuddhi-Magga* contains a chapter on "World Cycles."[5] "There are three destructions: the destruction by water, the destruction by fire, the destruction by wind." After the catastrophe of the deluge, "when now a long period has elapsed from the cessation of the rains, a second sun appeared." In the interim the world was enveloped in gloom. "When this second sun appears, there is no distinction of day and night," but "an incessant heat beats upon the world." When the fifth sun appeared, the ocean gradually dried up; when the sixth sun appeared, "the whole world became filled with smoke." "After the lapse of another long period, a seventh sun appears, and the whole world breaks into flames." This Buddhist book refers also to a more ancient "Discourse on the Seven Suns."[6]

The Brahmans called the epochs between two destructions "the great days."[7]

The Sibylline books recite the ages in which the world underwent destruction and regeneration. "The Sibyl told as follows: 'The nine suns are nine ages. . . . Now is the seventh sun.' " The Sibyl prophesied two ages yet to come—that of the eighth and of the ninth sun.[8]

[1] *Codex Vaticanus* A, plates vii-x.
[2] F. L. de Gómara, *Conquista de Mexico* (1870 ed.), II, 261.
[3] *Liber memorialis* ix.
[4] Brasseur, *Histoire des nations civilisées du Mexique*, I, 206.
[5] Warren, *Buddhism in Translations*, p. 322.
[6] *Ibid.*
[7] In the Talmud the "God's day" is equal to a millennium, so also in II Peter 3 : 8.
[8] J. Schleifer, "Die Erzählung der Sibylle. Ein Apokryph nach den karshunischen, arabischen und äthiopischen Handschriften zu London, Oxford, Paris und Rom," *Denkschrift der Kaiserl. Akadamie der Wiss., Philos.-hist. Klasse* (Vienna), LIII (1910).

The aborigines of British North Borneo, even today, declare that the sky was originally low, and that six suns perished, and at present the world is illuminated by the seventh sun.[1]

Seven solar ages are referred to in Mayan manuscripts, in Buddhist sacred books, in the books of the Sibyl. In all quoted sources the "suns" are explained (by the sources themselves) as signifying consecutive epochs, each of which went down in a great, general destruction.

Did the reason for the substitution of the word "sun" for "epoch" by the peoples of both hemispheres lie in the changed appearance of the luminary and in its changed path across the sky in each world age?

[1] Cf. Dixon, *Oceanic Mythology*, p. 178.

# PART I

## *Venus*

No book, or collection of books, in the history of mankind has had a more attentive reading, a wider circulation, or more diligent investigation than the Old Testament.

—R. H. Pfeiffer,
*Introduction to the Old Testament*

CHAPTER I

# THE MOST INCREDIBLE STORY

THE MOST INCREDIBLE story of miracles is told about Joshua ben Nun, who, when pursuing the Canaanite kings at Beth-horon, implored the sun and the moon to stand still. "And he said in the sight of Israel, Sun, stand thou still upon Gibeon; and thou, Moon, in the valley of Ajalon. And the sun stood still, and the moon stayed, until the people had avenged themselves upon their enemies. Is not this written in the book of Jasher? So the sun stood still in the midst of heaven, and hasted not to go down about a whole day" (Joshua 10 : 12, 13).

This story is beyond the belief of even the most imaginative or the most pious person. Waves of stormy sea may have drowned one host and been merciful to another. The earth could crack asunder and swallow up human beings. The Jordan could be blocked by a slice of its bank falling into the bed of the river. Jericho's walls—not by the blast of trumpets, but by an incidental earthquake—could have been breached.

But that the sun and the moon should halt in their movement across the firmament—this could be only the product of fancy, a poetic image, a metaphor;[1] a hideous implausibility when imposed as a subject for belief;[2] a matter for scorn—it manifests even a want of reverence for the Supreme Being.

According to the knowledge of our age—not of the age when the Book of Joshua or of Jasher was written—this could have happened if the earth had ceased for a time to roll along its prescribed path. Is such a disturbance conceivable? No record of the slightest confusion is registered in the present annals of the earth. Each year consists of 365 days, 5 hours, and 49 minutes.

[1] "Certainly one could not conceive a more effective flight of fancy, or one more fitted for the heights of one heroic and lyrical composition." G. Schiaparelli, *Astronomy in the Old Testament* (1905), p. 40.

[2] W. Whiston wrote in his *New Theory of the Earth* (6th ed., 1755), pp. 19-21, concerning the wonder of the sun standing still: "The Scripture did not intend to teach men philosophy, or accommodate itself to the true and Pythagoric system of the world." And again: "The prophets and holy penmen themselves . . . being seldom or never philosophers, were not capable of representing these things otherwise than they, with the vulgar, understood them."

A departure of the earth from its regular rotation is thinkable, but only in the very improbable event that our planet should meet another heavenly body of sufficient mass to disrupt the eternal path of our world.

It is true that aerolites or meteorites reach our earth continually, sometimes by the thousands and tens of thousands. But no dislocation of our precise turning round and round has ever been perceived.

This does not mean that a larger body, or a larger number of bodies, could not strike the terrestrial sphere. The large number of asteroids between the orbits of the planets Mars and Jupiter suggests that at some unknown time another planet revolved there; now only these meteorites follow approximately the path along which the destroyed planet circled the sun. Possibly a comet ran into it and shattered it.

That a comet may strike our planet is not very probable, but the idea is not absurd. The heavenly mechanism works with almost absolute precision; but unstable, their way lost, comets by the thousands, by the millions, revolve in the sky, and their interference may disturb the harmony. Some of these comets belong to our system. Periodically they return, but not at very exact intervals, owing to the perturbations caused by gravitation toward the larger planets when they fly too close to them. But innumerable other comets, often seen only through the telescope, come flying in from immeasurable spaces of the universe at very great speed, and disappear—possibly forever. Some comets are visible only for hours, some for days or weeks or even months.

Might it happen that our earth, the earth under our feet, would roll toward perilous collision with a huge mass of meteorites, a trail of stones flying at enormous speed around and across our solar system?

This probability was analysed with fervour during the last century. From the time of Aristotle, who asserted that a meteorite, which fell at Aegospotami when a comet was glowing in the sky, had been lifted from the ground by the wind and carried in the air and dropped over that place, until the year 1803 when, on April 26, a shower of meteorites fell at l'Aigle in France and was investigated by Biot for the French Academy of Sciences, the scholarly world—and in the meantime there lived Copernicus, Galileo Galilei, Kepler, Newton, and Huygens—did not believe that such a thing as a stone falling from the sky was possible at all. And this despite many occasions when stones fell before the eyes of a crowd, as did the aerolite in the presence of Emperor Maximilian and his court in Ensisheim, Alsace, on November 7, 1492.[1]

Only shortly before 1803, the Academy of Sciences of Paris refused to believe that, on another occasion, stones had fallen from the sky. The fall of meteorites on July 24, 1790 in southwest France was pronounced "un phénomène physiquement impossible."[2] Since the year 1803, however, scholars have believed that stones fall from the sky. If a stone can collide with the earth, and occasionally a shower of stones, too, cannot a full-sized comet fly into the face of the earth? It was calculated that such a possibility exists but that it is very unlikely to occur.[3]

If the head of a comet should pass very close to our path, so as to effect a distortion in the career of the earth, another phenomenon besides the disturbed movement of the planet would probably occur : a rain of meteorites would strike the earth and would increase to a torrent. Stones scorched by flying through the atmosphere would be hurled on home and head.

In the Book of Joshua, two verses before the passage about

[1] C. P. Olivier, *Meteors* (1925), p. 4.

[2] P. Bertholon, *Pubblicaziòni della specola astronomica Vaticana* (1913).

[3] D. F. Arago computed on some occasion that there is one chance in 280 million that a comet will hit the earth. Nevertheless, a hole one mile in diameter in Arizona is a sign of an actual headlong collision of the earth with a small comet or asteroid. On June 30, 1908, a calculated forty-thousand-ton mass of iron fell in Siberia at 60° 56′ north latitude and 101° 57′ east longitude. In 1946 the small Giacobini-Zinner comet passed within 131,000 miles of the point where the earth was eight days later.

While investigating whether an encounter between the earth and a comet had been the subject of a previous discussion, I found that W. Whiston, Newton's successor at Cambridge and a contemporary of Halley, in his *New Theory of the Earth* (the first edition of which appeared in 1696) tried to prove that the comet of 1680, to which he (erroneously) ascribed a period of 575½ years, caused the biblical Deluge on an early encounter.

G. Cuvier, who was unable to offer his own explanation of the causes of great cataclysms, refers to the theory of Whiston in the following terms: "Whiston fancied that the earth was created from the atmosphere of one comet, and that it was deluged by the tail of another. The heat which remained from its first origin, in his opinion, excited the whole antediluvian population, men and animals, to sin, for which they were all drowned in the deluge, excepting the fish, whose passions were apparently less violent."

I. Donnelly, author, reformer, and member of the United States House of Representatives, tried in his book *Ragnarok* (1883) to explain the presence of till and gravel on the rock substratum in America and Europe by hypothesizing an encounter with a comet, which rained till on the terrestrial hemisphere facing it at that moment. He placed the event in an indefinite period, but at a time when man already populated the earth. Donnelly did not show any awareness that Whiston was his predecessor. His assumption that there is till only in one half of the earth is arbitrary and wrong.

the sun that was suspended on high for a number of hours without moving to the occident, we find this passage:

"As they [the Canaanite kings] fled from before Israel, and were in the going down to Beth-horon . . . the Lord cast down *great stones* from heaven upon them unto Azekah, and they died: they were more which died with hail stones [stones of *barad*] than they whom the children of Israel slew with the sword."[1]

The author of the Book of Joshua was surely ignorant of any connection between the two phenomena. He could not be expected to have had any knowledge about the nature of aerolites, about the forces of attraction between celestial bodies, and the like. As these phenomena were recorded to have occurred together, it is improbable that the records were invented.

The meteorites fell on the earth in a torrent. They must have fallen in very great numbers for they struck down more warriors than the swords of the adversaries. To have killed persons by the hundreds or thousands in the field, a cataract of stones must have fallen. Such a torrent of great stones would mean that a train of meteorites or a comet had struck our planet.

The quotation in the Bible from the Book of Jasher is laconic and may give the impression that the phenomenon of the motionless sun and moon was local, seen only in Palestine between the valley of Ajalon and Gibeon. But the cosmic character of the prodigy is pictured in a thanksgiving prayer ascribed to Joshua

> Sun and moon stood still in heaven,
> and Thou didst stand in Thy wrath against our oppressors. . . .
>
> All the princes of the earth stood up,
> the kings of the nations had gathered themselves together. . . .
>
> Thou didst destroy them in Thy fury,
> and Thou didst ruin them in Thy rage.
>
> Nations raged from fear of Thee,
> kingdoms tottered because of Thy wrath. . . .
>
> Thou didst pour out Thy fury upon them. . . .
> Thou didst terrify them in Thy wrath. . . .

[1] Joshua 10 : 11.

The Earth quaked and trembled from the noise of Thy thunders.

Thou didst pursue them in Thy storm,
Thou didst consume them in the whirlwind. . . .

Their carcasses were like rubbish.[1]

The wide radius over which the heavenly wrath swept is emphasized in the prayer: "All the kingdoms tottered. . . ."

A torrent of large stones coming from the sky, an earthquake, a whirlwind, a disturbance in the movement of the earth—these four phenomena belong together. It appears that a large comet must have passed very near to our planet and disrupted its movement; a part of the stones dispersed in the neck and tail of the comet smote the surface of our earth a shattering blow.

Are we entitled, on the basis of the Book of Joshua, to assume that at some date in the middle of the second millennium before the present era the earth was interrupted in its regular rotation by a comet? Such a statement has so many implications that it should not be made thoughtlessly. To this I say that though the implications are great and many, the present research in its entirety is an interlinked sequence of documents and other evidence, all of which in common carry the weight of this and other statements in this book.

The problem before us is one of mechanics. Points on the outer layers of the rotating globe (especially near the equator) move at a higher linear velocity than points on the inner layers, but at the same angular velocity. Consequently, if the earth were suddenly stopped (or slowed down) in its rotation, the inner layers might come to rest (or their rotational velocity might be slowed) while the outer layers would still tend to go on rotating. This would cause friction between the various liquid or semifluid layers, creating heat ; on the outer-most periphery the solid layers would be torn apart, causing mountains and even continents to fall or rise.

As I shall show later, mountains fell and others rose from level ground; the earth with its oceans and continents became heated; the sea boiled in many places, and rock liquefied; volcanoes ignited and forests burned. Would not a sudden stop by the earth, rotating at a little over one thousand miles an hour at its equator, mean a complete destruction of the world? Since the world survived, there must have been a

[1] Ginzberg, *Legends*, IV, 11-12.

mechanism to cushion the slowing down of terrestrial rotation, if it really occurred, or another escape for the energy of motion besides transformation into heat, or both. Or if rotation persisted undisturbed, the terrestrial axis may have tilted in the presence of a strong magnetic field, so that the sun appeared to lose for hours its diurnal movement.[1] These problems are kept in sight and are faced in the Epilogue of this volume.

## ON THE OTHER SIDE OF THE OCEAN

The Book of Joshua, compiled from the more ancient Book of Jasher, relates the order of events. "Joshua . . . went up from Gilgal all night." In the early morning he fell upon his enemies unawares at Gibeon, and "chased them along the way that goes up to Beth-horon." As they fled, great stones were cast from the sky. That same day ("in the day when the Lord delivered up the Amorites") the sun stood still over Gibeon and the moon over the valley of Ajalon. It has been noted that this description of the position of the luminaries implies that the sun was in the forenoon position.[2] The Book of Joshua says that the luminaries stood in the midst of the sky.

Allowing for the difference in longitude, it must have been early morning or night in the Western Hemisphere.

We go to the shelf where stand books with the historical traditions of the aborigines of Central America.

The sailors of Columbus and Cortes, arriving in America, found there literate peoples who had books of their own. Most of these books were burned in the sixteenth century by the Dominican monks. Very few of the ancient manuscripts survived, and these are preserved in the libraries of Paris, the Vatican, the Prado, and Dresden; they are called codici, and their texts have been studied and partly read. However, among the Indians of the days of the conquest and also of the following century there were literary men who had access to the knowledge written in pictographic script by their forefathers.[3]

[1] This explanation was suggested to me by M. Abramovich of Tel Aviv.

[2] H. Holzinger, *Josua* (1901), p. 40, in "Hand-commentar zum Alten Testament," ed. K. Marti. R. Eisler, "Joshua and the Sun," *American Journal of Semitic Languages and Literature*, XLII (1926), 83: "It would have had no sense early in the morning of a battle, with a whole day ahead, to have prayed for the lengthening of the sunlight even into the night time."

[3] The Mayan tongue is still spoken by about 300,000 people, but of the Mayan hieroglyphics only the characters employed in the calendar are known for certain.

In the Mexican *Annals of Cuauhtitlan* [1]—the history of the empire of Culhuacan and Mexico, written in Nahua-Indian in the sixteenth century—it is related that during a cosmic catastrophe that occurred in the remote past, the night did not end for a long time.

The biblical narrative describes the sun as remaining in the sky for an additional day ("about a whole day"). The Midrashim, the books of ancient traditions not embodied in the Scriptures, relate that the sun and the moon stood still for thirty-six *itim*, or eighteen hours,[2] and thus from sunrise to sunset the day lasted about thirty hours.

In the Mexican annals it is stated that the world was deprived of light and the sun did not appear for a fourfold night. In a prolonged day or night time could not be measured by the usual means at the disposal of the ancients.[3]

Sahagun, the Spanish savant who came to America a generation after Columbus and gathered the traditions of the aborigines, wrote that at the time of one cosmic catastrophe the sun rose only a little way over the horizon and remained there without moving; the moon also stood still.[4]

I am dealing with the Western Hemisphere first, because the biblical stories were not known to its aborigines when it was discovered. Also, the tradition preserved by Sahagun bears no trace of having been introduced by the missionaries: in his version there is nothing to suggest Joshua ben Nun and his war against the Canaanite kings; and the position of the sun, only a very little above the eastern horizon, differs from the biblical text, though it does not contradict it.

We could follow a path around the earth and inquire into the various traditions concerning the prolonged night and prolonged day, with sun and moon absent or tarrying at different points along the zodiac, while the earth underwent a bombardment of stones in a world ablaze. But we must postpone this journey. There was more than one catastrophe when, according to the memory of mankind, the earth refused to

[1] Known also as *Codex Chimalpopoca.* "This manuscript contains a series of annals of very ancient date, many of which go back to more than a thousand years before the Christian era" (Brasseur).

[2] Sefer Ha-Yashar, ed. L. Goldschmidt (1923); Pirkei Rabbi Elieser (Hebrew sources differ as to how long the sun stood still); the Babylonian Talmud, Tractate Aboda Zara 25a; Targum Habakkuk 3 : 11.

[3] With the exception of the water clock.

[4] Bernardino de Sahagun (1499?-1590), *Historia general de las cosas de Nueva España,* new ed. 1938 (5 vols.) and 1946 (3 vols.). French transl. D. Jourdanet and R. Simeon (1880), p. 481.

play the chronometer by undisturbed rotation on its axis. First, we must differentiate the single occurrences of cosmic catastrophes, some of which took place before the one described here, some after it; some of which were of greater extent, and some of lesser.

## CHAPTER II

# FIFTY-TWO YEARS EARLIER

THE PRE-COLUMBIAN written traditions of Central America tell us that fifty-two years before the catastrophe that closely resembles that of the time of Joshua, another catastrophe of world dimensions had occurred.[1] It is therefore only natural to go back to the old Israelite traditions, as narrated in the Scriptures, to determine whether they contain evidence of a corresponding catastrophe.

The time of the Wandering in the Desert is given by the Scriptures as forty years. Then, for a number of years before the day of the disturbed movement of the earth, the protracted conquest of Palestine went on.[2] It seems reasonable, therefore, to ask whether a date fifty-two years before this event would coincide with the time of the Exodus.

In the work *Ages in Chaos*, I describe at some length the catastrophe that visited Egypt and Arabia. In that work it is explained that the Exodus took place amid a great natural upheaval that terminated the period of Egyptian history known as the Middle Kingdom. There I endeavour to show that contemporary Egyptian documents describe the same disaster accompanied by "the plagues of Egypt," and that the traditions of the Arabian Peninsula relate similar occurrences in this land and on the shores of the Red Sea. In that work I refer also to Beke's idea that Mt. Sinai was a smoking volcano. However, I reveal that "the scope of the catastrophe must have exceeded by far the measure of the disturbance which could be caused by one active volcano," and I promise to answer the question: "Of what nature and dimension was this catastrophe, or this series of catastrophes, accompanied by plagues?" and to publish an investigation into the nature of great catastrophes of the past. Both works—the reconstruction of history and the reconstruction of natural history—were conceived within the short interval of half a year; the desire to establish a correct historical

[1] These sources will be cited on subsequent pages.

[2] According to rabbinical sources, the war of conquest in Palestine lasted fourteen years.

chronology before fitting the acts of nature into the periods of human history impelled me to complete *Ages in Chaos* first.[1]

I shall employ some of the historical material from the first chapters of *Ages in Chaos*. There I use it for the purpose of synchronizing events in the histories of the countries around the eastern Mediterranean; here I shall use it to show that the same events took place all around the world, and to explain the nature of these events.

## THE RED WORLD

In the middle of the second millennium before the present era, as I intend to show, the earth underwent one of the greatest catastrophes in its history. A celestial body that only shortly before had become a member of the solar system—a new comet—came very close to the earth. The account of this catastrophe can be reconstructed from evidence supplied by a large number of documents.

The comet was on its way from its perihelion and touched the earth first with its gaseous tail. Later in this book I shall show that it was about this comet that Servius wrote: "Non igneo sed sanguineo rubore fuisse" (It was not of a flaming but of a bloody redness).

One of the first visible signs of this encounter was the reddening of the earth's surface by a fine dust of rusty pigment. In sea, lake, and river this pigment gave a bloody colouring to the water. Because of these particles of ferruginous or other soluble pigment, the world turned red.

The *Manuscript Quiché* of the Mayas tells that in the Western Hemisphere, in the days of a great cataclysm, when the earth quaked and the sun's motion was interrupted, the water in the rivers turned to blood.[2]

Ipuwer, the Egyptian eyewitness of the catastrophe, wrote his lament on papyrus:[3] "The river is blood," and this corresponds with the Book of Exodus (7 : 20): "All the waters that

[1] In order of publication it will follow the present volume.

[2] Brasseur, *Histoire des nations civilisées du Mexique*, I, 130.

[3] A. H. Gardiner, *Admonitions of an Egyptian Sage from a hieratic papyrus in Leiden* (1909). Its author was an Egyptian named Ipuwer. Hereafter the text will be cited as "Papyrus Ipuwer."

In *Ages in Chaos* I shall develop evidence to show that this papyrus describes events contemporaneous with the end of the Middle Kingdom in Egypt and the Exodus. It must have been composed shortly following the catastrophe.

were in the river were turned to blood." The author of the papyrus also wrote: "Plague is throughout the land. Blood is everywhere," and this, too, corresponds with the Book of Exodus (7 : 21): "There was blood throughout all the land of Egypt."

The presence of the hematoid pigment in the rivers caused the death of fish followed by decomposition and smell. "And the river stank" (Exodus 7 : 21). "And all the Egyptians digged round about the river for water to drink; for they could not drink of the water of the river" (Exodus 7 : 24). The papyrus relates: "Men shrink from tasting; human beings thirst after water," and "That is our water! That is our happiness! What shall we do in respect thereof? All is ruin."

The skin of men and of animals was irritated by the dust, which caused boils, sickness, and the death of cattle—"a very grievous murrain."[1] Wild animals, frightened by the portents in the sky, came close to the villages and cities.

The summit of mountainous Thrace received the name "Haemus," and Apollodorus related the tradition of the Thracians that the summit was so named because of the "stream of blood which gushed out on the mountain" when the heavenly battle was fought between Zeus and Typhon, and Typhon was struck by a thunderbolt.[3] It is said that a city in Egypt received the same name for the same reason.[4]

The mythology which personified the forces of the cosmic drama described the world as coloured red. In one Egyptian myth the bloody hue of the world is ascribed to the blood of Osiris, the mortally wounded planet god; in another myth it is the blood of Seth or Apopi; in the Babylonian myth the world was coloured red by the blood of the slain Tiamat, the heavenly monster.[5]

The Finnish epos of *Kalevala* describes how, in the days of the cosmic upheaval, the world was sprinkled with red milk.[6] The Altai Tartars tell of a catastrophe when "blood turns the whole world red," and a world conflagration follows.[7] The Orphic hymns refer to the time when the heavenly vault, "mighty Olympus, trembled fearfully . . . and the earth

[1] Exodus 9 : 3; cf. Papyrus Ipuwer 5 : 5.
[2] Ginzberg, *Legends*, V, 430.
[3] Apollodorus, *The Library* (transl. J. G. Frazer, 1921), VI.
[4] Frazer's comment to Apollodorus' *Library*, I, 50.
[5] *The Seven Tablets of Creation*, ed. L. W. King (1902).
[6] *Kalevala*, Rune 9.
[7] U. Holmberg, *Finno-Ugric, Siberian Mythology* (1927), p. 370.

around shrieked fearfully, and the sea was stirred [heaped], troubled with its purple waves."[1]

An old subject for debate is: Why is the Red Sea so named? If a sea is called Black or White, that may be due to the dark colouring of the water or to the brightness of the ice and snow. The Red Sea has a deep blue colour. As no better reason was found, a few coral formations or some red birds on its shores were proposed as explanations of its name.[2]

Like all the water in Egypt, the water on the surface of the Sea of the Passage was of a red tint. It appears that Raphael was not mistaken when, in painting the scene of the passage, he coloured the water red.

It was, of course, not this mountain or that river or that sea exclusively that was reddened, thus earning the name Red or Bloody, as distinguished from other mountains and seas. But crowds of men, wherever they were, who witnessed the cosmic upheaval and escaped with their lives, ascribed the name Haemus or Red to particular places.

The phenomenon of "blood" raining from the sky has also been observed in limited areas and on a small scale in more recent times. One of these occasions, according to Pliny, was during the consulship of Manius Acilius and Gaius Porcius.[3] Babylonians, too, recorded red dust and rain falling from the sky;[4] instances of "bloody rain" have been recorded in divers countries.[5] The red dust, soluble in water, falling from the sky in water drops, does not originate in clouds, but must come from volcanic eruptions or from cosmic spaces. The fall of meteorite dust is a phenomenon generally known to take place mainly after the passage of meteorites; this dust is found on the snow of mountains and in polar regions.[6]

[1] "To Minerva" in *Orphic Hymns* (transl. A. Buckley), ed. with the *Odyssey* of Homer (1861).

[2] H. S. Palmer, *Sinai* (1892). Probably at that time the mountainous land of Seir, upon which the Israelites wandered, received the name Edom (Red), and Erythrea (erythraios–red in Greek) its name; Erythrean Sea was in antiquity the name of the Arabian Gulf of the Indian Ocean, applied also to the Red Sea.

[3] Pliny, *Natural History*, ii, 57. Another instance, according to Plutarch, occurred in the reign of Romulus.

[4] F. X. Kugler, "Babylonische Zeitordnung" (Vol. II of his *Sternkunde und Sterndienst in Babel*) (1909-1910), p. 114.

[5] D. F. Arago, *Astronomie populaire* (1854-1857), IV, 209 f.; Abel-Rémusat, *Catalogue des bolides et des aérolithes observés à la Chine et dans les pays voisins* (1819), p. 6.

[6] It is estimated that approximately one ton of meteorite dust falls daily on the globe.

## THE HAIL OF STONES

Following the red dust, a "small dust," like "ashes of the furnace," fell "in all the land of Egypt" (Exodus 9 : 8), and then a shower of meteorites flew toward the earth. Our planet entered deeper into the tail of the comet. The dust was a forerunner of the gravel. There fell "a very grievous hail, such as has not been in Egypt since its foundations" (Exodus 9: 18). Stones of "barad," here translated "hail," is, as in most places where mentioned in the Scriptures, the term for meteorites. We are also informed by Midrashic and Talmudic sources that the stones which fell on Egypt were hot;[1] this fits only meteorites, not a hail of ice.[2] In the Scriptures it is said that these stones fell "mingled with fire" (Exodus 9 : 24), the meaning of which I shall discuss in the following section, and that their fall was accompanied by "loud noises" (*kolot*), rendered as "thunderings," a translation which is only figurative, and not literally correct, because the word for "thunder" is *raam*, which is not used here. The fall of meteorites is accompanied by crashes or explosion-like noises, and in this case they were so "mighty," that, according to the Scriptural narrative, the people in the palace were terrified as much by the din of the falling stones as by the destruction they caused (Exodus 9 : 28).

The red dust had frightened the people, and a warning to keep men and cattle under shelter had been issued: "Gather thy cattle and all that thou hast in the field; for upon every man and beast which shall be found in the field, and shall not be brought home, the hailstones shall come down upon them, and they shall die" (Exodus 9 : 19). "And he that regarded not the word of the Lord left his servants and his cattle in the field" (Exodus 9 : 21).

Similarly, the Egyptian eyewitness: "Cattle are left to stray, and there is none to gather them together. Each man fetches

[1] The Babylonian Talmud, Tractate Berakhot 54b; other sources in Ginzberg, *Legends*, VI, 178.

[2] In the Book of Joshua it is said that "great stones" fell from the sky, and then they are referred to as "stones of barad."

"The ancient Egyptian word for 'hail,' *ar*, is also applied to a driving shower of sand and stones; in the contest between Horus and Set, Isis is described as sending upon the latter *ar n sa*, 'a hail of sand.'" A. Macalister, "Hail," in Hastings, *Dictionary of the Bible* (1901-1904).

for himself those that are branded with his name."[1] Falling stones and fire made the frightened cattle flee.

Ipuwer also wrote: "Trees are destroyed," "No fruits, no herbs are found," "Grain has perished on every side," "That has perished which yesterday was seen. The land is left to its weariness like the cutting of flax."[2] In one day fields were turned to wasteland. In the Book of Exodus (9 : 25) it is written: "And the hail [stones of barad] smote every herb of the field, and brake every tree of the field."

The description of such a catastrophe is found in the *Visuddhi-Magga*, a Buddhist text on the world cycles. "When a world cycle is destroyed by wind . . . there arises in the beginning a cycle-destroying great cloud . . . There arises a wind to destroy the world cycle, and first it raises a fine dust, and then coarse dust, and then fine sand, and then coarse sand, and then grit, stones, up to boulders as large . . . as mighty trees on the hill tops." The wind "turns the ground upside down," large areas "crack and are thrown upwards," "all the mansions on earth" are destroyed in a catastrophe when "worlds clash with worlds."[3]

The Mexican *Annals of Cuauhtitlan* describe how a cosmic catastrophe was accompanied by a hail of stones; in the oral tradition of the Indians, too, the motif is repeated time and time again: In some ancient epoch the sky "rained, not water, but fire and red-hot stones,"[4] which is not different from the Hebrew tradition.

## NAPHTHA

Crude petroleum is composed of two elements, carbon and hydrogen. The main theories of the origin of petroleum are:

1. *The inorganic theory:* Hydrogen and carbon were brought together in the rock formations of the earth under great heat and pressure.

2. *The organic theory:* Both the hydrogen and carbon which compose petroleum come from the remains of plant and animal life, in the main from microscopic marine and swamp life.

[1] Papyrus Ipuwer 9 : 2-3.
[2] *Ibid.*, 4 : 14; 6 : 1; 6 : 3; 5 : 12.
[3] "World Cycles," *Visuddhi-Magga*, in Warren, *Buddhism in Translations*, p. 328.
[4] Alexander, *Latin American Mythology*, p. 72.

The organic theory implies that the process started after life was already abundant, at least at the bottom of the ocean.[1]

The tails of comets are composed mainly of carbon and hydrogen gases. Lacking oxygen, they do not burn in flight, but the inflammable gases, passing through an atmosphere containing oxygen, will be set on fire. If carbon and hydrogen gases, or vapour of a composition of these two elements, enter the atmosphere in huge masses, a part of them will burn, binding all the oxygen available at the moment; the rest will escape combustion, but in swift transition will become liquid. Falling on the ground, the substance, if liquid, would sink into the pores of the sand and into clefts between the rocks; falling on water, it would remain floating if the fire in the air is extinguished before new supplies of oxygen arrive from other regions.

The descent of a sticky fluid which came earthward and blazed with heavy smoke is recalled in the oral and written traditions of the inhabitants of both hemispheres.

*Popol-Vuh*, the sacred book of the Mayas, narrates: [2] "It was ruin and destruction . . . the sea was piled up . . . it was a great inundation . . . people were drowned in a sticky substance raining from the sky. . . . The face of the earth grew dark, and the gloomy rain endured days and nights. . . . And then there was a great din of fire above their heads." The entire population of the land was annihilated.

The *Manuscript Quiché* perpetuated the picture of the population of Mexico perishing in a downpour of bitumen: [3] "There descended from the sky a rain of bitumen and of a sticky substance. . . . The earth was obscured and it rained day and night. And men ran hither and thither and were as if seized by madness; they tried to climb to the roofs, and the houses crashed down; they tried to climb the trees, and the trees cast them far away; and when they tried to escape in caves and caverns, these were suddenly closed."

[1]Even before Plutarch the problem of the origin of petroleum was much discussed. Speaking of the visit of Alexander to the petroleum sources of Iraq, Plutarch said: "There has been much discussion about the origin of [this naphtha]." But in the extant text of Plutarch a sentence containing one of two rival views is missing. The remaining text reads: ". . . or whether rather the liquid substance that feeds the flame flows out from the soil which is rich and productive of fire." Plutarch, *Lives* (transl. B. Perrin, 1919), "The Life of Alexander," xxv.
[2] *Popol-Vuh, le livre sacré*, ed. Brasseur (1861), Chap. III, p. 25.
[3] Brasseur; *Histoire des nations civilisées du Mexique*, I, 55.

A similar account is preserved in the *Annals of Cuauhtitlan.*[1] The age which ended in the rain of fire was called *Quiauhtonatiuh,* which means "the sun of fire-rain." [2]

And far away, in the other hemisphere, in Siberia, the Voguls carried down through the centuries and millennia this memory: "God sent a sea of fire upon the earth. . . . The cause of the fire they call 'the fire-water.' " [3]

Half a meridian to the south, in the East Indies, the aboriginal tribes relate that in the remote past *Sengle-Das* or "water of fire" rained from the sky; with very few exceptions, all men died.[4]

The eighth plague as described in the Book of Exodus was "*barad* [meteorites] and fire mingled with the *barad,* very grievous, such as there was none like it in all the land of Egypt since it became a nation" (Exodus 9 : 24). There were "thunder [correct: loud noises] and *barad,* and the fire ran along the ground" (Exodus 9 : 23).

The Papyrus Ipuwer describes this consuming fire: "Gates, columns, and walls are consumed by fire. The sky is in confusion."[5] The papyrus says that this fire almost "exterminated mankind."

The Midrashim, in a number of texts, state that naphtha, together with hot stones, poured upon Egypt. "The Egyptians refused to let the Israelites go, and He poured out naphtha over them, burning blains [blisters]." It was a "stream of hot naphtha." [6] Naphtha is petroleum in Aramaic and Hebrew.

The population of Egypt was "pursued with strange rains and hails and showers inexorable, and utterly consumed with fire: for what was more marvellous of all, in the water which quencheth all things the fire wrought yet more mightily," [7] which is the nature of burning petroleum; in the register of the plagues in Psalms 105 it is referred to as "flaming fire," and in Daniel (7 : 10) as "river of fire" or "fiery stream."

In the Passover Haggadah it is said that "mighty men of

[1] Brasseur, *Sources de l'histoire primitive du Mexique,* p. 28.
[2] E. Seler, *Gesammelte Abhandlungen zur amerikanischen Sprach- und Altertumsgeschichte* (1902-1923), II, 798.
[3] Holmberg, *Finno-Ugric, Siberian Mythology,* p. 368.
[4] *Ibid.,* p. 369. Also A. Nottrott, *Die Gosnerische Mission unter den Kohls* (1874), p. 25. See R. Andree, *Die Flutsagen* (1891).
[5] Papyrus Ipuwer 2 : 10; 7 : 1; 11 : 11; 12 : 6.
[6] Midrash Tanhuma, Midrash Psikta Raboti, and Midrash Wa-Yosha. For other sources see Ginzberg, *Legends,* II, 342-343, and V, 426.
[7] *The Wisdom of Solomon* (transl. Holmes, 1913) in *The Apocrypha and Pseudepigrapha of the Old Testament,* ed. R. H. Charles.

Pul and Lud [Lydia in Asia Minor] were destroyed with consuming conflagration on the Passover."

In the valley of the Euphrates the Babylonians often referred to "the rain of fire," vivid in their memory.[1]

All the countries whose traditions of fire-rain I have cited actually have deposits of oil: Mexico, the East Indies, Siberia, Iraq, and Egypt.

For a span of time after the combustive fluid poured down, it may well have floated upon the surface of the seas, soaked the surface of the ground, and caught fire again and again. "For seven winters and summers the fire has raged . . . it has burnt up the earth," narrate the Voguls of Siberia.[2]

The story of the wandering in the desert contains a number of references to fire springing out of the earth. The Israelites travelled three days' journey away from the Mountain of the lawgiving, and it happened that "the fire of the Lord burnt among them, and consumed them that were in the uttermost parts of the camp" (Numbers 11 : 1). The Israelites continued on their way. Then came the revolt of Korah and his confederates. "And the earth opened her mouth, and swallowed them up. . . . And all Israel that were round about them fled at the cry of them. . . . And there came out a fire from the Lord, and consumed the two hundred and fifty men that offered incense."[3] When they kindled the fire of incense, the vapours which rose out of the cleft in the rock caught the flame and exploded.

Unaccustomed to handling this oil, rich in volatile derivatives, the Israelite priests fell victims to the fire. The two elder sons of Aaron, Nadab and Abihu, "died before the Lord, when they offered strange fire before the Lord, in the wilderness of Sinai."[4] The fire was called strange because it had not been known before and because it was of foreign origin.

If oil fell on the desert of Arabia and on the land of Egypt and burned there, vestiges of conflagration must be found in some of the tombs built before the end of the Middle Kingdom, into which the oil or some of its derivatives might have seeped.

[1] See A. Schott, "Die Vergleiche in den Akkadischen Königsinschriften," *Mitt. d. Vorderasiat. Ges.*, XXX (1925), 89, 106.
[2] Holmberg, *Finno-Ugric, Siberian Mythology*, p. 369.
[3] Numbers 16 : 32-35. Cf. Psalms 106 : 17-18.
[4] Numbers 3 : 4. Cf. Numbers 26 : 61.

We read in the description of the tomb of Antefoker, vizier of Sesostris I, a pharaoh of the Middle Kingdom: "A problem is set us by a conflagration, clearly deliberate, which has raged in the tomb, as in many another. . . . The combustible material must not only have been abundant, but of a light nature; for a fierce fire which speedily spent itself seems alone able to account for the fact that tombs so burnt remain absolutely free from blackening, except in the lowest parts; nor are charred remains found as a rule. The conditions are puzzling."[1]

"And what does natural history tell us?" asked Philo in his *On the Eternity of the World*,[2] and answered: "Destructions of things on earth, destructions not of all at once but of a very large number, are attributed by it to two principal causes, the tremendous onslaughts of fire and water. These two visitations, we are told, descend in turns after very long cycles of years. When the agent is the conflagration, a stream of heaven-sent fire pours out from above and spreads over many places and overruns great regions of the inhabited earth."

The rain of fire-water contributed to the earth's supply of petroleum; rock oil in the ground appears to be, partly at least, "star oil" brought down at the close of world ages, notably the age that came to its end in the middle of the second millennium before the present era.

The priests of Iran worshipped the fire that came out of the ground. The followers of Zoroastrianism or Mazdaism are also called fire worshippers. The fire of the Caucasus was held in great esteem by all the inhabitants of the adjacent lands. Connected with the Caucasus and originating there is the legend of Prometheus.[3] He was chained to a rock for bringing fire to man. The allegorical character of this legend gains meaning when we consider Augustine's words that Prometheus was a contemporary of Moses.[4]

Torrents of petroleum poured down upon the Caucasus and were consumed. The smoke of the Caucasus fire was still in the imaginative sight of Ovid, fifteen centuries later, when he described the burning of the world.

The continuing fires in Siberia, the Caucasus, in the Arabian

[1] N. de Garis Davies, *The Tomb of Antefoker, Vizier of Sesostris I* (1920), p. 5.
[2] *On the Eternity of the World*, Vol. IX of Philo (transl. F. H. Colson, 1941), Sect. 146-147.
[3] See A. Olrik, *Ragnarok* (German ed., 1922).
[4] *The City of God*, Bk. XVIII, Chap. 8 (transl. M. Dods, ed. P. Schaff, 1907).

desert, and everywhere else were blazes that followed the great conflagration of the days when the earth was caught in vapours of carbon and hydrogen.

In the centuries that followed, petroleum was worshipped, burned in holy places; it was also used for domestic purposes. Then many ages passed when it was out of use. Only in the middle of the last century did man begin to exploit this oil, partly contributed by the comet of the time of the Exodus. He utilized its gifts, and today his highways are crowded with vehicles propelled by oil. Into the heights rose man, and he accomplished the age-old dream of flying like a bird; for this, too, he uses the remnants of the intruding star that poured fire and sticky vapour upon his ancestors.

## THE DARKNESS

The earth entered deeper into the tail of the onrushing comet and approached its body. This approach, if one is to believe the sources, was followed by a disturbance in the rotation of the earth. Terrific hurricanes swept the earth because of the change or reversal of the angular velocity of rotation and because of the sweeping gases, dust, and cinders of the comet.

Numerous rabbinical sources describe the calamity of darkness; the material is collated as follows:[1]

An exceedingly strong wind endured seven days. All the time the land was shrouded in darkness. "On the fourth, fifth, and sixth days, the darkness was so dense that they [the people of Egypt] could not stir from their place." "The darkness was of such a nature that it could not be dispelled by artificial means. The light of the fire was either extinguished by the violence of storm, or else it was made invisible and swallowed up in the density of the darkness. . . . Nothing could be discerned. . . . None was able to speak or to hear, nor could anyone venture to take food, but they lay themselves down . . . their outward senses in a trance. Thus they remained, overwhelmed by the affliction."

The darkness was of such kind that "their eyes were blinded by it and their breath choked"; [2] it was "not of ordinary earthy

[1] Ginzberg, *Legends*, II, 360.

[2] Josephus, *Jewish Antiquities* (transl. H. St. J. Thackeray, 1930), Bk. II, xiv. 5.

kind."[1] The rabbinical tradition, contradicting the spirit of the Scriptural narrative, states that during the plague of darkness the vast majority of the Israelites perished and that only a small fraction of the original Israelite population of Egypt was spared to leave Egypt. Forty-nine out of every fifty Israelites are said to have perished in this plague.[2]

A shrine of black granite found at el-Arish on the border of Egypt and Palestine bears a long inscription in hieroglyphics. It reads: "The land was in great affliction. Evil fell on this earth. . . . There was a great upheaval in the residence. . . . Nobody could leave the palace [there was no exit from the palace] during nine days, and during these nine days of upheaval there was such a tempest that neither men nor gods [the royal family] could see the faces of those beside them."[3]

This record employs the same description of the darkness as Exodus 10 : 22: "And there was a thick darkness in all the land of Egypt three days. They saw not one another, neither rose any from his place for three days."

The difference in the number of the days (three and nine) of the darkness is reduced in the rabbinical sources, where the time is given as seven days. The difference between seven and nine days is negligible if one considers the subjectivity of the time estimation under such conditions. Appraisal of the darkness with respect to its impenetrability is also subjective; rabbinical sources say that for part of the time there was a very slight visibility, but for the rest (three days) there was no visibility at all.

It should be kept in mind that, as in the case I have already discussed, a day and a night of darkness or light can be described as one day or as two days.

That both sources, the Hebrew and the Egyptian, refer to the same event can be established by another means also. Following the prolonged darkness and the hurricane, the pharaoh, according to the hieroglyphic text of the shrine, pursued the "evil-doers" to "the place called Pi-Khiroti." The same place is mentioned in Exodus 14 : 9: "But the Egyptian pursued after them, all the horses and chariots of Pharaoh . . .

[1] Ginzberg, *Legends*, II, 359.

[2] Targum Yerushalmi, Exodus 10 : 23; *Mekhilta d'rabbi Simon ben Jokhai* (1905), p. 38.

[3] F. L. Griffith, *The Antiquities of Tel-el-Yahudiyeh and Miscellaneous Work in Lower Egypt in 1887-88* (1890); G. Goyon, "Les Travaux de Chou et les tribulations de Geb d'après Le Naos 2248 d'Ismailia," *Kemi, Revue de Philol. et d'arch. égypt.* (1936).

and overtook them encamping by the sea, beside Pi-ha-khiroth." [1]

The inscription on the shrine also narrates the death of the pharaoh during this pursuit under exceptional circumstances: "Now when the Majesty fought with the evil-doers in this pool, the place of the whirlpool, the evil-doers prevailed not over his Majesty. His Majesty leapt into the place of the whirlpool." This is the same apotheosis described in Exodus 15 : 19: "For the horse of Pharaoh went in with his chariots and with his horsemen into the sea, and the Lord brought again the waters of the sea upon them."

If "the Egyptian darkness" was caused by the earth's stasis or tilting of its axis, and was aggravated by a thin cinder dust from the comet, then the entire globe must have suffered from the effect of the two concurring phenomena; in either the eastern or the western parts of the world there must have been a very extended, gloomy day.

Nations and tribes in many places of the globe, to the south, to the north, and to the west of Egypt, have old traditions about a cosmic catastrophe during which the sun did not shine; but in some parts of the world the traditions maintain that the sun did not set for a period of time equal to a few days.

Tribes of the Sudan to the south of Egypt refer in their tales to a time when the night would not come to an end.[2]

*Kalevala*, the epos of the Finns, tell of a time when hailstones of iron fell from the sky, and the sun and the moon disappeared (were stolen from the sky) and did not appear again; in their stead, after a period of darkness, a new sun and a new moon were placed in the sky.[3] Caius Julius Solinus writes that "following the deluge which is reported to have occurred in the days of Ogyges, a heavy night spread over the globe." [4]

In the manuscripts of Avila and Molina, who collected the traditions of the Indians of the New World, it is related that the sun did not appear for five days; a cosmic collision of stars preceded the cataclysm; people and animals tried to escape to mountain caves. "Scarcely had they reached there when the sea, breaking out of bounds following a terrifying shock, began

[1] The syllable *ha* is the definite article in Hebrew and in this case belongs between "Pi" and "Khiroth."

[2] L. Frobenius, *Dichten und Denken im Sudan* (1925), p. 38.

[3] *Kalevala* (transl. J. M. Crawford, 1888), p. xiii.

[4] Caius Julius Solinus, *Polyhistor*. French transl. by M. A. Agnant, 1847, Chap. xi, reads: "a heavy night spread over the globe for nine consecutive days." Other translators render: "nine consecutive months."

to rise on the Pacific coast. But as the sea rose, filling the valleys and the plains around, the mountain of Ancasmarca rose, too, like a ship on the waves. During the five days that this cataclysm lasted, the sun did not show its face and the earth remained in darkness."[1]

Thus the traditions of the Peruvians describe a time when the sun did not appear for five days. In the upheaval, the earth changed its profile, and the sea fell upon the land.[2]

East of Egypt, in Babylonia, the eleventh tablet of the *Epic of Gilgamesh* [Gilgamish] refers to the same events. From out the horizon rose a dark cloud and it rushed against the earth; the land was shrivelled by the heat of the flames. "Desolation . . . stretched to heaven; all that was bright was turned into darkness. . . . Nor could a brother distinguish his brother. . . . Six days . . . the hurricane, deluge, and tempest continued sweeping the land . . . and all human back to its clay was returned."[3]

The Iranian book *Anugita* reveals that a threefold day and threefold night concluded a world age,[4] and the book *Bundahis*, in a context that I shall quote later and that shows a close relation to the events of the cataclysm I describe here, tells of the world being dark at midday as though it were in deepest night: it was caused, according to the *Bundahis*, by a war between the stars and the planets.[5]

A protracted night, deepened by the onrushing dust sweeping in from interplanetary space, enveloped Europe, Africa, and America, the valleys of the Euphrates and the Indus also. If the earth did not stop rotating but slowed down or was tilted, there must have been a longitude where a prolonged day was followed by a prolonged night. Iran is so situated that, if one is to believe the Iranian tradition, the sun was absent for a threefold day, and then it shone for a threefold day. Farther to the east there must have been a protracted day corresponding to the protracted night in the west.

According to "Bahman Yast," at the end of a world age in eastern Iran or in India the sun remained ten days visible in the sky.

In China, during the reign of the Emperor Yahou, a great

[1] Brasseur, *Sources de l'histoire primitive du Mexique*, p. 40.
[2] Andree, *Die Flutsagen*, p. 115.
[3] *The Epic of Gilgamish* (transl. R. C. Thompson, 1928).
[4] "The Anugita" (transl. K. T. Telang, 1882) in Vol. VIII of *The Sacred Books of the East*.
[5] "The Bundahis" in *Pahlavi Texts* (transl. E. W. West) (*The Sacred Books of the East*, V [1880]), Pt. I, p. 17.

catastrophe brought a world age to a close. For ten days the sun did not set.[1] The events of the time of the Emperor Yahou deserve close examination; I shall return to the subject shortly.[2]

## EARTHQUAKE

The earth, forced out of its regular motion, reacted to the close approach of the body of the comet: a major shock convulsed the lithosphere, and the area of the earthquake was the entire globe.

Ipuwer witnessed and survived this earthquake. "The towns are destroyed. Upper Egypt has become waste. . . . All is ruin." "The residence is overturned in a minute."[3] Only an earthquake could have overturned the residence in a minute. The Egyptian word for "to overturn" is used in the sense of "to overthrow a wall."[4]

This was the tenth plague. "And Pharaoh rose up in the night, he, and all his servants, and all the Egyptians; and there was a great cry in Egypt; for there was not a house where there was not one dead" (Exodus 12 : 30). Houses fell, smitten by one violent blow. "[The angel of the Lord] passed over the houses of the children of Israel in Egypt, when he smote the Egyptians, and delivered our houses" (Exodus 12 : 27). *Nogaf*, meaning "smote," is the word used for a very violent blow, as, for instance, goring by the horns of an ox. The Passover Haggadah says: "The firstborn of the Egyptians didst Thou crush at midnight."

The reason why the Israelites were more fortunate in this plague than the Egyptians probably lies in the kind of material of which their dwellings were constructed. Occupying a marshy district and working on clay, the captives must have lived in huts made of clay and reeds, which are more resilient than brick or stone. "The Lord will pass over the door, and will not suffer the destroyer to come and smite your houses."[5] An

[1] Cf. "Yao," *Universal Lexicon* (1732-1754), Vol. LX.
[2] The way the Egyptians estimated the time the sun was not in the sky must have been similar to the Chinese method of estimation. It is very probable that these peoples reckoned the disturbance as lasting five days and five nights (because a ninefold or tenfold period elapsed from one sunrise or sunset to the other).
[3] Papyrus Ipuwer 2 : 11; 3 : 13.
[4] Gardiner's commentary to Papyrus Ipuwer.
[5] Exodus 12 : 23. The King James Version, "will not suffer the destroyer to come *in* unto your houses to smite *you*," is not correct.

example of the selective action of a natural agent upon various kinds of construction is narrated also in Mexican annals. During a catastrophe accompanied by hurricane and earthquake, only the people who lived in small log cabins remained uninjured; the larger buildings were swept away. "They found that those who lived in small houses had escaped, as well as the newly-married couples, whose custom it was to live for a few years in cabins in front of those of their fathers-in-law." [1]

In *Ages in Chaos* (my reconstruction of ancient history), I shall show that "first-born" (*bkhor*) in the text of the plague is a corruption of "chosen" (*bchor*). All the flower of Egypt succumbed in the catastrophe.

"Forsooth: The children of princes are dashed against the walls . . . the children of princes are cast out in the streets"; "the prison is ruined," wrote Ipuwer,[2] and this reminds us of princes in palaces and captives in dungeons who were victims in the disaster (Exodus 12 : 29).

To confirm my interpretation of the tenth plague as an earthquake, which should be obvious from the expression, "to smite the houses," I find a corroborating passage of Artapanus in which he describes the last night before the Exodus, and which is quoted by Eusebius: There was "hail and earthquake by night, so that those who fled from the earthquake were killed by the hail, and those who sought shelter from the hail were destroyed by the earthquake. And at that time all the houses fell in, and most of the temples." [3]

Also, Hieronymus (St. Jerome) wrote in an epistle that "in the night in which Exodus took place, all the temples of Egypt were destroyed either by an earthshock or by the thunderbolt." [4] Similarly in the Midrashim: "The seventh plague, the plague of barad [meteorites]: earthquake, fire, meteorites." [5] It is also said that the structures which were erected by the Israelite slaves in Pithom and Ramses collapsed or were swallowed by the earth.[6] An inscription which dates from the beginning of the New Kingdom refers to a temple of the Middle Kingdom

[1] Diego de Landa, *Yucatan, before and after the Conquest* (transl. W. Gates, 1937), p. 18.
[2] Papyrus Ipuwer 5 : 6; 6 : 12.
[3] Eusebius, *Preparation for the Gospel* (transl. E. H. Gifford, 1903), Bk. IX, Chap. xxvii.
[4] Cf. S. Bochart, *Hierozoicon* (1675), I, 344.
[5] The Mishna of Rabbi Eliezer, ed. H. G. Enelow (1933).
[6] Ginzberg, *Legends*, II, 241. Pithom was excavated by E. Naville (*The Store-City of Pithom and the Route of the Exodus* [1885]), but he did not dig beneath the alyer of the New Kingdom.

that was "swallowed by the ground" at the close of the Middle Kingdom.[1]

The head of the celestial body approached very close, breaking through the darkness of the gaseous envelope, and according to the Midrashim, the last night in Egypt was as bright as the noon on the day of the summer solstice.[2]

The population fled. "Men flee. . . . Tents are what they make like the dwellers of hills," wrote Ipuwer.[3] The population of a city destroyed by an earthquake usually spends the nights in the fields. The Book of Exodus describes a hurried flight from Egypt on the night of the tenth plague; a "mixed multitude" of non-Israelites left Egypt together with the Israelites, who spent their first night in Sukkoth (huts).[4]

"The lightnings lightened the world: the earth trembled and shook. . . . Thou leddest thy people like a flock by the hand of Moses and Aaron."[5] They were brought out of Egypt by a portent which looked like a stretched arm—"by a stretched out arm and by great terrors," or "with a mighty hand, and with an outstretched arm, and with great terribleness, and with signs, and with wonders."[6]

## "13"

"At midnight" all the houses of Egypt were smitten; "there was not a house where there was not one dead." This happened on the night of the fourteenth of the month Aviv (Exodus 12 : 6; 13 : 4). This is the night of Passover. It appears that the Israelites originally celebrated Passover on the eve of the fourteenth of Aviv.

The month Aviv is called "the first month" (Exodus 12 : 18). Thout was the name of the first month of the Egyptians. What, for the Israelites, became a feast, became a day of sadness and fasting for the Egyptians. "The thirteenth day of the month Thout [is] a very bad day. Thou shalt not do anything on this day. It is the day of the combat which Horus waged with Seth."[7]

[1] The inscription of Queen Hatshepsut at Speos Artemidos, J. Breasted, *Ancient Records of Egypt*, Vol. II, Sec. 300.
[2] Zohar ii, 38a-38b.
[3] Papyrus Ipuwer 10 : 2.
[4] Exodus 12 : 37-38.
[5] Psalms 77 : 18, 20.
[6] Deuteronomy 4 : 34; 26 : 8.
[7] W. Max Müller, *Egyptian Mythology* (1918), p. 126.

The Hebrews counted (and still count) the beginning of the day from sunset;[1] the Egyptians reckoned from sunrise.[2] As the catastrophe took place at midnight, for the Israelites it was the fourteenth day of the (first) month; for the Egyptians it was the thirteenth day.

An earthquake caused by contact or collision with a comet must be felt simultaneously all around the world. An earthquake is a phenomenon that occurs from time to time; but an earthquake accompanying an impact in the cosmos would stand out and be recalled as a memorable date by survivors.

In the calendar of the Western Hemisphere, on the thirteenth day of the month, called *olin*, "motion" or "earthquake,"[3] a new sun is said to have initiated another world age.[4] The Aztecs, like the Egyptians, reckoned the day from sunrise.[5]

Here we have, *en passant*, the answer to the open question concerning the origin of the superstition which regards the number 13, and especially the thirteenth day, as unlucky and inauspicious. It is still the belief of many superstitious persons, unchanged through thousands of years and even expressed in the same terms: "The thirteenth day is a very bad day. You shall not do anything on this day."

I do not think that any record of this belief can be found dating from before the time of the Exodus. The Israelites did not share this superstition of the evil-working number thirteen (or fourteen).

[1] Leviticus 23 : 32.
[2] K. Sethe, "Die ägyptische Zeitrechnung" (*Göttingen Ges. d. Wiss.*, 1920), pp. 130 ff.
[3] See *Codex Vaticanus No.* 3773 (B), elucidated by E. Seler (1902-1903).
[4] Seler, *Gesammelte Abhandlungen*, II, 798, 800.
[5] L. Ideler, *Historische Untersuchungen uber die astronomischen Beobachtungen der Alten* (1806), p. 26.

## CHAPTER III

# THE HURRICANE

THE SWIFT SHIFTING of the atmosphere under the impact of the gaseous parts of the comet, the drift of air attracted by the body of the comet, and the rush of the atmosphere resulting from inertia when the earth stopped rotating or shifted its poles, all contributed to produce hurricanes of enormous velocity and force and of worldwide dimensions.

*Manuscript Troano* and other documents of the Mayas describe a cosmic catastrophe during which the ocean fell on the continent, and a terrible hurricane swept the earth.[1] The hurricane broke up and carried away all towns and all forests.[2] Exploding volcanoes, tides sweeping over mountains, and impetuous winds threatened to annihilate humankind, and actually did annihilate many species of animals. The face of the earth changed, mountains collapsed, other mountains grew and rose over the onrushing cataract of water driven from oceanic spaces, numberless rivers lost their beds, and a wild tornado moved through the debris descending from the sky. The end of the world age was caused by Hurakan, the physical agent that brought darkness and swept away houses and trees and even rocks and mounds of earth. From this name is derived "hurricane," the word we use for a strong wind. Hurakan destroyed the major part of the human race. In the darkness swept by wind, resinous stuff fell from the sky and participated with fire and water in the destruction of the world.[3] For five days, save for the burning naphtha and burning volcanoes, the world was dark, since the sun did not appear.

The theme of a cosmic hurricane is reiterated time and again in the Hindu *Vedas* and in the Persian *Avesta*,[4] and *diluvium venti*, the deluge of wind, is a term known from many ancient

[1] Brasseur, *Manuscrit Troano* (1869), p. 141.
[2] In the documents of the collection of Kingsborough, the writings of Gómara, Mitolinia, Sahagun, Landa, Cogolludo, and other authors of the early postconquest time, the cataclysm of deluge, hurricane, and volcanoes is referred to in numerous passages. See, e.g., *Gómara, Conquista de México*, II, pp. 261 ff.
[3] *Popol-Vuh*, Chap. III.
[4] Cf. A. J. Carnoy, *Iranian Mythology* (1917).

authors.[1] In the Section, "The Darkness," I quoted rabbinical sources on the "exceedingly strong west wind" that endured for seven days when the land was enveloped in darkness, and the hieroglyphic inscription from el-Arish about "nine days of upheaval" when "there was such a tempest" that nobody could leave the palace or see the faces of those beside him, and the eleventh tablet of the *Epic of Gilgamesh* which says that "six days and a night . . . the hurricane, deluge, and tempest continued sweeping the land," and mankind perished almost altogether. In the battle of the planet-god Marduk with Tiamat, "he [Marduk] created the evil wind, and the tempest, and the hurricane, and the fourfold wind, and the sevenfold wind, and the whirlwind, and the wind which had no equal."[2]

The Maoris narrate [3] that amid a stupendous catastrophe "the mighty winds, the fierce squalls, the clouds, dense, dark, fiery, wildly drifting, wildly bursting," rushed on creation, in their midst Tawhirima-tea, father of winds and storms, and swept away giant forests and lashed the waters into billows whose crests rose high like mountains. The earth groaned terribly, and the ocean fled.

"The earth was submerged in the ocean but was drawn by Tefaafanau," relate the aborigines of Paumotu in Polynesia. The new isles "were bated by a star." In the month of March the Polynesians celebrate a god, Taafanua.[4] "In Arabic, Tyfoon is a whirlwind and Tufan is the Deluge; and the same word occurs in Chinese as Ty-fong."[5] It appears as though the noise of the hurricane was overtoned by a sound not unlike the name Typhon, as if the storm were calling him by name.

The cosmic upheaval proceeded with a "mighty strong west wind," [6] but before the climax, in the simple words of the Scriptures, "the Lord caused the sea to go back by a strong east wind all that night, and made the sea dry land, and the waters were divided." [7]

The Israelites were on the shore of the Sea of Passage at the climax of the cataclysm. The name Jam Suf is generally rendered as Red Sea; the Passage is supposed to have taken

[1] Cf. Eisler, *Weltmantel und Himmelszelt*, II, 453. The Talmud also occasionally uses the notion of "cosmic wind." The Babylonian Talmud, Tractate Berakhot, 13.
[2] *Seven Tablets of Creation*, the fourth tablet.
[3] E. B. Tylor, *Primitive Culture* (1929), I, 322 ff.
[4] Williamson, *Religious and Cosmic Beliefs of Central Polynesia*, I, 36, 154, 237.
[5] G. Rawlinson, *The History of Herodotus* (1858-1862), II, 225 note.
[6] Exodus 10 : 19.
[7] Exodus 14 : 21.

place either at the Gulf of Suez or at Akaba Gulf of the Red Sea, but sometimes the site of the Passage is identified as one of the inner lakes on the route from Suez to the Mediterranean. It is argued that *suf* means "reed" (papyrus reed), and since papyrus reed does not grow in salt water, Jam Suf must have been a lagoon of fresh water.[1] We will not enter here into a discussion where the Sea of the Passage was. The inscription on the shrine found in el-Arish may provide some indication where the Pharaoh was engulfed by the whirlpool;[2] in any event, the topographical distribution of sea and land did not remain the same as before the cataclysm of the days of the Exodus. But the name of the Sea of the Passage—Jam Suf—is derived not from "reed," but from "hurricane," *suf*, *sufa*, in Hebrew. In Egyptian the Red Sea is called *shari*, which signifies the sea of percussion (*mare percussionis*) or the sea of the stroke or of the disaster.[3]

The Haggadah of Passover says: "Thou didst sweep the land of Moph and Noph . . . on the Passover."[4]

The hurricane that brought to an end the Middle Kingdom in Egypt—"the blast of heavenly displeasure" in the language of Manetho—swept through every corner of the world. In order to distinguish, in the traditions of the peoples, this *diluvium venti* of cosmic dimensions from local disastrous storms, other cosmic disturbances like disappearance of the sun or change of sky must be found accompanying the hurricane.

In the Japanese cosmogonical myth, the sun goddess hid herself for a long time in a heavenly cave in fear of the storm god. "The source of light disappeared, the whole world became dark," and the storm god caused monstrous destruction. Gods made terrible noise so that the sun should reappear, and from their tumult the earth quaked.[5] In Japan and in the vast extent of the ocean hurricanes and earthquakes are not rare occurrences; but they do not disturb the day-night succession, nor is there any resulting permanent change in the sky and its luminaries. "The sky was low," relate the Polynesians of Takaofo Island, and "then the winds and waterspouts and the hurricanes came, and carried up the sky to its present height."[6]

[1] Cf. Isaiah 19 : 6.
[2] See p. 60.
[3] Akerblad, *Journal asiatique*, XIII (1834), 349; F. Fresnel, *ibid.*, 4e Série, XI (1848); cf. Peyron, *Lexicon linguae copticae* (1835), p. 304.
[4] Moph and Noph refer to Memphis.
[5] Nihongi, "Chronicles of Japan from the Earliest Times" (transl. W. G. Aston), *Transactions and Proceedings of the Japanese Society*, I (1896), 37 f., 47.
[6] Williamson, *Religious and Cosmic Beliefs of Central Polynesia*, I, 44.

"When a world cycle is destroyed by wind," says the Buddhist text on the "World Cycles," the wind also turns "the ground upside down, and throws it into the sky," and "areas of one hundred leagues in extent, two hundred, three hundred, five hundred leagues in extent, crack and are thrown upward by the force of the wind" and do not fall again but are "blown to powder in the sky and annihilated." "And the wind throws up also into the sky the mountains which encircle the earth . . . [they] are ground to powder and destroyed." The cosmic wind blows and destroys "a hundred thousand times ten million worlds." [1]

## THE TIDE

The ocean tides are produced by the action of the sun and to a larger extent by that of the moon. A larger body than the moon or one nearer to the earth would act with greater effect. A comet with a head as large as the earth, passing sufficiently close, would raise the waters of the oceans miles high.[2] The slowing down or stasis of the earth in its rotation would cause a tidal recession of water toward the poles,[3] but the celestial body near by would disturb this poleward recession, drawing the water toward itself.

The traditions of many peoples persist that seas were torn apart and their water heaped high and thrown upon the continents. In order to establish that these traditions refer to one and the same event, or at least to an event of the same order, we must keep to this guiding sequence: the great tide followed a disturbance in the motion of the earth.

The Chinese annals, which I have mentioned and which I intend to quote more extensively in a subsequent section, say that in the time of Emperor Yahou the sun did not go down for ten days. The world was in flames, and "in their vast extent" the waters "over-topped the great heights, threatening the heavens with their floods." The water of the ocean was heaped up and cast upon the continent of Asia; a great tidal wave swept over the mountains and broke in the middle of the

[1] Warren, "World Cycles," *Buddhism*, p. 328.
[2] Cf. J. Lalande, *Abrégé d'astronomie* (1795), p. 340, who computed that a comet with a head as large as the earth, at a distance of 13,290 *lieues*, or about four diameters of the earth, would raise ocean tides 2,000 *toises* or about four kilometres high.
[3] P. Kirchenberg, *La Théorie de la relativité* (1922), pp. 131-132.

Chinese Empire. The water was caught in the valleys between the mountains, and the land was flooded for decades.

The traditions of the people of Peru tell that for a period of time equal to five days and five nights the sun was not in the sky, and then the ocean left the shore and with a terrible din broke over the continent; the entire surface of the earth was changed in this catastrophe.[1]

The Choctaw Indians of Oklahoma relate: "The earth was plunged in darkness for a long time." Finally a bright light appeared in the north, "but it was mountain-high waves, rapidly coming nearer." [2]

In these traditions there are two concurrent elements: a complete darkness that endured a number of days (in Asia, prolonged day) and, when the light broke through, a mountain-high wave that brought destruction.

The Hebrew story of the passage of the sea contains the same elements. There was a prolonged and complete darkness (Exodus 10 : 21). The last day of the darkness was at the Red Sea.[3] When the world plunged out of darkness, the bottom of the sea was uncovered, the waters were driven apart and heaped up like walls in a double tide.[4] The Septuagint translation of the Bible says that the water stood "as a wall," and the Koran, referring to this event, says "like mountains." In the old rabbinical literature it is said that the water was suspended as if it were "glass, solid and massive." [5]

The commentator Rashi, guided by the grammatical structure of the sentence in the Book of Exodus, explained in accordance with Mechilta: "The water of all oceans and seas was divided." [6]

The Midrashim contain the following description: "The waters were piled up to the height of sixteen hundred miles, and they could be seen by all the nations of the earth." [7] The figure in this sentence intends to say that the heap of water was tremendous. According to the Scriptures, the waters

[1] Andree, *Die Flutsagen*, p. 115.
[2] H. S. Bellamy, *Moons, Myths and Man* (1938), p. 277.
[3] Exodus 14 : 20; Ginzberg, *Legends*, II, 359.
[4] "The waters were a wall unto them on their right hand, and on their left." Exodus 14 : 22.
[5] A. Calmet, *Commentaire, l'Exode* (1708), p. 159: "Les eaux demeurent suspendues, comme une glace solide et massive."
[6] *Rashi's Commentary to Pentateuch* (English transl. by M. Rosenbaum and A. M. Silberman, 1930).
[7] Ginzberg, *Legends*, III, 22; Targum Yerushalmi, Exodus 14 : 22.

climbed the mountains and stood above them, and they mounted to the heavens.[1]

A sea rent apart was a marvellous spectacle and could not have been forgotten. It is mentioned in numerous passages in the Scriptures. "The pillars of heaven tremble. . . . He divideth the sea with his power." [2] "Marvellous things did he in the sight of their fathers. . . . He divided the sea, and caused them to pass through; and he made the waters to stand as a heap." [3] "He gathereth the waters of the sea together as a heap . . . let all the inhabitants of the world stand in awe of him." [4]

Then the Great Sea (the Mediterranean) broke into the Red Sea in an enormous tidal wave.[5]

It was an unusual event, and because it was unusual, it became the most impressive recollection in the very long history of this people. All peoples and nations were blasted by the same fire and shattered in the same fury. The tribes of Israel on the shore of a sea found in this annihilation their salvation from bondage. They escaped destruction but their oppressors perished before their eyes. They extolled the Creator, took upon themselves the burden of moral rules, and considered themselves chosen for a great destiny.

When the Spaniards conquered Yucatan, Indians versed in their ancient literature related to the conquerors the tradition handed down to them by their ancestors: their forefathers were delivered from pursuit by some other people when the Lord opened for them a way in the midst of the sea.[6]

This tradition is so similar to the Jewish tradition of the Passage that some friars who came to America believed that the Indians of America were of Jewish origin. Friar Diego de Landa wrote: "Some old men of Yucatan say that they have heard from their ancestors that this country was peopled by a certain race who came from the east, whom God delivered by opening for them twelve roads through the sea. If this is true, all the inhabitants of the Indies must be of Jewish descent." [7]

It may have been an echo of what happened at the Sea of

[1] Psalms 104 : 6-8; 107 : 25-26.
[2] Job 26 : 11-12.
[3] Psalms 78 : 12-13.
[4] Psalms 33 : 7-8.
[5] Mekhilta Beshalla 6, 33a; other sources in Ginzberg, *Legends*, VI, 10.
[6] Antonio de Herrera, *Historia general de las Indias Occidentales*, Vol. IV, Bk. 10, Chap. 2; Brasseur, *Histoire des nations civilisées du Mexique*, I, 66.
[7] De Landa, *Yucatan*, p. 8.

Passage, or a description of a similar occurrence at the same time but in another place.

According to the Lapland cosmogonic story,[1] "when the wickedness increased among the human beings," the midmost of the earth "trembled with terror so that the upper layers of the earth fell away and many of the people were hurled down into those caved-in places to perish." "And Jubmel, the heaven-lord himself, came down. . . . His terrible anger flashed like red, blue, and green fire-serpents, and people hid their faces, and the children screamed with fear. . . . The angry god spoke: 'I shall reverse the world. I shall bid the rivers flow upward; I shall cause the sea to gather together itself up into a huge towering wall which I shall hurl upon your wicked earth-children, and thus destroy them and all life.' "

> Jubmel set a storm-wind blowing,
> and the wild air-spirits raging. . . .
> Foaming, dashing, rising sky-high
> came the sea-wall, crushing all things.
> Jubmel, with one strong upheaval,
> made the earth-lands all turn over;
> then, the world again he righted.
> Now the mountains and the highlands
> could no more be seen by Beijke [sun].
> Filled with groans of dying people,
> was the fair earth, home of mankind.
> No more Beijke shone in heaven.

According to the Lapland epic, the world was overwhelmed by the hurricane and the sea, and almost all human beings perished. After the sea-wall fell on the continent, gigantic waves continued to roll and dead bodies were dashed about in dark waters.

The great earthquake and the chasms that opened in the ground, the appearance of a celestial body with serpentlike flashes, rivers flowing upward, a sea-wall that crushed everything, mountains that became levelled or covered with water, the world that was turned over and then righted, the sun that no more shone in the sky—all these are motifs which we found in the description of the calamities of the time of the Exodus.

In many places of the world, and especially in the north, large boulders are found in a position which proves that a

[1] Leonne de Cambrey, *Lapland Legends* (1926).

great force must have lifted them up and carried them long distances before depositing them where they are found today. Sometimes these large loose rocks are of entirely different mineral composition than the local rocks, but are akin to formations many miles away. Thus, occasionally an erratic boulder of granite perches on top of a high ridge of dolerite, whereas the nearest outcrops of granite lie far away. These erratic boulders may weigh as much as ten thousand tons, about as much as one hundred thirty thousand people.[1]

To explain these facts, the scholars of the first half of the nineteenth century assumed that enormous tides had swept over the continents and carried with them masses of stone. The transfer of the rocks was explained by the tides, but what could have caused those billows to rise high over the continents?

"It was conceived that somehow and somewhere in the far north a series of gigantic waves was mysteriously propagated. These waves were supposed to have precipitated themselves upon the land, and then swept madly on over mountain and valley alike, carrying along with them a mighty burden of rocks and stones and rubbish. Such deluges were styled 'waves of translation'; and the till was believed to represent the materials which they hurried along with them in their wild course across the country."[2] The stones and boulders on the hilltops and the mounds of sand and gravel in the lowlands were explained by this theory. Critics, however, maintained that "it was unfortunate for this view that it violated at the very outset the first principles of science, by assuming the former existence of a cause which there was little in nature to warrant . . . spasmodic rushes of the sea across a whole country had fortunately never been experienced within the memory of man."[3] That the correctness of the last sentence is questionable is shown by references to the traditions of a number of peoples.

Wherever possible, the movement of stones was attributed to the progress of the ice sheet in the glacial ages and to glaciers on the mountain slopes.

Agassiz, in 1840, assumed that just as the Alpine moraines

[1] The Madison boulder near Conway, New Hampshire, measures 90 by 40 by 38 feet, and weighs almost 10,000 tons. "It is composed of granite, quite unlike the bedrock beneath it; hence the boulder is typically 'erratic.'" Daly, *The Changing World of the Ice Age*, p. 16.
[2] J. Geikie, *The Great Ice Age and Its Relation to the Antiquity of Man* (1894), pp. 25-26.
[3] *Ibid.*

were left behind by the retreating glaciers, so the moraines in the flatlands of northern Europe and America could have been caused by the movement of great continental ice sheets (and thus introduced the theory of ice ages). Although this is correct to some extent, the analogy is not exact, as the glaciers of the Alps push the stones down, not up the slope. Meeting an upward motion of the ice, large boulders would probably sink into the ice.

The problem of the migration of the stones must be regarded as only partially connected with the progress and retreat of the ice sheet, if at all. Billows miles high travelled over the land, originating in causes described in this book.

It can be established by the extent of denudation of the rocks under the erratic boulders that the latter were deposited at their places during human history. So, for instance, in Wales and Yorkshire, where this effect was evaluated in terms of time, the "amount of denudation of limestone rocks on which boulders lie" is a "proof that a period of no more than six thousand years has elapsed since the boulders were left in their positions."[1]

The fact that accumulations of stones were transferred from the equator toward the higher latitudes, an enigmatic problem in the ice theory, can be explained by the poleward recession of the equatorial waters at the moment the velocity of rotation of the earth was reduced or its poles were shifted. In the Northern Hemisphere, in India, the moraines were carried from the equator not only toward higher latitudes, but also toward the Himalaya Mountains, and in the Southern Hemisphere from the equatorial regions of Africa toward the higher latitudes, across the prairies and deserts and forests of the black continent.

## THE BATTLE IN THE SKY

At the same time that the seas were heaped up in immense tides, a pageant went on in the sky which presented itself to the horrified onlookers on earth as a gigantic battle. Because this battle was seen from almost all parts of the world, and because it impressed itself very strongly upon the imagination of the peoples, it can be reconstructed in some detail.

[1] Upham, *The Glacial Lake Agassiz* (1895), p. 239.

When the earth passed through the gases, dust, and meteorites of the tail of the comet, disturbed in rotation, it proceeded on a distorted orbit. Emerging from the darkness, the Eastern Hemisphere faced the head of the comet. This head only shortly before had passed close to the sun and was in a state of candescence. The night the great earthquake shook the globe was, according to rabbinical literature, as bright as the day of the summer solstice. Because of the proximity of the earth, the comet left its own orbit and for a while followed the orbit of the earth. The great ball of the comet retreated, then again approached the earth, shrouded in a dark column of gases which looked like a pillar of smoke during the day and of fire at night, and the earth once more passed through the atmosphere of the comet, this time at its neck. This stage was accompanied by violent and incessant electrical discharges between the atmosphere of the tail and the terrestrial atmosphere. There was an interval of about six days between these two close approaches. Emerging from the gases of the comet, the earth seems to have changed the direction of its rotation, and the pillar of smoke moved to the opposite horizon.[1] The column looked like a gigantic moving serpent.

When the tidal waves rose to their highest point, and the seas were torn apart, a tremendous spark flew between the earth and the globe of the comet, which instantly pushed down the miles-high billows. Meanwhile, the tail of the comet and its head, having become entangled with each other by their close contact with the earth, exchanged violent discharges of electricity. It looked like a battle between the brilliant globe and the dark column of smoke. In the exchange of electrical potentials, the tail and the head were attracted one to the other and repelled one from the other. From the serpentlike tail extensions grew, and it lost the form of a column. It looked now like a furious animal with legs and with many heads. The discharges tore the column to pieces, a process that was accompanied by a rain of meteorites upon the earth. It appeared as though the monster were defeated by the brilliant globe and buried in the sea, or wherever the meteorites fell. The gases of the tail subsequently enveloped the earth.

The globe of the comet, which lost a large portion of its atmosphere as well as much of its electrical potential, withdrew from the earth but did not break away from its attraction.

[1] Cf. Exodus 14 : 19.

Apparently, after a six-week interval, the distance between the earth and the globe of the comet again diminished. This new approach of the globe could not be readily observed because the earth was shrouded in the clouds of dust left by the comet on its former approach as well as by dust ejected by the volcanoes. After renewed discharges, the comet and the earth parted.

This behaviour of the comet is of great importance in problems of celestial mechanics. That a comet, encountering a planet, can become entangled and drawn away from its own path, forced into a new course, and finally liberated from the influence of the planet is proved by the case of Lexell's comet, which in 1767 was captured by Jupiter and its moons. Not until 1779 did it free itself from this entanglement. A phenomenon that has not been observed in modern times is an electrical discharge between a planet and a comet and also between the head of a comet and its trailing part.

The events in the sky were viewed by the peoples of the world as a fight between an evil monster in the form of a serpent and the light-god who engaged the monster in battle and thus saved the world. The tail of the comet, leaping back and forth under the discharges of the flaming globe, was regarded as a separate body, inimical to the globe of the comet.

A full survey of the religious and folklore motifs which mirror this event would require more space than is at my disposal here; it is difficult to find a people or tribe on the earth that does not have the same motif at the very focus of its religious beliefs.[1]

Since the descriptions of the battle between Marduk and Tiamat, the dragon, or Isis and Seth, or Vishnu and the serpent, or Krishna and serpent, or Ormuzd and Ahriman follow an almost identical pattern and have many details in common with the battle of Zeus and Typhon, I shall give here Apollodorus' description of this battle.[2]

Typhon "out-topped all the mountains, and his head often brushed the stars. One of his hands reached out to the west and the other to the east, and from them projected a hundred dragons' heads. From the thighs downward he had huge coils of vipers which . . . emitted a long hissing. . . . His body was all winged . . . and fire flashed from his eyes. Such and

[1] I intend to handle a portion of this material in an essay on The Dragon.
[2] Apollodorus, *The Library*, Epitome II (transl. Frazer).

so great was Typhon when, hurling kindled rocks, he made for the very heaven with hissing and shouts, spouting a great jet of fire from his mouth." To the sky of Egypt Zeus pursued Typhon "rushing at heaven." "Zeus pelted Typhon at a distance with thunderbolts, and at close quarters struck him down with an adamantine sickle, and as he fled pursued him closely as far as Mount Casius, which overhangs Syria. There, seeing the monster sore wounded, he grappled with him. But Typhon twined about him and gripped him in his coils. . . ." "Having recovered his strength Zeus suddenly from heaven riding in a chariot of winged horses, pelted Typhon with thunderbolts. . . . So being again pursued he [Typhon] came to Thrace and in fighting at Mount Haemus he heaved whole mountains . . . a stream of blood gushed out on the mountain, and they say that from that circumstance the mountain was called Haemus [bloody]. And when he started to flee through the Sicilian sea, Zeus cast Mount Etna in Sicily upon him. That is a huge mountain, from which down to this day they say that blasts of fire issue from the thunderbolts that were thrown."

The struggle left deep marks on the entire ancient world. Some districts were specially associated with the events of this cosmic fight. The Egyptian shore of the Red Sea was called Typhonia.[1] Strabo narrates also that the Arimi (Aramaeans or Syrians) were terrified witnesses of the battle of Zeus with Typhon. And Typhon, "who, they add, was a dragon, when struck by the bolts of lightning, fled in search of a descent underground," [2] and not only did he cut furrows into the earth and form the beds of the rivers, but descending underground, he made fountains break forth.

Similar descriptions come from various places of the ancient world, in which the nations relate the experience of their ancestors who witnessed the great catastrophe of the middle of the second millennium.

At that time the Israelites had not yet arrived at a clear monotheistic concept and, like other peoples, they saw in the great struggle a conflict between good and evil. The author of the Book of Exodus, suppressing the conception of the ancient Israelites, presented the portent of fire and smoke moving in a column as an angel or messenger of the Lord. However, many

[1] Strabo, *The Geography* (transl. H. L. Jones, 1924), vii, 3, 8.
[2] *Ibid.*

passages in other books of the Scriptures preserved the picture as it impressed itself upon eyewitnesses. *Rahab* is the Hebrew name for the contester with the Most High. "O Lord God of hosts, who is a strong Lord like unto thee? . . . Thou hast broken Rahab in pieces. . . . The heavens are thine, the earth also is thine: as for the world and the fulness thereof, thou hast founded them. The north and the south thou hast created them."[1] Deutero-Isaiah prayed: "Awake, awake, put on strength, O arm of the Lord; awake as in the ancient days, in the generations of old. Art thou not it that hath cut Rahab, and wounded the dragon? Art thou not it which hath dried the sea, the waters of the great deep; that hath made the depths of the sea a way for the ransomed to pass over?"[2] From these passages it is clear that the battle of the Lord with Rahab was not a primeval battle before Creation, as some scholars think.[3]

Isaiah prophesied for the future: "In that day the Lord with his sore and great and strong sword shall punish leviathan the piercing serpent, even leviathan that crooked serpent; and he shall slay the dragon that is in the sea."[4]

The "crooked serpent" is shown in many ancient pictures from China to India, to Persia, to Assyria, to Egypt, to Mexico. With the rise of the monotheistic concept, the Israelites regarded this crooked serpent, the contester with the Most High, as the Lord's own creation.

"He stretcheth out the north over the empty place, and hangeth the earth upon nothing. . . . The pillars of heaven tremble. . . . He divideth the sea with his power . . . his hand hath formed the crooked serpent."[5] The Psalmist also says:[6] "God is my King of old. . . . Thou didst divide the sea by thy strength. . . . Thou brakest the heads of leviathan in pieces. . . . Thou didst cleave the fountain and the flood: Thou driedst up mighty rivers."

The sea was cleft, the earth was cut with furrows, great rivers disappeared, others appeared. The earth rumbled for many years, and the peoples thought that the fiery dragon that had been struck down had descended underground and was groaning there.

[1] Psalms 89 : 10-12.
[2] Isaiah 51 : 9-10.
[3] See S. Reinach, *Cults, Myths and Religion* (1912), pp. 42 ff; H. Gunkel, *Schöpfung und Chaos in Urzeit und Endzeit* (1895); J. Pedersen, *Israel, Its Life and Culture* (1926), pp. 472 ff.
[4] Isaiah 27 : 1.
[5] Job 26 : 7-13.
[6] Psalms 74 : 12-15.

# THE COMET OF TYPHON

> Of all the mysterious phenomena which accompanied the Exodus, this mysterious Pillar seems the first to demand explanation.
>
> —W. PHYTHIAN-ADAMS
> *The Call of Israel*

One of the places of the heavenly combat between elementary forces of nature—as narrated by Apollodorus and Strabo—was on the way from Egypt to Syria.[1] According to Herodotus, the final act of the fight between Zeus and Typhon took place at Lake Serbon on the coastal route from Egypt to Palestine.[2] On the way from Egypt to Palestine the Israelites, after a night of terror and strong east wind, witnessed the upheaval of the day of the Passage. These parallel circumstances lead to a conclusion that will sound somewhat strange. Typhon (Typheus) lies on the bottom of the sea where the spell-bound Israelites saw the upheaval of nature: darkness, hurricane, mountains of water, fire and smoke, recorded in the Greek legend as the circumstances in which the battle of Zeus with the dragon Typhon was fought. In the same pit of the sea lie the pharaoh and his hosts.[3]

Up to now I have identified Rahab-Typhon as a comet. But if Typhon lies on the bottom of the sea, is he not the pharaoh? This would mean that in the legend of Typhon two elements were welded together: the pharaoh, who perished in the catastrophe, and the outrageous rebel against Zeus, the lord of the sky.[4]

In Pliny's *Natural History*, the ninety-first section of the

[1] Mount Casius, mentioned by Apollodorus, is the name of Mount Lebanon as well as of Mount Sinai. Cf. Pomponius Mela *De situ orbis*.

[2] Herodotus iii, 5. Also Apollonius Rhodius in the *Argonautica*, Bk. ii, says that Typhon "smitten by the bolt of Zeus . . . lies whelmed beneath the waters of the Serbonian lake."

[3] In *Ages in Chaos*, evidence will be presented to identify the pharaoh of the Exodus as Taui Thom, the last king of the Middle Kingdom. He is Tau Timaeus (Tutimaeus) of Manetho, in whose days "a blast of God's displeasure" fell upon Egypt and terminated the period at present known as the Middle Kingdom. The name of his queen is given in the naos of el-Arish as Tephnut.

Ra-uah-ab is a name met among the Egyptian kings of that period (W.M.F. Petrie, *A History of Egypt*, I, 227); it could have served as origin for the Hebrew word for dragon, *Rahab*. See note 4.

[4] Actually, "dragon" became the appellation of Egyptian pharaohs in the prophetic literature. Cf. Ezekiel 32 : 2.

second book reads: [1] "A terrible comet was seen by the people of Ethiopia and Egypt, to which Typhon, the king of that period, gave his name; it had a fiery appearance and was twisted like a coil, and it was very grim to behold: it was not really a star so much as what might be called a ball of fire."

The visit of a disastrous comet, so many times referred to in this book, is told in plain words, not in disguise. However, I must find support for my assumption that the comet of the days of King Typhon was the comet of the days of the Exodus.

I investigated the writings of the old chronographers, and in *Cometographia* of Hevelius (1668) I found references to the works of Calvisius, Helvicus, Herlicius, and Rockenbach, all of whom used manuscripts for the most part and not printed sources, as they lived only a little over one century after the invention of movable characters and the printing press.

Hevelius wrote (in Latin): "In the year of the world 2453 (1495 B.C.), according to certain authorities, a comet was seen in Syria, Babylonia, India, in the sign Jo, in the form of a disc, at the very time when the Israelites were on their march from Egypt to the Promised Land. So Rockenbach. The Exodus of the Israelites is placed by Calvisius in the year of the world 2453, or 1495 B.C." [2]

I was fortunate enough to locate one copy of Rockenbach's *De cometis tractatus novus methodicus* in the United States.[3] This book was published in Wittenberg in 1602. Its author was professor of Greek, mathematics and law, and dean of philosophy at Frankfort. He wrote his book using old sources which he did not name: "*ex probatissimis & antiquissimis veterum scriptoribus*" (from the most trustworthy and the most ancient of the early writers). As a result of his diligent gathering of ancient material, he made the following entry:

"In the year of the world two thousand four hundred and fifty-three—as many trustworthy authors, on the basis of many conjectures, have determined—a comet appeared which Pliny also mentioned in his second book. It was fiery, of irregular circular form, with a wrapped head; it was in the shape of a globe and was of terrible aspect. It is said that King Typhon ruled at that time in Egypt. . . . Certain [authorities] assert that the comet was seen in Syria, Babylonia, India, in the sign

[1] Pliny, *Natural History*, ii, 91 (transl. Rackham, 1938).
[2] J. Hevelius, *Cometographia* (1668), pp. 794 f.
[3] In the library of the American Antiquarian Society, Worcester, Mass.

of Capricorn, in the form of a disc, at the time when the children of Israel advanced from Egypt toward the Promised Land, led on their way by the pillar of cloud during the day and by the pillar of fire at night." [1]

Rockenbach did not draw any conclusion on the relation of the comet of the days of Exodus to the natural phenomena of that time; his intent was only to fix the date of the comet of Typhon.

Among the early authors, Lydus, Servius (who quotes Avienus), Hephaestion, and Junctinus, in addition to Pliny, mention the Typhon comet.[2] It is depicted as an immense globe (*globus immodicus*) of fire, also as a sickle, which is a description of a globe illuminated by the sun, and close enough to be observed thus. Its movement was slow, its path was close to the sun. Its colour was bloody: "It was not of fiery, but of bloody redness." It caused destruction "in rising and setting." Servius writes that this comet caused many plagues, evils, and hunger.

To discover what were the manuscript sources of Abraham Rockenbach that led him to the same conclusion at which we have arrived, namely, that the Typhon comet appeared in the time of the Exodus, is a task not yet accomplished. Servius says that more information about the calamities caused by this comet is to be found in the writings of the Roman astrologer Campester and in the works of the Egyptian astrologer Petosiris.[3] It is possible that copies of works of some authors containing citations from the writings of these ancient astrologers, preserved in the libraries of Europe, were Rockenbach's manuscript sources.

[1] "Anno mundi, bis millesimo, quadrigentesimo quinquagesimo tertio, Cometa (ut multi probati autores de tempore hoc statuunt, ex conjecturis multis) cuius Plinius quoque lib. 2 cap. 25 mentionem facit, igneus, formam imperfecti circuli, in se convoluti caputq; globi repraesentans, aspectu terribilis apparuit, Typhonq; a rege, tunc temporis ex Aegypto imperium tenente, dictus est, qui rex, ut homines fide digni asserunt, auxilio gigantum, reges Aegyptoru devicit. Visus quoq; est, ut aliqui volut, in Siria, Babylonia, India, in signo capricorni, sub forma rotae, eo tempore, quando filii Israel ex Aegypto in terram promissam, duce ac viae monstratore, per diem columna nubis, noctu vero columna ignis, ut cap. 7.8.9.10 legitur profecti sunt."

[2] Johannis Laurentii Lydi *Liber de ostentis et calendaria Graeca omnia* (ed. by C. Wachsmuth, 1897),.p. 171. In this work Wachsmuth also printed excerpts from Hephaestion, Avienus apud Servium, and Junctinus.

[3] The time when Campester flourished is not known, but it is assumed to have been in the third or fourth century of the present era. See Pauly-Wissowa, *Real-Encyclopädie der classischen Altertumswissenschaft*, *s.v.* The time of Petosiris is tentatively dated in the second pre-Christian era (Pauly-Wissowa, *s.v.*). But he is mentioned in *The Danaïdes* of Aristophanes (−448 to −388). See also E. Riess, *Nechepsonis et Petosiridis fragmenta magica* (1890).

Campester, as quoted by Lydus, was certain that should the comet Typhon again meet the earth, a four-day encounter would suffice to destroy the world.[1] This implies also that the first encounter with the comet Typhon brought the earth to the brink of destruction.

But even without this sombre prognostication of Campester, we have a very imposing and quite inexhaustible array of references to Typhon and its destructive action against the world: almost every Greek author referred to it. The real nature of Typhon being that of a comet, as explained by Pliny and others, all references to the disasters caused by Typhon must be understood as descriptions of natural catastrophes in which the earth and the comet were involved. As is known, Pallas of the Greeks was another name for Typhon; also Seth of the Egyptians was an equivalent of Typhon.[2] Thus the number of references to the comet Typhon can be enlarged by references to Pallas and Seth.

It was not only Abraham Rockenbach who synchronized the appearance of the comet Typhon with the Exodus of the Israelites from Egypt. Looking for authors who might have done likewise, I found that Samuel Bochart, a scholarly writer of the seventeenth century, in his book *Hierozoicon*,[3] has a passage in which he maintains that the plagues of the days of the Exodus resemble the calamities that Typhon brought in his train, and that therefore "the flight of Typhon is the Exodus of Moses from Egypt."[4] In this he actually follows the passage transmitted by Plutarch.[5] But since Typhon, according to Pliny and others, was a comet, Samuel Bochart was close to the conclusions at which we arrive, travelling along another route.

1 Campester in Lydus *Liber de ostentis*; cf. *Handwörterbuch des deutschen Aberglaubens* (1932-1933), Vol. V, *s.v.* "Komet."

2 "The Egyptians regularly call Typhon 'Seth'; it means 'overmastering' and 'overpowering,' and in very many instances 'turning back,' and again 'over-passing'." Plutarch, *Isis and Osiris* (transl. F. C. Babbitt, 1936), 41 and 49.

3 Bochart, *Hierozoicon*, I, 343.

4 "Fuga Typhonis est Mosis ex Egypto excessus." *Ibid.*, p. 341.

5 "Those who relate that Typhon's flight from the battle [with Horus] was made on the back of an ass and lasted seven days, and that after he had made his escape, he became the father of sons, Hierosolymus [Jerusalem] and Judaeus, are manifestly, as the very names show, attempting to drag the Jewish traditions into the legend." Plutarch, *Isis and Osiris*, 32.

A phenomenon of great significance took place. The head of the comet did not crash into the earth, but exchanged major electrical discharges with it. A tremendous spark sprang forth at the moment of the nearest approach of the comet, when the waters were heaped at their highest above the surface of the earth and before they fell down, followed by a rain of debris torn from the very body and tail of the comet.

"And the Angel of God, which went before the camp of Israel, removed and went behind them; and the pillar of the cloud went from before their face, and stood behind them . . . and it was a cloud and darkness but it gave light by night." An exceedingly strong wind and lightnings rent the cloud. In the morning the waters rose as a wall and moved away. "And the children of Israel went into the midst of the sea upon the dry ground: and the waters were a wall unto them on their right hand, and on their left. And the Egyptians pursued . . . And it came to pass, that in the morning watch the Lord looked unto the host of the Egyptians through the pillar of fire and of the cloud, and troubled the host of the Egyptians, and took off their chariot wheels . . . and the waters returned, and covered the chariots, and the horsemen, and all the host of Pharaoh that came into the sea after them; there remained not so much as one of them." [1]

The immense tides were caused by the presence of a celestial body close by; they fell when a discharge occurred between the earth and the other body.

Artapanus, the author of the no longer extant *De Judaeis*, apparently knew that the words, "The Lord looked unto the host of the Egyptians through the pillar of fire and of the cloud," refer to a great lightning. Eusebius quotes Artapanus: "But when the Egyptians . . . were pursuing them, a fire, it is said, shone out upon them from the front, and the sea overflowed the path again, and the Egyptians were all destroyed by the fire and the flood." [2]

[1] Exodus 14 : 19 ff.
[2] Eusebius, *Preparation for the Gospel* (transl. Gifford), Bk. ix, Chap. xxvii. Calmet, *Commentaire, l'Exode*, p. 154, correctly understood the passage in Artapanus because he paraphrases it as follows: "Artapanus dans Eusèbe dit que les Egyptiens fureut frappés de la foudre, et abbatus par le feu du ciel dans le même temps que l'ean de la mer vint tomber sur eux."

The great discharges of interplanetary force are commemorated in the traditions, legends, and mythology of all the peoples of the world. The god—Zeus of the Greeks, Odin of the Icelanders, Ukko of the Finns, Perun of the Russian pagans, Wotan (Woden) of the Germans, Mazda of the Persians, Marduk of the Babylonians, Shiva of the Hindus—is pictured with lightning in his hand and described as the god who threw his thunderbolt at the world overwhelmed with water and fire.

Similarly, many psalms of the Scriptures commemorate the great discharges. "Then the earth shook and trembled; the foundations also of the hills moved and were shaken. . . . He bowed the heavens also, and came down . . . he did fly upon the wings of the wind. . . . At the brightness that was before him his thick clouds passed, hail stones and coals of fire. The Lord also thundered in the heavens, and the Highest gave his voice; hail stones and coals of fire . . . and he shot out lightnings . . . Then the channels of waters were seen, and the foundations of the world were discovered."[1] "The voice of the Lord is powerful. . . . The voice of the Lord breaketh the cedars. . . . The voice of the Lord divideth the flames of fire. The voice of the Lord shaketh the wilderness; the Lord shaketh the wilderness of Kadesh."[2] "The kingdoms were moved; he uttered his voice, the earth melted."[3] "The waters saw thee; they were afraid: the depths also were troubled . . . the skies sent out a sound: thine arrows also went abroad. The voice of thy thunder was in the heaven; the lightnings lightened the universe: the earth trembled and shook."[4] "Clouds and darkness are round about him . . . a fire goeth before him and burneth up his enemies round about. . . . His lightnings enlightened the world: the earth saw, and trembled."[5]

Nothing is easier than to add to the number of such quotations from other parts of the Scriptures—Job, the Song of Deborah, the Prophets.

With the fall of the double wall of water, the Egyptian host was swept away. The force of the impact threw Pharaoh's army into the air. "Come and see the works of God: he is terrible in his doing toward the children of men. He turned

[1] Psalms 18 : 7-15.
[2] Psalms 29 : 4-8.
[3] Psalms 46 : 6.
[4] Psalms 77 : 16-19. *Tevel* is the universe, but the King James Version translates "world"; world is *olam*.
[5] Psalms 97 : 2-4.

the sea into dry land: they went through the flood on foot. . . . Thou hast caused men to ride over our heads; we went through fire and through water." [1]

This tossing of the Egyptian host into the air by an avalanche of water is referred to also in the Egyptian source I quoted before: on the shrine found in el-Arish the story is told of a hurricane and of a prolonged darkness when nobody could leave the palace, and of the pursuit by the pharaoh Taoui-Thom of the fleeing slaves whom he followed to Pi-khiroti, which is the biblical Pi-ha-khiroth. "His Majesty leapt into the place of the whirlpool." Then it is said that he was "lifted by a great force." [2]

Although the larger part of the Israelite fugitives were already out of the reach of the falling tidal waves, a great number of them perished in this disaster, as in the previous ones of fire and hurricane of cinders. That Israelites perished at the Sea of Passage is implied in Psalm 68 where mention is made of "my people" that remained in "the depths of the sea."

These tidal waves also overwhelmed entire tribes who inhabited Tehama, the thousand-mile-long coastal region of the Red Sea.

"God sent against the Djorhomites swift clouds, ants, and other signs of his rage, and many of them perished. . . . In the land of Djohainah an impetuous torrent carried off all of them in a night. The scene of this catastrophe is known by the name of Idam (fury)." The author of this passage, Masudi, an Arab author of the tenth century, quotes an earlier author, Omeyah, son of Abu-Salt: "In days of yore the Djorhomites settled in Tehama, and a violent flood carried all of them away." [3]

Likewise the tradition related in Kitab Alaghani [4] is familiar with the plague of insects (ants of the smallest variety) that forced the tribe to migrate from Hedjaz to their native land, where they were destroyed by "Toufan"—a deluge. In my reconstruction of ancient history, I endeavour to establish the synchronism of these events and the Exodus.

[1] Psalms 66 : 5-12. On cosmic discharges see *infra* the Sections, "Ignis e Coelo" and "Synodos."
[2] Griffith, *The Antiquities of Tel-el-Yahudiyeh*; Goyon, "Les travaux de Chou et les tribulations de Geb," *Kemi* (1936).
[3] El-Maçoudi, *Les Prairies d'or* (transl. C. Barbier and P. de Courteille, 1861), III, Chap. 39. An English translation is by A. Sprenger (1841): El-Mas'udi, *Meadows of Gold and Mines of Gems.*
[4] F. Fresnel, "Sur l'Histoire des Arabes avant l'Islamisme (Kitab alaghaniyy)," *Journal asiatique* (1838).

## THE COLLAPSED SKY

The rain of meteorites and fire from the sky, the clouds of dust of exogenous origin that drifted low, and the displacement of the world quarters created the impression that the sky had collapsed.

The ancient peoples of Mexico referred to a world age that came to its end when the sky collapsed and darkness enshrouded the world.[1]

Strabo relates, in the name of Ptolemaeus, the son of Lagus, a general of Alexander and founder of the Egyptian dynasty called by his name, that the Celti who lived on the shores of the Adriatic were asked by Alexander what it was they most feared, to which they replied that they feared no one, but only that the sky might collapse.[2]

The Chinese refer to the collapse of the sky which took place when the mountains fell.[3] Because mountains fell or were levelled at the same time when the sky was displaced, ancient peoples, not only the Chinese, thought that mountains support the sky.

"The earth trembled, and the heavens dropped . . . the mountains melted," says the Song of Deborah.[4] "The earth shook, the heavens also dropped at the presence of God: even Sinai itself was moved," says the psalmist.[5]

The tribes of Samoa in their legends refer to a catastrophe when "in days of old the heavens fell down." The heavens or the clouds were so low that the people could not stand erect without touching them.[6]

The Finns tell in their *Kalevala* that the support of the sky gave way and then a spark of fire kindled a new sun and a new moon.[7] The Lapps make offerings accompanied by the prayer that the sky should not lose its support and fall down.[8] The Eskimos of Greenland are afraid that the support of the sky may fail and the sky fall down and kill all human beings; a

[1] Seler, *Gesammelte Abhandlungen*, II, 798.
[2] Strabo, *The Geography*, vii, 3, 8.
[3] A. Forke, *The World Conception of the Chinese* (1925), p. 43.
[4] Judges 5 : 4-5.
[5] Psalms 68 : 8. On periodic collapses of the firmament see also Rashi's commentary on Genesis 11 : 1, referred to in the Section, "World Ages."
[6] Williamson, *Religious and Cosmic Beliefs of Central Polynesia*, I, 41.
[7] See Section, "The Darkness," note 8.
[8] Olrik, *Ragnarok* (German ed.), p. 446.

darkening of the sun and the moon will precede such a catastrophe.[1]

The primitives of Africa, in eastern as well as western provinces of the continent, tell about the collapse of the sky in the past. The Ovaherero tribesmen say that many years ago "the Greats of the sky" (Eyuru) let the sky fall on the earth; almost all the people were killed, only a few remained alive. The tribes of Kanga and Loanga also have a tradition of the collapse of the sky which annihilated the human race. The Wanyoro in Unyoro likewise relate that the sky fell on the earth and killed everybody: the god Kagra threw the firmament upon the earth to destroy mankind.[2]

The tradition of the Cashinaua, the aborigines of western Brazil, is narrated as follows: "The lightnings flashed and the thunders roared terribly and all were afraid. Then the heaven burst and the fragments fell down and killed everything and everybody. Heaven and earth changed places. Nothing that had life was left upon the earth."[3]

In this tradition are included the same elements: the lightnings and thunderings, "the bursting of heaven," the fall of meteorites. About the change of places between heaven and earth there is more to say, and I shall not postpone the subject for long.

[1] Olrik, *Ragnarok*, (German ed.), p. 406. The tradition was told by the Eskimos to P. Egede (1734-1740).
[2] L. Frobenius, *Die Weltanschauung der Naturvölker* (1898), pp. 355-357.
[3] Bellamy, *Moons, Myths and Man*, p. 80.

CHAPTER IV

# BOILING EARTH AND SEA

TWO CELESTIAL BODIES were driven near to each other. The interior of the terrestrial globe pushed toward the exterior. The earth, disturbed in its rotation, developed heat. The land surface became hot. Various sources of many peoples describe the melting of the earth's surface and the boiling of the sea.

The earth burst and lava flowed. The Mexican sacred book, *Popol-Vuh*, the *Manuscript Cakchiquel*, the *Manuscript Troano* all record how the mountains in every part of the Western Hemisphere simultaneously gushed lava. The volcanoes that opened along the entire chain of the Cordilleras and in other mountain ranges and on flat land vomited fire, vapour, and torrents of lava. These and other Mexican sources relate how, at the closing hours of the age that was brought to an end by the rain of fire, mountains swelled under the pressure of molten masses and new ridges rose; new volcanoes sprang out of the earth, and streams of lava flowed out of the cleft earth.[1]

Events underlying Greek and Mexican traditions are narrated in the Scriptures. "The mountains shake with the swelling . . . the earth melted."[2] "Clouds and darkness . . . fire . . . the earth saw and trembled. The hills melted like wax."[3] "He looketh on the earth, and it trembleth: he toucheth the hills, and they smoke."[4] "The earth trembled . . . the mountains melted . . . even that Sinai."[5] "He rebuketh the sea, and maketh it dry, and drieth up all the rivers. . . . The mountains quake at him, and the hills melt, and the earth is burned . . . yea, the world, and all that dwell therein."[6]

The rivers steamed, and even the bottom of the sea boiled here and there. "The sea boiled, and all the shores of the ocean boiled, all the middle of it boiled," says the *Zend-Avesta*. The star Tistrya made the sea boil.[7]

[1] See Seler, *Gesammelte Abhandlungen*, II, 798.
[2] Psalms 46 : 3-6.
[3] Psalms 97 : 2-5.
[4] Psalms 104 : 32.
[5] Song of Deborah, Judges 5 : 4-5.
[6] Nahum 1 : 4-5.
[7] *The Zend-Avesta* (Pt. II, p. 95 of J. Darmesteter's translation, 1883); Carnoy, *Iranian Mythology*, p. 268.

The traditions of the Indians retain the memory of this boiling of the water in river and sea. The tribes of British Columbia tell: "Great clouds appeared . . . such a great heat came, that finally the water boiled. People jumped into the streams and lakes to cool themselves, and died." [1] On the North Pacific coast of America the tribes insist that the ocean boiled: "It grew very hot . . . many animals jumped into the water to save themselves, but the water began to boil." [2] The Indians of the Southern Ute tribe in Colorado recorded in their legends that the rivers boiled.[3]

Jewish tradition, as preserved in the rabbinical sources, declares that the mire at the bottom of the Sea of Passage was heated. "The Lord fought against the Egyptians with the pillar of cloud and fire. The mire was heated to the boiling point by the pillar of fire." [4] The rabbinical sources say also that the pillar of fire and smoke levelled mountains.[5]

Hesiod in his *Theogony*, relating the upheaval caused by a celestial collision, says: "The huge earth groaned. . . . A great part of the huge earth was scorched by the terrible vapour and melted as tin melts when heated by man's art . . . or as iron, which is hardest of all things, is softened by glowing fire in mountain glens." [6]

According to the traditions of the New World, the profile of the land changed in a catastrophe, new valleys were formed, mountain ridges were torn apart, new gulfs were cut out, ancient heights were overturned and new ones sprang up. The few survivors of the ruined world were enveloped in darkness, "the sun in some way did not exist," and in intervals in the light of blazing fires they saw the silhouettes of new mountains.

The Mayan sacred book *Popol-Vuh* says that the god "rolled mountains" and "removed mountains," and "great and small mountains moved and shaked." Mountains swelled with lava. Coniraya-Viracocha, the god of the Incas, raised mountains from the flat land and flattened other mountains.[7]

And similarly, "When Israel went out of Egypt . . . the sea

[1] "Kaska Tales" collected by J. A. Teit, *Journal of American Folk-lore*, XXX (1917), 440.
[2] S. Thompson, *Tales of the North American Indians* (1929); H. B. Alexander, *North American Mythology* (1916), p. 255.
[3] R. H. Lowie, "Southern Ute," *Journal of American Folk-lore*, XXXVII (1924).
[4] Ginzberg, *Legends*, III, 49.
[5] *Ibid.*, II, 375; III, 316; VI, 116. Tractate Berakhot, 59a-59b.
[6] Hesiod, *Theogony* (transl. Evelyn-White), ll. 856 ff.
[7] Brasseur, *Sources de l'histoire primitive du Mexique*, pp. 30, 35, 37, 47.

saw and fled . . . the mountains skipped like rams, and the little hills like lambs. . . . Tremble, thou earth, at the presence of the Lord."[1]

"Which removeth the mountains . . . which overturneth them in his anger; which shaketh the earth out of her place . . . which commandeth the sun and it riseth not . . . which alone spreadeth out the heavens, and treadeth upon the waves of the sea."[2]

## MOUNT SINAI

Along the eastern shore of the Red Sea there stretches a mountainous crest with a number of volcanic craters, at present extinguished; some, however, were active not many centuries ago. One of these volcanoes is usually described as the Mount of the Lawgiving: In the seventies of the last century a scholar, Charles Beke, suggested that Mount Sinai was a volcano in the Arabian Desert.[3] The Book of Deuteronomy (4: 11) says "the mountain burned with fire unto the midst of heaven, with darkness, clouds, and thick darkness." Beke's idea was rejected by his contemporaries and ultimately by himself.[4] Modern scholars, however, agree with his original theory, and for this reason they look for the Mount of the Lawgiving among the volcanoes of Mount Seir and not on the traditional Sinai Peninsula where there are no volcanoes. Thus the claims of the rival peaks of the Sinai Peninsula for the honour of being the Mount of the Lawgiving[5] are silenced by new contestants.

It is true that it is stated "the mountains melted . . . even that Sinai,"[6] but this melting of summits does not necessarily mean an opening up of craters. Rocks turned into a flowing mass.

The plateau of the Sinai Peninsula is covered with formations of basalt lava;[7] wide stretches of the Arabian desert also glisten with lava.[8] Lava formations, interspersed with extinguished

1 Psalms 114 : 1-7.
2 Job 9 : 5-8.
3 Beke, *Mount Sinai, a Volcano* (1873).
4 *The Late Dr. Charles Beke's Discoveries of Sinai in Arabia and of Midian* (1878), pp. 436, 561.
5 Cf. Palmer, *Sinai: From the Fourth Egyptian Dynasty to the Present Day.*
6 Song of Deborah, Judges 5 : 5.
7 W. M. Flinders Petrie, "The Metals in Egypt," *Ancient Egypt* (1915), refers to "the enormous eruption of ferruginous basalt . . . which probably burnt up forests in its outflow."
8 N. Glueck, *The Other Side of the Jordan* (1940), p. 34.

volcanoes, stretch from the vicinity of Palmyra southward into Arabia as far as Mecca.[1] Only a few thousand years ago the deserts glowed with the beacons of many volcanoes, mountains melted, and lava flowed over the ground from numerous fissures.

The celestial body that the great Architect of nature sent close to the earth, made contact with it in electrical discharges, retreated, and approached again. If we are to believe the Scriptural data, there elapsed seven weeks, or by another computation, about two months [2] from the day of the Exodus to the day of the revelation at Mount Sinai.

"There were thunders and lightnings, and a thick cloud upon the mount, and the voice of the trumpet exceeding loud; so that all the people that was in the camp trembled. . . . And mount Sinai was altogether on a smoke . . . and the smoke thereof ascended as the smoke of a furnace, and the whole mount quaked greatly. And when the voice of the trumpet sounded long, and waxed louder and louder, Moses spake, and God answered him by a voice." [3]

The Talmud and Midrashim describe the Mountain of the Lawgiving as quaking so greatly that it appeared as if it were lifted up and shaken above the heads of the people; and the people felt as if they were no longer standing securely on the ground, but were held up by some invisible force.[4] The presence of a heavenly body overhead caused this phenomenon and this feeling.

"Then the earth shook and trembled; the foundations also of the hills moved and were shaken, because he was wroth. . . . He bowed the heavens also, and came down: and darkness was under his feet. . . . At the brightness that was before him his thick clouds passed, hail stones and coals of fire. The Lord also thundered in the heavens . . . hail stones and coals of fire. . . . He shot out lightnings. . . . Then the channels of waters were seen, and the foundations of the world were discovered." [5]

Earth and heaven participated in the cosmic convulsion. In the Fourth Book of Ezra the occurrences witnessed at Mount

[1] C. P. Grant, *The Syrian Desert* (1937), p. 9.
[2] Exodus 19 : 1.
[3] Exodus 19 : 16-19.
[4] Cf. Ginzberg, *Legends*, II, 92, 95.
[5] Psalms 18 : 7-15. An identical text is found in 2 Samuel 22.

Sinai are described in these words: "Thou didst bow down the heavens, didst make the earth quake, and convulsed the world. Thou didst cause the deeps to tremble and didst alarm the spheres." [1]

The approach of a star toward the earth in the days of the revelation at Sinai is implied by the text of the Tractate Shabbat: Although the ancestors of the later proselytes were not present at the Mountain of the Lawgiving, their star was there close by.[2]

An author of the first century of the present era, whose work on biblical antiquities has been ascribed to Philo, the Alexandrian philosopher, thus describes the commotion on the earth below and in the sky above: "The mountain [Sinai] burned with fire and the earth shook and the hills were removed and the mountains overthrown; the depths boiled, and all the inhabitable places were shaken . . . and flames of fire shone forth and thunderings and lightnings were multiplied, and winds and tempests made a roaring: the stars were gathered together [collided]." [3] Referring to the verse, "He bowed the heavens also, and came down" (Psalms 18), Pseudo-Philo describes the events of Mount Sinai and says that the Lord "impeded the course of the stars." [4] "The earth was stirred from her foundation, and the mountains and the rocks trembled in their fastenings, and the clouds lifted up their waves against the flame of the fire that it should not consume the world . . . and all the waves of the sea came together." [5]

The Hindus depict the cosmic catastrophe at the end of a world age: "The whole world breaks into flames. So also a hundred thousand times ten million worlds. All the peaks of Mount Sineru, even those which are hundreds of leagues in height, crumble and disappear in the sky. The flames of fire rise up and envelop the heaven."[6] The sixth sun or sun age ended. Similarly, in the Jewish tradition, with the revelation at Sinai the sixth world age was terminated and the seventh began.[7]

[1] IV Ezra (transl. Box), in *The Apocrypha and Pseudepigrapha of the Old Testament*, ed. R. H. Charles.
[2] The Babylonian Talmud, Tractate Shabbat 146a. According to Midrash Shir (15a-15b) the pharaoh warned the Israelites not to leave Egypt, because they would meet the bloody star Ra (in Hebrew "Evil").
[3] *The Biblical Antiquities of Philo* (transl. M. R. James, 1917), Chap. XI.
[4] *Ibid.*, Chap. XXIII.
[5] *Ibid.*, Chap. XXXII.
[6] Warren, *Buddhism*, p. 323.
[7] Midrash Rabba, Bereshit.

Earthquakes are often accompanied by a roaring noise that comes from the bowels of the earth. This phenomenon was known to early geographers. Pliny [1] wrote that earthquakes are "preceded or accompanied by a terrible sound." Vaults supporting the ground give way and it seems as though the earth heaves deep sighs. The sound was attributed to the gods and called theophany.

The eruptions of volcanoes are also accompanied by loud noises. The sound produced by Krakatoa in the East Indies, during the eruption of 1883, was so loud that it was heard as far as Japan, 3,000 miles away, the farthest distance travelled by sound recorded in modern annals. [2]

In the days of the Exodus, when the world was shaken and rocked, and all volcanoes vomited lava and all continents quaked, the earth groaned almost unceasingly. At an initial stage of the catastrophe, according to Hebrew tradition, Moses heard in the silence of the desert the sound which he interpreted to mean, "I am that I am." [3] "I am Yahweh," heard the people on the frightful night at the Mountain of the Lawgiving.[4] "The whole mount quaked greatly" and "the voice of the trumpet sounded long." [5] "And all the people saw the roars, and the torches, and the noise of the trumpet, and the mountain smoking: and when the people saw it, they trembled, and stood afar off."[6]

It was a perfect setting for hearing words in the voice of nature in an uproar. An inspired leader interpreted the voice he heard, ten long, trumpetlike blasts. The earth groaned: for weeks now all its strata had been disarranged, its orbit distorted, its world quarters displaced, its oceans thrown upon its continents, its seas turned into deserts, its mountains upheaved, its islands submerged, its rivers running upstream—a world flowing with lava, shattered by meteorites, with yawning chasms, burning naphtha, vomiting volcanoes, shaking ground,

[1] Pliny, *Natural History*, ii, 82.
[2] G. J. Symons (ed.), *The Eruption of Krakatoa: Report of the Krakatoa Committee of the Royal Society* (of London) (1888).
[3] Exodus 3 : 14.
[4] Exodus 20 : 1.
[5] Exodus 19 : 18-19.
[6] Exodus 20 : 18; "the thunderings and the lightnings" of the King James Version is not an exact translation of *Kolot* and *Lapidim*.

a world enshrouded in an atmosphere filled with smoke and vapour.

Twisting of strata and building of mountains, earthquakes and rumbling of volcanoes joined in an infernal din. It was a voice not only in the desert of Sinai; the entire world must have heard it. "The sky and the earth resounded . . . mountains and hills were moved," said the Midrash. "Loud did the firmament roar, and earth with echo resounded," says the epic of Gilgamesh.[1] In Hesiod "the huge earth groaned" when Zeus lashed Typhon with his bolts—"the earth resounded terribly, and the wide heaven above."[2]

The approach of two charged globes toward each other could also produce trumpetlike sounds, varying as the distance between them increased or lessened.[3] It appears that this phenomenon is described by Pseudo-Philo as "testimony of the trumpets between the stars and their Lord."[4] Here we can trace the origin of the Pythagorean notion of the "music of the spheres" and the idea that stars make music. In Babylonia the spheres of the planets were called "voices" and they were supposed to produce music.[5] According to Midrashic literature, the trumpet sounding at Mount Sinai had seven different pitches (or notes), and the rabbinical literature speaks of "the heavenly music" heard at the revelation. "At the first sound the sky and the earth moved, the seas and the rivers turned to flight, mountains and hills were loosened in their foundations."[6]

Homer depicts a similar occurrence in these words: "The wide earth rang, and round about great heaven pealed as with a trumpet."[7] "The world all burns at the blast of the horn," is said in the Völuspa.[8]

According to the Hebrew tradition, all the nations heard the roaring of the lawgiving. It appears that at Mount Sinai the sound that "sounded long" rose ten times; in this roaring the Hebrews heard the Decalogue.

"Thou shalt not kill" (*Lo tirzah*); "Thou shalt not commit adultery" (*Lo tin'af*); "Thou shalt not steal" (*Lo tignov*). . . .

[1] *Epic of Gilgamish* (transl. Thompson).
[2] *Theogony*, ll. 820 ff., 852 ff.
[3] This phenomenon of sound between two charged bodies changing with distance is utilized for musical effect by Theremin.
[4] *The Biblical Antiquities of Philo*, Chap. XXXII.
[5] E. F. Weidner, *Handbuch der Babylonischen Astronomie* (1915), I, 75.
[6] Sefer Pirkei Rabbi Elieser.
[7] *The Iliad*, xxi, 385 ff. (transl. A. T. Murray, 1924).
[8] Cf. W. Bousset, *The Antichrist Legend* (transl. A. H. Keane, 1896), p. 113.

"These words [of the Decalogue] . . . were not heard by Israel alone, but by the inhabitants of all the earth. The Divine voice divided itself into the seventy tongues of men, so that all might understand it. . . . The souls of the heathens almost fled from them when they heard it."[1]

The din caused by the groaning earth repeated itself again and again, but not so loud, as subterranean strata readjusted themselves after being dislocated; earthquakes incessantly shook the ground for years. The Papyrus Ipuwer calls these years "years of noise." "Years of noise. There is no end to noise," and again, "Oh, that the earth would cease from noise, and tumult (uproar) be no more."[2]

The sound probably had the same pitch all over the world as it came from the deep interior of the earth, all of whose strata were dislocated when it was thrown from its orbit and forced from its axis.

The great king-lawgiver of China, in whose time a dreadful cataclysm took place and the order of nature was disturbed, bore the name Yahou.[3] In the Preface to the *Shu King*, attributed to Confucius, it is written: "Examining into antiquity, we find that the Emperor Yaou was called Fang-heun."[4] Yahou was a surname given to him in the time following the flood, apparently inspired by the sound of the earth's groaning.

The same sound was heard in those years in the Western Hemisphere or wherever the ancestors of the Indians then lived. They relate that once when the heavens were very close to the earth, all mankind lifted the sky little by little at the repeated shouting "Yahu," which rang all over the world.[5]

In Indonesia an oath is accompanied by the invocation of the heavenly bodies. An arrow is shot toward the sky, "while

[1] Ginzberg, *Legends*, III, 97; the Babylonian Talmud, Tractate Shabbat 88b.

[2] Papyrus Ipuwer 4 : 2, 4-5.

[3] For the Chinese pronunciation of this name see R. van Bergen, *Story of China* (1902), p. 112 : "At the time of the flood, the Emperor of China was named Yau (Yah-oo)."

[4] *Shoo-king, the Canon of Yaou* (transl. James Legge), Vol. III, Pt. 1 of *The Chinese Classics* (Hongkong, 1865). In this edition Legge used this spelling of the name of the book and of the name of the king; his later spelling is different.

In Volume LX of the *Universal Lexicon* (Leipzig and Halle, 1732-1754), *s.v.* "Yao," it is said that some call Yao by the name Tam and also Tao. This is curious because in my reconstruction of ancient history I come to the conclusion that the name of the pharaoh of the Exodus was Taui Thom (Greek "Tau Timaeus") of the Thirteenth Dynasty, the last of the Middle Kingdom. He was a contemporary of this Chinese king.

[5] F. Shelton, "Mythology of Puget Sound: Origin of the Exclamation 'Yahu,'" *Journal of American Folk-lore*, XXXVII (1924).

all present raise a cry of 'ju ju huwe.' "[1] The same sound is heard in the very name Jo, Jove (Jupiter). The name Yahweh is preserved in shorter forms, as well, Yahou and Yo,[2] as the name of the Deity in the Bible.[3] Diodorus wrote of Moses that he had received his laws from the God invoked by the name Iao.[4]

In Mexico, Yao or Yaotl is the god of war; the similarity of sound has already been noted.[5]

*Nihongi*, chronicles of Japan from the earliest times, begins with a reference to the time when "of old, heaven and earth were not yet separated, and the In and Yo not yet divided." Yo is the earth. The time when the sky touched the earth is the time when the heavy dust and vapour-charged clouds of the comet enveloped the globe and lay very close to the ground.

## EMPEROR YAHOU

The history of China is commonly supposed to extend back to grey antiquity. But in reality the sources of the ancient period of the Chinese past are very scanty, for they were destroyed by the Emperor Tsin-chi-hoang (246—209 before the present era). He ordered all books on history and astronomy, as well as works of classic literature, to be burned. Search for these books was made throughout the empire for this purpose. The story persists that a few remnants of the old literature were again put into writing from the memory of an old man; some were said to have been found hidden in the sepulchre of Confucius, and are ascribed to his pen.

Of these few remains of the old lore, the most cherished are those which tell of the Emperor Yahou and his times. His personality and his period are considered as "the most auspicious in the Chinese annals." [6] The history of China preceding his reign is ascribed to the mythical period of the Chinese past.

[1] J. G. Frazer, *The Worship of Nature* (1926), p. 665. F. Boas, *Kwakiutl Culture as Reflected in Mythology* (1935), p. 130, tells of *Yuwe gendayusens na lax* ("the wind edge of our world"), from where also come "death-bringing arrows that set mountains on fire."

[2] Psalms 68 : 4.

[3] Cf. R. A. Bowman, "Yahweh the Speaker," *Journal of Near Eastern Studies*, III (1944). H. Torczyner, *Die Bundeslade und die Anfänge der Religion Israels* (1930), p. iii, sees a connection between the name *jhwh* and the Arab word *wahwa*, to roar.

[4] Diodorus of Sicily, *Library of History*, I, 94.

[5] Brasseur, *Quatre lettres sur le Mexique*, p. 374.

[6] H. Murray, J. Crawfurd, and others, *An Historical and Descriptive Account of China.*

In the days of Yahou the event occurred which separates the almost obliterated and very dim past of China from the period that is considered historical: China was overwhelmed by an immense catastrophe.

"At that time the miracle is said to have happened that the sun during a span of ten days did not set, the forests were ignited, and a multitude of abominable vermin was brought forth." [1] "In the lifetime of Yao [Yahou] the sun did not set for ten full days and the entire land was flooded." [2]

An immense wave "that reached the sky" fell down on the land of China. "The water was well up on the high mountains, and the foothills could not be seen at all." [3] (This recalls Psalms 104: "The waters stood above the mountains . . . they go up by the mountains" and Psalms 107: "The waves mount up to the heaven.")

"Destructive in their overflow are the waters of the inundation," said the emperor. "In their vast extent they embrace the hills and overtop the great heights, threatening the heavens with their floods." The emperor ordered that all efforts be made to open outlets for the waters that were caught in the valleys between the mountains. For many years the population laboured, trying to free the plains and valleys of the waters of the flood by digging channels and draining the fields. For a considerable number of years all efforts were in vain. The minister who was in charge of this urgent and immense work, Khwan, was sentenced to death because of his failure—"For nine years he laboured, but the work was unaccomplished" [4]—and only his son Yu succeeded in draining the land. This achievement was so highly rated that Yu became emperor of China after King Shun, first successor to Yahou. This Yu was the founder of the new and notable dynasty called by his name.

The chronicles of modern China preserve records of one million lives lost in a single overflow of the Yellow River.[5] Another natural catastrophe—the earthquake—also caused

[1] "Yao," *Universal Lexicon*, Vol. LX (1749).
[2] J. Hübner, *Kurze Fragen aus der politischen Historie* (1729).
[3] *The Shu King, the Cànon of Yao* (transl. Legge, 1879). See also C. L. J. de Guignes, *Le Chou-king* (1770), Pt. 1, Chap. 1, and J. Moryniac, *Histoire générale de la Chine* (1877), I, 53.
[4] *The Shu King.*
[5] Andree, *Die Flutsagen*, p. 36; C. Deckert, "Der Hoangho und seine Stromlaufänderung," *Globus, Zeitschrift für Länder- und Völkerkunde*, LIII (1888), 129, concerning the flood of 1887.

great devastation in China at various times: it is estimated that in the year 1556 the quaking earth took 830,000 lives and 3,000,000 in 1662.[1] Was not the catastrophe of the time of Yahou one of the major inundations of rivers, as modern scholars suppose it to have been? But the fact that this catastrophe has been vivid in traditions for thousands of years, while neither the overflow of the Yellow River, when a million people perished, nor the great earthquakes, play a conspicuous part in the recollections of the nation, is an argument against the established interpretation.

Rivers do not overflow in the form of a sky-high wave. The overflowing rivers of China subside in a few weeks, and the water does not remain in the plains until the following spring, but flows away, and the ground dries in a few more weeks. The flood of Yahou required draining for many years, and during all this period water covered the lower part of the country.

Yahou's reign is remembered for the following undertaking: This emperor sent scholars to different parts of China, and even to Indo-China, to find out the location of north, west, east, and south by observing the direction of the sun's rising and setting and the motion of the stars. He also charged his astronomers to find out the duration of seasons, and to draw up a new calendar. *The Shu King* is called the oldest book of Chinese chronicles, rewritten from memory or from some hidden manuscript after the burning of books by Tsin-chi-hoang. In its oldest section, the canon of Yaou [Yahou], it is written:

"Thereupon Yaou [Yahou] commanded He and Ho, in reverent accordance with the wide heavens, to calculate and delineate the movements and appearances of the sun, the moon, the stars, and the zodiacal spaces; and to deliver respectfully the seasons to the people."[2]

The necessity, soon after the flood, of finding anew the four directions and learning anew the movements of the sun and the moon, of delineating the zodiacal signs, of compiling the calendar, of informing the population of China of the sequence of the seasons, creates the impression that during the catastrophe the orbit of the earth and the year, the inclination of the axis and the seasons, the orbit of the moon and the month, changed.

[1] Daly, *Our Mobile Earth*, p. 3.
[2] *The Shoo-king* (Hong Kong edition).

We are not told what caused the cataclysm, but it is written in ancient annals that during the reign of Yahou "a brilliant star issued from the constellation Yin." [1]

According to the old Tibetan traditions, the highlands of Tibet, too, were flooded in a great cataclysm.[2] The traditions of the Tibetans speak also of terrifying comets that caused great upheavals.[3]

Calculations were undertaken to establish the dates of the Emperor Yahou. On the basis of a remark that the constellation Niao, thought to be the constellation Hydra, culminated at sunset on the day of the vernal equinox in the time of Yahou, it was reckoned that the flood occurred in the twenty-third century before the present era, but this date has been questioned by many. Sometimes it has also been supposed that the "Flood of Yahou" was the Chinese story of the universal flood, but this point of view has been abandoned. The story of the deluge of Noah has its parallel in a Chinese tradition about a universal flood in prehistoric times, in the days of Fo-hi, who alone of all the country was saved. The flood of Yahou is sometimes regarded as simultaneous with the flood of Ogyges.

The flood of Ogyges did not occur in the third millennium, but in the middle of the second millennium before this era. In the section entitled "The Floods of Deucalion and Ogyges," the synchronism of these devastations with the catastrophes of the days of Moses and Joshua will be demonstrated and supported by ancient and chronological sources.

When we summarize what has been told about the time of Yahou, we have the following data: the sun did not set for a number of days, the forests were set on fire, vermin filled the country, a high wave "reaching the sky" poured over the face of the land and swept water over the mountain peaks and filled the valleys for many years; in the days of Yahou the four quarters of the heaven were established anew, and observations of the duration of the year and month and of the order of the seasons were made. The history of China in the period before this catastrophe is quite obliterated.

All these data are in accord with the traditions of the Jewish people about the events connected with the Exodus: the sun

[1] *The Annals of the Bamboo Books*, Vol. 3, Pt. 1 of *The Chinese Classics* (transl. Legge), p. 112.

[2] Andree, *Die Flutsagen*, quoting S. Turner, *An Account of an Embassy to the Court of the Teshoo Lama in Tibet* (1800).

[3] Eckstein, *Sur les Sources de la cosmogonie du Sanchoniathon* (1860), p. 227.

disappeared for a number of days; the land was filled with vermin; gigantic sky-high tidal waves divided the sea; the world burned. As we shall see, the Hebrew sources, too, reveal that a new calendar was established reckoning from the days of the catastrophe and that the seasons and the four quarters of the heaven were no longer the same.

CHAPTER V

# EAST AND WEST

OUR PLANET ROTATES from west to east. Has it always done so? In this rotation from west to east, the sun is seen to rise in the east and set in the west. Was the east the primeval and only place of the sunrise?

There is testimony from all parts of the world that the side which is now turned toward the evening once faced the morning.

In the second book of his history, Herodotus relates his conversations with Egyptian priests on his visit to Egypt some time during the second half of the fifth century before the present era. Concluding the history of their people, the priests told him that the period following their first king covered three hundred and forty-one generations, and Herodotus calculated that, three generations being equal to a century, the whole period was over eleven thousand years. The priests asserted that within historical ages and since Egypt became a kingdom, "four times in this period (so they told me) the sun rose contrary to his wont; twice he rose where he now sets, and twice he set where he now rises."[1]

This passage has been the subject of exhaustive commentaries, the authors of which tried to invent every possible explanation of the phenomenon, but failed to consider the meaning which was plainly stated by the priests of Egypt, and their efforts through the centuries have remained fruitless.

The famous chronologist of the sixteenth century, Joseph Scaliger, weighed the question whether the Sothis period, or time reckoning by years of 365 days which, when compared with the Julian calendar, accumulated an error of a full year in 1,461 years, was hinted at by this passage in Herodotus, and remarked: "*Sed hoc non fuerit occasum et orientem mutare*" (No reversal of sunrise and sunset takes place in a Sothis period).[2]

Did the words of the priests to Herodotus refer to the slow change in the direction of the terrestrial axis during a period of approximately 25,800 years, which is brought about by its

[1] Herodotus, Bk. ii, 142 (transl. A. D. Godley, 1921).

[2] Joseph Scaliger, *Opus de emendatione temporum* (1629), III, 198.

spinning or by the slow movement of the equinoctial points of the terrestrial orbit (precession of the equinoxes)? So thought Alexander von Humboldt of "the famous passage of the second book of Herodotus which so strained the sagacity of the commentators." [1] But this is also a violation of the meaning of the words of the priests, for during the period of spinning, orient and occident do not exchange places.

One may doubt the trustworthiness of the priests' statements, or of Egyptian tradition in general, or attack Herodotus for ignorance of the natural sciences,[2] but there is no way to reconcile the passage with present-day natural science. It remains "a very remarkable passage of Herodotus that has become the despair of commentators." [3]

Pomponius Mela, a Latin author of the first century, wrote: "The Eygptians pride themselves on being the most ancient people in the world. In their authentic annals . . . one may read that since they have been in existence, the course of the stars has changed direction four times, and that the sun has set twice in that part of the sky where it rises today." [4]

It should not be deduced that Mela's only source for this statement was Herodotus. Mela refers explicitly to Egyptian written sources. He mentions the reversal in the movement of the stars as well as of the sun; if he had copied Herodotus, he would probably not have mentioned the reversal in the movement of the stars (*sidera*). At a time when the movement of the sun, planets, and stars was not yet regarded as the result of the movement of the earth, the change in the direction of the sun was not necessarily connected in Mela's mind with a similar change in the movement of all heavenly bodies.[5]

If, in Mela's time, there were Egyptian historical records which referred to the rising of the sun in the west, we ought to investigate the old Egyptian literary sources extant today.

The Magical Papyrus Harris speaks of a cosmic upheaval of fire and water when "the south becomes north, and the Earth turns over." [6]

[1] Humboldt, *Vues des Cordillères*, II, 131 (*Researches*, II, 30).
[2] A. Wiedemann, *Herodots zweites Buch* (1890), p. 506: "Tiefe Stufe seiner naturwissenschaftlichen Kenntnisse."
[3] P. M. de la Faye in *Histoire de l'art égyptien* by Prisse d'Avennes (1879), p. 41.
[4] Pomponius Mela *De situ orbis*. i. 9. 8.
[5] Mela, differing from Herodotus, computed the length of Egyptian history as equal to 330 generations until Amasis (died -525) and figured it at more than thirteen thousand years.
[6] H. O. Lange, "Der Magische Papyrus Harris," *K. Danske Videnskabernes Selskab* (1927), p. 58.

In the Papyrus Ipuwer it is similarly stated that "the land turns round [over] as does a potter's wheel" and the "Earth turned upside down." [1] This papyrus bewails the terrible devastation wrought by the upheaval of nature. In the Ermitage Papyrus (Leningrad, 1116b recto) also, reference is made to a catastrophe that turned the "land upside down; happens that which never (yet) had happened." [2] It is assumed that at that time—in the second millennium—people were not aware of the daily rotation of the earth, and believed that the firmament with its luminaries turned around the earth; therefore, the expression, "the earth turned over," does not refer to the daily rotation of the globe.

Nor do these descriptions in the papyri of Leiden and Leningrad leave room for a figurative explanation of the sentence, especially if we consider the text of the Papyrus Harris—the turning over of the earth is accompanied by the interchange of the south and north poles.

Harakhte is the Egyptian name for the western sun. As there is but one sun in the sky, it is supposed that Harakhte means the sun at its setting. But why should the sun at its setting be regarded as a deity different from the morning sun? The identity of the rising and the setting sun is seen by everyone. The inscriptions do not leave any room for misunderstanding: "Harakhte, he riseth in the west." [3]

The texts found in the pyramids say that the luminary "ceased to live in the occident, and shines, a new one, in the orient." [4]

After the reversal of direction, whenever it may have occurred, the words "west" and "sunrise" were no longer synonyms, and it was necessary to clarify references by adding: "the west which is at the sun-setting." It was not mere tautology, as the translator of this text thought.[5]

Inasmuch as the hieroglyphs were deciphered in the nineteenth century, it would be only reasonable to expect that since then the commentaries on Herodotus and Mela would have been written after consulting the Egyptian texts.

[1] Papyrus Ipuwer 2 : 8. Cf. Lange's (German) translation of the papyrus (*Sitzungsberichte d. Preuss. Akad. der Wissenschaften* [1903], pp. 601-610).
[2] Gardiner, *Journal of Egyptian Archaeology*, I (1914); *Cambridge Ancient History*, I, 346.
[3] Breasted, *Ancient Records of Egypt*, III, Sec. 18.
[4] L. Speelers, *Les Textes des Pyramides* (1923), I.
[5] K. Piehl, *Inscriptions hiéroglyphiques*, seconde série (1892), p. 65: "l'ouest qui est à l'occident."

In the tomb of Senmut, the architect of Queen Hatshepsut, a panel on the ceiling shows the celestial sphere with the signs of the zodiac and other constellations in "a reversed orientation" of the southern sky.[1]

The end of the Middle Kingdom antedated the time of Queen Hatshepsut by several centuries. The astronomical ceiling presenting a reversed orientation must have been a venerated chart, made obsolete a number of centuries earlier.

"A characteristic feature of the Senmut ceiling is the astronomically objectionable orientation of the southern panel." The centre of this panel is occupied by the Orion-Sirius group, in which Orion appears west of Sirius instead of east. "The orientation of the southern panel is such that the person in the tomb looking at it has to lift his head and face north, not south." "With the reversed orientation of the south panel, Orion, the most conspicuous constellation of the southern sky, appeared to be moving eastward, i.e., in the wrong direction."[2]

The real meaning of "the irrational orientation of the southern panel" and the "reversed position of Orion" appears to be this: the southern panel shows the sky of Egypt as it was before the celestial sphere interchanged north and south, east and west. The northern panel shows the sky of Egypt as it was on some night of the year in the time of Senmut.

Was there no autochthonous tradition in Greece about the reversals of the revolution of the sun and stars?

Plato wrote in his dialogue, "The Statesman" (*Politicus*): "I mean the change in the rising and setting of the sun and the other heavenly bodies, how in those times they used to set in the quarter where they now rise, and used to rise where they now set . . . the god at the time of the quarrel, you recall, changed all that to the present system as a testimony in favour of Atreus." Then he proceeded: "At certain periods the universe has its present circular motion, and at other periods it revolves in the reverse direction. . . . Of all the changes which take place in the heavens this reversal is the greatest and most complete." [3]

Plato continued his dialogue, using the above passage as the introduction to a fantastic philosophical essay on the reversal

[1] A. Pogo, "The Astronomical Ceiling Decoration in the Tomb of Senmut (XVIIIth Dynasty)," *Isis* (1930), p. 306.
[2] *Ibid.*, pp. 306, 315, 316.
[3] Plato, *The Statesman* or *Politicus* (transl. H. N. Fowler, 1925), pp. 49, 53.

of time. This minimizes the value of the quoted passage despite the categorical form of his statement.

The reversal of the movement of the sun in the sky was not a peaceful event; it was an act of wrath and destruction. Plato wrote in *Politicus*: "There is at that time great destruction of animals in general, and only a small part of the human race survives."

The reversal of the movement of the sun was referred to by many Greek authors before and after Plato. According to a short fragment of an historical drama by Sophocles (*Atreus*), the sun rises in the east only since its course was reversed. "Zeus . . . changed the course of the sun, causing it to rise in the east and not in the west." [1]

Euripides wrote in *Electra*: "Then in his anger arose Zeus, turning the stars' feet back on the fire-fretted way; yea, and the sun's car splendour-burning, and the misty eyes of the morning grey. And the flash of his chariot-wheels back-flying flushed crimson the face of the fading day. . . . The sun . . . turned backward . . . with the scourge of his wrath in affliction repaying mortals." [2]

Many authors in later centuries realized that the story of Atreus described some event in nature. But it could not have been an eclipse. Strabo was mistaken when he tried to rationalize the story by saying that Atreus was an early astronomer who "discovered that the sun revolves in a direction opposite to the movement of the heavens." [3] During the night the stars move from east to west two minutes faster than the sun which moves in the same direction during the day.[4]

Even in poetical language such a phenomenon would not have been described as follows: "And the sun-car's winged speed from the ghastly strife turned back, changing his westering track through the heavens unto where blush-burning dawn rose," as Euripides wrote in another work of his.[5]

Seneca knew more than his older contemporary Strabo.

[1] *The Fragments of Sophocles*, ed. by A. C. Pearson (1917), III, 5, Fragment 738; see also *ibid.*, I, 93. Those of the Greek authors who ascribed a permanent change in the direction of the sun to the time of the Argive tyrant Atreus, confused two events and welded them into one: a lasting reversal of west and east in earlier times and a temporary retrograde movement of the sun in the days of the Argive tyrants.

[2] Euripides, *Electra* (transl. A. S. Way), ll. 727 ff.

[3] Strabo, *The Geography*, i, 2, 15.

[4] Every night stars rise four minutes earlier: the earth rotates 366¼ times in a year in relation to the stars, but 365¼ times in relation to the sun.

[5] Euripides, *Orestes* (transl. A. S. Way), ll. 1001 ff.

In his drama *Thyestes*, he gave a powerful description of what happened when the sun turned backward in the morning sky, which reveals much profound knowledge of natural phenomena. When the sun reversed its course and blotted out the day in mid-Olympus (noon), and the sinking sun beheld Aurora, the people, smitten with fear, asked: "Have we of all mankind been deemed deserving that heaven, its poles uptorn, should overwhelm us? In our time has the last day come?"[1]

The early Greek philosophers, and especially Pythagoras, would have known about the reversal of the revolution of the sky, if it actually occurred, but as Pythagoras and his school kept their knowledge secret, we must depend upon the authors who wrote about the Pythagoreans. Aristotle says that the Pythagoreans differed between the right- and the left-hand motion of the sky ("the side from which the stars rise" is heaven's right, "and where they set . . . its left" [2]), and in Plato we find: "A direction from left to right—and that will be from west to east."[3] The present sun moves in the opposite direction.

In the language of a symbolic and philosophical astronomy, probably of Pythagorean origin, Plato describes in *Timaeus* the effects of a collision of the earth "overtaken by a tempest of winds" with "alien fire from without, or a solid lump of earth," or waters of "the immense flood which foamed in and streamed out": the terrestrial globe engages in all motions, "forwards and backwards, and again to right and to left, and upwards and downwards, wandering every way in all the six directions." [4]

As the result of such a collision, described in a not easily understandable text which represents the earth as possessing a soul, there was a "violent shaking of the revolutions of the Soul," "a total blocking of the course of the same," "shaking of the course of the other," which "produced all manner of twistings, and caused in their circles fractures and disruptures of every possible kind, with the result that, as they [the earth and the "perpetually flowing stream"?] barely held together one with another, they moved indeed but more irrationally,

[1] Seneca, *Thyestes* (transl. F. J. Miller), ll. 794 ff.
[2] Aristotle, *On the Heavens*, II, ii (transl. W. K. C. Guthrie, 1939). Cf. also Plutarch, who, in his *The Opinions of the Philosophers*, wrote that according to Pythagoras, Plato, and Aristotle, "east is the right side, and west is the left side."
[3] Plato, *Laws* (transl. R. G. Bury, 1926), Bk. iv, ll. 760 D.
[4] Plato, *Timaeus* (transl. Bury, 1929), 43 B and C.

being at one time reversed, at another oblique, and again upside down."[1] In Plato's terminology, "revolution of the same" is from east to west, and "revolution of the other" is from west to east.[2] In *The Statesman*, Plato put this symbolic language into very simple terms, speaking of the reversal of the quarters in which the sun rises and sets.

I shall return later to some other Greek references to the sun setting in the east.[3]

Caius Julius Solinus, a Latin author of the third century of the present era, wrote of the people living on the southern borders of Egypt: "The inhabitants of this country say that they have it from their ancestors that the sun now sets where it formerly rose."[4]

The traditions of peoples agree in synchronizing the changes in the movement of the sun with great catastrophes which terminated world ages. The changes in the movement of the sun in each successive age make the use by many peoples of the term "sun" for "age" understandable.

"The Chinese say that it is only since a new order of things has come about that the stars move from east to west."[5] "The signs of the Chinese zodiac have the strange peculiarity of proceeding in a retrograde direction, that is, against the course of the sun."[6]

In the Syrian city Ugarit (Ras Shamra) was found a poem dedicated to the planet-goddess Anat, who "massacred the population of the Levant" and who "exchanged the two dawns and the position of the stars."[7]

The hieroglyphics of the Mexicans describe four movements of the sun, *nahui ollin tonatiuh*. "The Indian authors translate *ollin* by 'motions of the sun.' When they find the number *nahui* added, they render *nahui ollin* by the words 'sun (tonatiuh) in his four motions.' "[8] These "four motions" refer "to four prehistoric suns" or "world ages," with shifting cardinal points.[9]

[1] Cf. Bury's comments to *Timaeus*, notes, pp. 72, 80.
[2] Plato, *Timaeus*, 43 D and E.
[3] See for literature Frazer's note to Epitome II in his translation of Apollodorus; Wiedemann, *Herodots zweites Buch*, p. 506; Pearson, *The Fragments of Sophocles*, III, note to Fragment 738.
[4] Solinus, *Polyhistor*, xxxii.
[5] Bellamy, *Moons, Myths and Man*, p. 69.
[6] *Ibid.*
[7] C. Virolleaud, "La déesse Anat," *Mission de Ras Shamra*, Vol. IV (1938).
[8] Humboldt, *Researches*, I, 351. See also by the same author, *Examen critique de l'histoire de la géographie du nouveau continent* (1836-1839), II, 355.
[9] Seler, *Gesammelte Abhandlungen*, II, 799.

The sun that moves toward the east, contrary to the present sun, is called by the Indians Teotl Lixco.[1] The people of Mexico symbolized the changing direction of the sun's movement as a heavenly ball game, accompanied by upheavals and earthquakes on the earth.[2]

The reversal of east and west, if combined with the reversal of north and south, would turn the constellations of the north into constellations of the south, and show them in reversed order, as in the chart of the southern sky on the ceiling of Senmut's tomb. The stars of the north would become stars of the south; this is what seems to be described by the Mexicans as the "driving away of the four hundred southern stars." [3]

The Eskimos of Greenland told missionaries that in an ancient time the earth turned over and the people who lived then became antipodes.[4]

Hebrew sources on the present problem are numerous.[5] In Tractate Sanhedrin of the Talmud it is said: "Seven days before the deluge, the Holy One changed the primeval order and the sun rose in the west and set in the east." [6]

*Tevel* is the Hebrew name for the world in which the sun rose in the west.[7] *Arabot* is the name of the sky where the rising point was in the west.[8]

Hai Gaon, the rabbinical authority who flourished between 939 and 1038, in his *Responses* refers to the cosmic changes in which the sun rose in the west and set in the east.[9]

The Koran speaks of the Lord "of two easts and of two wests," [10] a sentence which presented much difficulty to the exegetes. Averrhoes, the Arab philosopher of the twelfth century, wrote about the eastward and westward movements of the sun.[11]

References to the reversal of the movement of the sun that

[1] Seler, perplexed by the statement of the old Mexican sources that the sun moved toward the east, writes: "The travelling toward the east and the disappearance in the east . . . must be understood literally. . . . However, one cannot imagine the sun as wandering eastward: the sun and the entire firmament of the fixed stars travel westward." "Einiges über die natürlichen Grundlagen mexicanischer Mythen" (1907) in *Gesammelte Abhandlungen*, Vol. III.

[2] *Ibid.* Also Brasseur, *Histoire des nations civilisées du Mexique*, I, 123.

[3] Seler, "Ueber die natürlichen Grundlagen," *Gesammelte Abhandlungen*, III, 320.

[4] Olrik, *Ragnarok*, p. 407.

[5] See M. Steinschneider, *Hebräische Bibliographie* (1877), Vol. XVIII.

[6] Tractate Sanhedrin 108b.

[7] Steinschneider, *Hebräische Bibliographie*, Vol. XVIII, pp. 61 ff.

[8] Ginzberg, *Legends*, I, 69.

[9] Taam Zekenim 55b, 58b.

[10] Koran, Sura LV.

[11] Steinschneider, *Hebräische Bibliographie*, Vol. XVIII.

have been gathered here do not refer to one and the same time: the Deluge, the end of the Middle Kingdom, the days of the Argive tyrants, were separated by many centuries. The tradition heard by Herodotus in Egypt speaks of four reversals. Later in this book and again in the book that will deal with earlier catastrophes, I shall return to this subject. At this point, I leave historical and literary evidence on the reversal of earth's cardinal points for the testimony of the natural sciences on the reversal of the magnetic poles of the earth.

## THE REVERSED POLARITY OF THE EARTH

A thunderbolt, on striking a magnet, reverses the poles of the magnet. The terrestrial globe is a huge magnet. A short circuit between it and another celestial body could result in the north and south magnetic poles of the earth exchanging places.

It is possible to detect in the geological records of the earth the orientation of the terrestrial magnetic field in past ages.

"When lava cools and freezes following a volcanic outburst, it takes up a permanent magnetization dependent upon the orientation of the Earth's magnetic field at the time. This, because of small capacity for magnetization in the Earth's magnetic field after freezing, may remain practically constant. If this assumption be correct, the direction of the originally acquired permanent magnetization can be determined by tests in the laboratory, provided that every detail of the orientation of the mass tested is carefully noted and marked when it is removed."[1]

We would expect to find a full reversal of magnetic direction. Although repeated heating of lava and rocks can change the picture, there must have remained rocks with inverted polarity. Another author writes:

"Examination of magnetization of some igneous rocks reveals that they are polarized oppositely from the prevailing present direction of the local magnetic field and many of the older rocks are less strongly magnetized than more recent ones. On the assumption that the magnetization of the rocks occurred

[1] J. A. Fleming, "The Earth's Magnetism and Magnetic Surveys" in *Terrestrial Magnetism and Electricity*, ed. by J. A. Fleming (1939), p. 32.

when the magma cooled and that the rocks have held their present positions since that time, this would indicate that the polarity of the Earth has been completely reversed within recent geological times."[1]

Because the physical facts seemed entirely inconsistent with every cosmological theory, the author of the above passage was cautious not to draw further conclusions from them.

The reversed polarity of lava indicates that in recent geological times the magnetic poles of the globe were reversed; when they had a very different orientation, abundant flows of lava took place.

Additional problems, and of a large scope, are: whether the position of the magnetic poles has anything to do with the direction of rotation of the globe, and whether there is an interdependence in the direction of the magnetic poles of the sun and of the planets.

## THE QUARTERS OF THE WORLD DISPLACED

The traditions gathered in the section before last refer to various epochs; actually, Herodotus and Mela say that according to Egyptian annals, the reversal of the west and east recurred: the sun rose in the west, then in the east, once more in the west, and again in the east.

Was the cosmic catastrophe that terminated a world age in the days of the fall of the Middle Kingdom and of the Exodus one of these occasions, and did the earth change the direction of its rotation at that time? If we cannot assert this much, we can at least maintain that the earth did not remain on the same orbit, nor its poles stay in their places, nor was the direction of the axis the same as before. The position of the globe and its course were not settled when the earth first came into contact with the onrushing comet; in Plato's terms, already partly quoted, the motion of the earth was changed by "blocking of the course" and went through "shaking of the revolutions" with "disruptures of every possible kind," so that the position of the earth became "at one time reversed, at another oblique, and again upside down," and it wandered "every way in all six directions."

[1] A. McNish, "On Causes of the Earth's Magnetism and Its Changes" in *Terrestrial Magnetism and Electricity*, ed. by Fleming, p. 326.

The Talmud and other ancient rabbinical sources tell of great disturbances in the solar movement at the time of the Exodus and the Passage of the Sea and the Lawgiving.[1] In old Midrashim it is repeatedly narrated that four times the sun was forced out of its course in the few weeks between the day of the Exodus and the day of the Lawgiving.[2]

The prolonged darkness (and prolonged day in the Far East) and the earthshock (i.e., the ninth and the tenth plagues) and the world conflagration were the result of one of these disturbances in the motion of the earth. A few days later, if we follow the biblical narration, immediately before the hurricane changed its direction, "the pillar of cloud went from before their faces and stood behind them"; this means that the column of fire and smoke turned about and appeared from the opposite direction. Mountainous tides uncovered the bottom of the sea; a spark sprang between two celestial bodies; and "at the turning of the morning," [3] the tides fell in a cataclysmic avalanche.

The Midrashim speak of a disturbance in the solar movement on the day of the Passage: the sun did not proceed on its course.[4] On that day, according to the Psalms (76 : 8), "the earth feared and was still." It is possible that Amos (8 : 8-9) is reviving the memory of this event when he mentions the "flood of Egypt," at the time "the earth was cast out of the sea, and dry land was swallowed by the sea," and "the sun was brought down at noon," although, as I show later on, Amos might have referred to a cosmic catastrophe of a more recent date.

Also, the day of the Lawgiving, when the worlds collided again, was, according to numerous rabbinical sources, a day of unusual length: the motion of the sun was disturbed.[5]

On this occasion, and generally in the days and months following the Passage, the gloom, the heavy and charged clouds, the lightning, and the hurricanes, aside from the devastation by earthquake and flood, made observation very difficult, if not impossible. "They walk on in darkness: all the foundations of

[1] See, e.g., the Babylonian Talmud, Tractate Taanit 20: Tractate Avoda Zara 25a.

[2] Pirkei Rabbi Elieser 41; Ginzberg, *Legends*, VI, 45-46.

[3] Rashi, the commentator, is surprised by the combination of the words, "at the turning of the morning" (*lifnot haboker*). The word *lifnot* (from *pana*), when used with reference to time, means "to turn away" or "to go down." The word is applied here, not to "day," which goes down, but to the morning, which rises, changes to day, but does not go down.

[4] Midrash Psikta Raboti; Likutim Mimidrash Ele Hadvarim (ed. Buber, 1885).

[5] Ginzberg, *Legends*, III, 109.

the earth are out of course" (Psalms 82 : 5) is a metaphor used by the Psalmist.

The Papyrus Ipuwer, which says that "the earth turned over like a potter's wheel" and "the earth is upside down," was written by an eyewitness of the plagues and the Exodus.[1] The change is described also in the words of another papyrus (Harris) which I have quoted once before: "The south becomes north, and the earth turns over."

Whether there was a complete reversal of the cardinal points as a result of the cosmic catastrophe of the days of the Exodus, or only a substantial shift, is a problem not solved here. The answer was not apparent even to contemporaries, at least for a number of decades. In the gloom that endured for a generation, observations were impossible and very difficult when the light began to break through.

The *Kalevala* relates that "dreaded shades" enveloped the earth, and "the sun occasionally steps from his accustomed path." [2] Then Ukko-Jupiter struck fire from the sun to light a new sun and a new moon, and a new world age began.

In *Völuspa* (Poetic Edda) of the Icelanders we read:

> No knowledge she [the sun] had where her home should be,
> The moon knew not what was his,
> The stars knew not where their stations were.

Then the gods set order among the heavenly bodies.

The Aztecs related: "There had been no sun in existence for many years. . . . [The chiefs] began to peer through the gloom in all directions for the expected light, and to make bets as to what part of heaven he [the sun] should first appear in. Some said 'Here', and some said 'There'; but when the sun rose, they were all proved wrong, for not one of them had fixed upon the east." [3]

Similarly, the Mayan legend tells that "it was not known from where the new sun would appear." "They looked in all directions, but they were unable to say where the sun would rise. Some thought it would take place in the north and their glances were turned in that direction. Others thought it would

[1] See the Section, "The Red World," note 2.

[2] J. M. Crawford in the Preface to his translation of *Kalevala*.

[3] Quoted by I. Donnelly, *Ragnarok*, p. 215, from Andres de Olmos. Donnelly thought that this tradition signified that "in the long-continued darkness they had lost all knowledge of the cardinal points"; he did not consider that it might refer to the displacement of the cardinal points.

be in the south. Actually, their guesses included all directions because the dawn shone all around. Some, however, fixed their attention on the orient, and maintained that the sun would come from there. It was their opinion that proved to be correct."[1]

According to the *Compendium* of Wong-shi-Shing (1526–1590), it was in the "age after the chaos, when heaven and earth had just separated, that is, when the great mass of cloud just lifted from the earth," that the heaven showed its face.[2]

In the Midrashim it is said that during the wandering in the desert the Israelites did not see the face of the sun because of the clouds. They were also unable to orient themselves on their march.[3]

The expression repeatedly used in the Books of Numbers and Joshua, "the east, to the sunrising,"[4] is not tautology, but a definition, which, by the way, testifies to the ancient origin of the literary materials that served as sources for these books; it is an expression that has its counterpart in the Egyptian "the west which is at the sun-setting."

The cosmological allegory of the Greeks has Zeus, rushing on his way to engage Typhon in combat, steal Europa (Erev, the evening land) and carry her to the west. Arabia (also Erev) kept its name, "the evening land,"[5] though it lies to the east of the centres of civilization—Egypt, Palestine, Greece. Eusebius, one of the Fathers of the Church, assigned the Zeus-Europa episode to the time of Moses and the Deucalion Flood, and Augustine wrote that Europa was carried by the king of Crete to his island in the west, "betwixt the departure of Israel out of Egypt and the death of Joshua."[6]

The Greeks, like other peoples, spoke of the reversal of the quarters of the earth and not merely in allegories but in literal terms.

The reversal of the earth's rotation, referred to in the written and oral sources of many peoples, suggests the relation of one of these events to the cataclysm of the day of the Exodus. Like the quoted passage from *Visuddhi-Magga*, the Buddhist text, and

[1] Sahagun, *Historia general de las cosas de Nueva España*, Bk. VII, Chap. 2.
[2] Quoted by Donnelly, *Ragnarok*, p. 210.
[3] Exodus 14 : 3 ; Numbers 10 : 31.
[4] Numbers 2 : 3; 34 : 15; Joshua 19 : 12.
[5] Cf. Isaiah 21 : 13. In Jeremiah 25 : 20 the name "Arab" is used to denote "a mingled people."
[6] Eusebius *Werke*, Vol. V, *Die Chronik* (transl. J. Karst, 1911), "Chronikon Kanon"; St. Augustine, *The City of God*, Bk. XVIII, Chap. 12.

the cited tradition of the Cashinaua tribe in western Brazil, the versions of the tribes and peoples of all five continents include the same elements, familiar to us from the Book of Exodus: lightning and "the bursting of heaven," which caused the earth to be turned "upside down," or "heaven and earth to change places." On the Andaman Islands the natives are afraid that a natural catastrophe will cause the world to turn over.[1] In Greenland also the Eskimos fear that the earth will turn over.[2]

Curiously enough, the cause of such perturbation is revealed in beliefs like that of the people of Flanders in Belgium. Thus we read: "In Menin (Flanders) the peasants say, on seeing a comet: 'The sky is going to fall; the earth is turning over!' "[3]

## CHANGES IN THE TIMES AND THE SEASONS

Many agents collaborated to change the climate. Insolation was impaired by heavy clouds of dust, and the radiation of heat from the earth was equally hindered.[4] Heat was generated by the earth's contacts with another celestial body; the earth was removed to an orbit farther from the sun; the polar regions were displaced; oceans and seas evaporated and the vapours precipitated as snow on new polar regions and in the higher latitudes in a long Fimbul-winter and formed new ice sheets; the axis on which the earth rotated pointed in a different direction, and the order of the seasons was disturbed.

Spring follows winter and fall follows summer because the earth rotates on an axis inclined toward the plane of its revolution around the sun. Should this axis become perpendicular to that plane, there would be no seasons on the earth. Should it change its direction, the seasons would change their intensity and their order.

The Egyptian papyrus known as Papyrus Anastasi IV contains a complaint about gloom and the absence of solar

[1] Hastings, "Eschatology," *Encyclopedia of Religion and Ethics*.
[2] Olrik, *Ragnarok*, p. 406.
[3] *Revue des traditions populaires*, XVII (1902-1903), 571.
[4] Cf. the works of Arrhenius on the influence of carbon dioxide in the atmosphere on the temperature, and J. Tyndall (*Heat a Mode of Motion*, 6th ed., pp. 417-418) on the influence on the climate of a theoretical layer of olefiant gas surrounding our earth at a short distance above its surface.

light; it says also: "The winter is come as (instead of) summer, the months are reversed and the hours are disordered." [1]

"The breath of heaven is out of harmony. . . . The four seasons do not observe their proper times," we read in the *Texts of Taoism.*[2]

In the historical memoirs of Se-Ma Ts'ien, as in the annals of the Shu King which we have already quoted, it is said that Emperor Yahou sent astronomers to the Valley of Obscurity and to the Sombre Residence to observe the new movements of the sun and of the moon and the syzygies or the orbital points of the conjunctions, also "to investigate and to inform the people of the order of the seasons." [3] It is also said that Yahou introduced a calendar reform: he brought the seasons into accord with the observations; he did the same with the months; and he "corrected the days." [4]

Plutarch gives the following description of a derangement of seasons: "The thickened air concealed the heaven from view, and the stars were confused with a disorderly huddle of fire and moisture and violent fluxions of winds. The sun was not fixed to an unwandering and certain course, so as to distinguish orient and occident, nor did he bring back the seasons in order." [5]

In another work of his, Plutarch ascribes these changes to Typhon, "the destructive, diseased and disorderly," who caused "abnormal seasons and temperatures." [6]

It is characteristic that in the written traditions of the peoples of antiquity the disorder of the seasons is directly connected with the derangement in the motion of the heavenly bodies.

The oral traditions of primitive peoples in various parts of the world also retain memories of this change in the movement of the heavenly bodies, the seasons, the flow of time, during a period when darkness enveloped the world. As an example I quote the tradition of the Oraibi in Arizona. They say that the firmament hung low and the world was dark, and no sun,

[1] A. Erman, *Egyptian Literature* (1927), p. 309. Cf. also J. Vandier, *La Famine dans l'Egypte ancienne* (1936), p. 118: "Les mois sont à l'envers, et les heures se confondent" (Papyrus Anastasi IV, 10), and R. Weill, *Bases, méthodes, et résultats de la chronologie égyptienne* (1926), p. 55.

[2] *Texts of Taoism* (transl. Legge), I, 301.

[3] *Les Mémoires historiques de Se-ma Ts'ien* (transl. E. Chavannes, 1895), p. 47.

[4] *Ibid.*, p. 62.

[5] Plutarch, "Of Eating of Flesh," *Morals* (transl. "by several hands," revised by W. Goodwin, ed. 1898).

[6] Plutarch, *Isis and Osiris*, 49.

no moon, nor stars were seen. "The people murmured because of the darkness and the cold." Then the planet god Machito "appointed times, and seasons, and ways for the heavenly bodies." [1]

Among the Incas the "guiding power in regulating the seasons and the courses of the heavenly bodies" was Uira-cocha. "The sun, the moon, the day, the night, spring, winter, are not ordained in vain by thee, O Uira-cocha." [2]

The American sources, which speak of a world coloured red, of a rain of fire, of world conflagration, of new rising mountains, of frightening portents in the sky, of a twenty-five year gloom, imply also that "the order of the seasons was altered at that epoch." "The astronomers and geologists whose concern is all this . . . should judge of the causes which could effect the derangement of the day and could cover the earth with tenebrosity," wrote a clergyman who spent many years in Mexico and in the libraries of the Old World which store ancient manuscripts of the Mayas and works of early Indian and Spanish authors about them.[3] It did not occur to him that the biblical narrative of the time of the Exodus contains the same elements.

With the end of the Middle Kingdom in Egypt, when the Israelites left that country, the old order of seasons came to an end and a new world age began. The Fourth Book of Ezra, which borrows from some earlier sources, refers to the "end of the seasons" in these words: "I sent him [Moses] and led my people out of Egypt, and brought them to Mount Sinai, and held him by me for many days. I told him many wondrous things, showed him the secrets of the times, declared to him the end of the seasons."

Because of various simultaneous changes in the movement of the earth and the moon, and because observation of the sky was hindered when it was hidden in smoke and clouds, the calendar could not be correctly computed; the changed lengths of the year, the month, and the day required prolonged, unobstructed observation. The words of the Midrashim, that Moses was unable to understand the new calendar, refer to

[1] Donnelly, *Ragnarok*, p. 212.
[2] C. Markham, *The Incas of Peru*, pp. 97-98.
[3] Brasseur, *Sources de l'histoire primitive du Mexique*, pp. 28-29. In his later work *Quatre lettres sur Mexique* (1868), Brasseur came to the conclusion that a stupendous catastrophe occurred in America and that migrating tribes carried the echo of this catastrophe to many peoples of the world.
[4] IV Ezra 14 : 4.

this situation; "the secrets of the calendar" (*sod ha-avour*), or more precisely, "the secret of the transition" from one time-reckoning to another, was revealed to Moses, but he had difficulty in comprehending it. Moreover, it is said in rabbinical sources that in the time of Moses the course of the heavenly bodies became confounded.[1]

The month of the Exodus, which occurred in the spring, became the first month of the year: "This month shall be unto you the beginning of months: it shall be the first month of the year to you."[2] Thus, the strange situation was created in the Jewish calendar that the New Year is observed in the seventh month of the year: the beginning of the calendar year was moved to a point about half a year away from the New Year in the autumn.

With the fall of the Middle Kingdom and the Exodus, one of the great world ages came to its end. The four quarters of the world were displaced, and neither the orbit nor the poles nor, probably, the direction of rotation remained the same. The calendar had to be adjusted anew. The astronomical values of the year and the day could not be the same before and after an upheaval in which, as the quoted Papyrus Anastasi IV says, the months were reversed and "the hours disordered."

The length of the year during the Middle Kingdom is not known from any contemporaneous document. Because in the Pyramid texts dating from the Old Kingdom there is mention of "five days," it was erroneously concluded that in that period a year of 365 days was already known.[3] But no inscription of the Old or Middle Kingdom has been found in which mention is made of a year of 365 days or even 360 days. Neither is any reference to a year of 365 days or to "five days" found in the very numerous inscriptions of the New Kingdom prior to the dynasties of the seventh century.[4] Thus the inference that "the five days" of the Pyramid Texts of the Old Kingdom signify the five days over 360 is not well founded.

There exists a direct statement found as a gloss on a manuscript of *Timaeus* that a calendar of a solar year of three hundred and sixty days was introduced by the Hyksos after the fall of

[1] Pirkei Rabbi Elieser 8; Leket Midrashim 2a; Ginzberg, *Legends*, VI, 24.
[2] Exodus 12 : 2.
[3] Breasted, *A History of Egypt*, p. 14.
[4] The table of the dynasties in Egypt and their chronological order are the subject of the forthcoming *Ages in Chaos*.

the Middle Kingdom;[1] the calendar year of the Middle Kingdom apparently had fewer days.

The fact I hope to be able to establish is that from the fifteenth century to the eighth century before the present era the astronomical year was equal to 360 days; neither before the fifteenth century, nor after the eighth century was the year of this length. In a later chapter of this work extensive material will be presented to demonstrate this point.

The number of days in a year during the Middle Kingdom was less than 360; the earth then revolved on an orbit somewhat closer than the present orbit of Venus. An investigation into the length of the astronomical year during the periods of the Old and Middle Kingdoms is reserved for that part of this work which will deal with the cosmic catastrophes that occurred before the beginning of the Middle Kingdom of Egypt.

Here I give space to an old Midrashic source which, taking issue with a contradiction in the scriptural texts referring to the length of time the Israelites sojourned in Egypt, maintains that "God hastened the course of the planets during Israel's stay in Egypt," so that the sun completed 400 revolutions during the space of 210 regular years.[2] These figures must not be taken as correct, since the intention was to reconcile two biblical texts, but reference to the different motion of the planets in the period of the Israelites' stay in Egypt during the Middle Kingdom is worth mentioning.

In Midrash Rabba,[3] it is said on the authority of Rabbi Simon that a new world order came into being with the end of the sixth world age at the revelation on Mount Sinai. "There was a weakening (*metash*) of the creation. Hitherto world time was counted, but henceforth we count it by a different reckoning." Midrash Rabba refers also to "the greater length of time taken by some planets."[4]

[1] See Bissing, *Geschichte Aegyptens* (1904), pp. 31, 33; Weill, *Chronologie égyptienne*, p. 32. But cf. also "The Book of Sothis" of Pseudo-Manetho in *Manetho* (transl. Waddell), Loeb Classical Library; there the introduction of the reform of adding five days to a year of 360 days is ascribed to the Hyksos King Aseth, who also introduced the worship of the bull calf Apis.

[2] An unknown Midrash quoted in Shita Mekubetzet, Nedarim 31b; see Ginzberg, *Legends*, V, 420.

[3] Midrash Rabbah, Bereshit (ed. Freeman and Simon), ix, 14.

[4] *Ibid.*, p. 73, footnote of the editors.

CHAPTER VI

# THE SHADOW OF DEATH

AN ENTIRE YEAR after the eruption of Krakatoa in the East Indies in 1883, sunset and sunrise in both hemispheres were very colourful. Lava dust suspended in the air and carried around the globe accounted for this phenomenon.[1]

In 1783, after the eruption of Skaptar-Jökull in Iceland, the world was darkened for months; records of this phenomenon are found in many contemporary authors. One German contemporary compared the gloomy world of the year 1783 with the Egyptian plague of darkness.[2]

The world was gloomy in the year of Caesar's death, —44. "After the murder of Caesar the dictator and during the Antonine war," there was "almost a whole year's continuous gloom," wrote Pliny.[3] Virgil described this year in these words: "The sun . . . veiled his shining face in dusky gloom, and godless age feared everlasting night. . . . Germany heard the clash of arms through all the sky; the Alps rocked with unwonted terrors . . . and spectres, pale in wondrous wise, were seen at evening twilight." [4]

On September 23, —44, a short while after the death of Caesar, on the very day when Octavian performed the rites in honour of the deceased, a comet became visible at daytime; it was very bright and moved from north to west. It was seen for only a few days and vanished while still in the north.[5]

It appears that the gloom which enveloped the world the year after Caesar's death was caused by the dust of the comet dispersed in the atmosphere. The "clash of arms" heard "through all the sky" was probably the sound that accompanied the entrance of the gases and dust into the earth's atmosphere.

[1] *The Eruption of Krakatoa: Report*, ed. by G. J. Symons, pp. 40 f.
[2] *Ibid.*, p. 393; W. J. Phythian-Adams, *The Call of Israel* (1934), p. 165.
[3] *Natural History*, Bk. ii, 30.
[4] Virgil, *Georgics* (transl. H. R. Fairclough, 1920), i, 466.
[5] Dio Cassius, *Roman History*, xlv. 7; Pliny ii. 71. 93; Suetonius *Caesar* 88; Plutarch *Caesar* 69. 3. It is remarkable that a new world age was proclaimed by an Etruscan diviner named Voclanius as having begun with the approach of the comet of –44. Cf. "Komet," by Stegemann in *Handwörterbuch des deutschen Aberglaubens* (1927).

If the eruption of a single volcano can darken the atmosphere over the entire globe, a simultaneous and prolonged eruption of thousands of volcanoes would blacken the sky. And if the dust of the comet of —44 had a darkening effect, contact of the earth with a great cinder-trailing comet of the fifteenth century before this era could likewise cause the blackening of the sky. As this comet activated all the volcanoes and created new ones, the cumulative action of the eruptions and of the comet's dust must have saturated the atmosphere with floating particles.

Volcanoes vomit water vapour as well as cinders. The heating effect of the contact of the globe with the comet must have caused a great evaporation from the surface of the seas and rivers. Two kinds of clouds—water vapour and dust—were formed. The clouds obscured the sky, and drifting very low, hung as a fog. The veil left by the gaseous trail of the hostile star and the smoke of the volcanoes caused darkness, not complete, but profound. This condition prevailed for decades, and only very gradually did the dust subside and the water vapours condense.

"A vast night reigned over all the American land, of which tradition speaks unanimously: in a sense the sun no longer existed for this ruined world which was lighted up at intervals only by frightful conflagrations, revealing the full horror of their situation to the small number of human beings that had escaped from these calamities." [1]

"Following the cataclysm caused by the waters, the author of the *Codex Chimalpopoca*, in his history of the suns, shows us terrifying celestial phenomena, twice followed by darkness that covered the face of the earth, in one instance for a period of twenty-five years." "This fact is mentioned in the *Codex Chimalpopoca* and in most of the traditions of Mexico." [2]

Gómara, the Spaniard who came to the Western Hemisphere in the middle of the sixteenth century, shortly after the conquest, wrote: [3] "After the destruction of the fourth sun, the world plunged in darkness during the space of twenty-five years. Amid this profound obscurity, ten years before the appearance of the fifth sun, mankind was regenerated."

In the years of this gloom, when the world was covered with clouds and shrouded in mist, the Quiché tribe migrated to

[1] Brasseur, *Sources de l'histoire primitive du Mexique*, p. 47.
[2] *Ibid.*, pp. 28-29.
[3] Gómara, *Conquista de Mexico*, II, 261. See Humboldt, *Researches*, II, 16.

Mexico, crossing a sea enveloped in a sombre fog.[1] In the so-called *Manuscript Quiché* it is also narrated that there was "little light on the surface of the earth . . . the faces of the sun and of the moon were covered with clouds." [2]

In the Ermitage Papyrus in Leningrad previously mentioned there are lamentations about a terrible catastrophe, when heaven and earth turned upside down ("I show thee the land upside down; it happened that which never had happened"). After this catastrophe, darkness covered the earth: "The sun is veiled and shines not in the sight of men. None can live when the sun is veiled by clouds. . . . None knoweth that midday is there; the shadow is not discerned. . . . Not dazzled is the sight when he [the sun] is beheld; he is in the sky like the moon." [3]

In this description the light of the sun is compared to the light of the moon; but even in the light of the moon objects cast a shadow. If the midday could not be discerned, the disc of the sun was not clearly visible, and only its diffused light made the day different from the night. The gloom gradually lifted with the passing years as the clouds became less thick; little by little the sky and the sun appeared less and less veiled.

The years of darkness in Egypt are described in a number of other documents. The Papyrus Ipuwer, which contains the story of the plagues of Egypt, says that the land is without light [dark].[4] In the Papyrus Anastasi IV the years of misery are described, and it is said: "The sun, it hath come to pass that it riseth not." [5]

It was the time of the wandering of the Israelites in the desert.[6] Is there any indication that the desert was dark? Jeremiah says (2 : 6): "Neither said they, Where is the Lord that brought us up out of the land of Egypt, that led us through the wilderness, through a land of deserts and of pits, through a land of drought, and of the shadow of death, through a land that no man passed through, and where no man dwelt?"

The "shadow of death" is related to the time of the wanderings in the desert after the Exodus from Egypt. The sinister meaning of the words "shadow of death" corresponds with the

[1] Brasseur, *Histoire des nations civilisées du Mexique*, I, 11.
[2] *Ibid.*, p. 113.
[3] Papyrus 1116b recto, published by Gardiner, *Journal of Egyptian Archaeology*, I (1914).
[4] Papyrus Ipuwer 9 : 8.
[5] Erman, *Egyptian Literature*, p. 309.
[6] See p. 60, note 3.

description of the Ermitage Papyrus: "None can live when the sun is veiled by clouds."

At intervals the earth was lighted by conflagrations in the desert.[1]

The phenomenon of gloom enduring for years impressed itself on the memory of the Twelve Tribes and is mentioned in many passages of the Bible: "Thou hast . . . covered us with the shadow of death" (Psalms 44 : 19); "The people that walked in darkness . . . in the land of the shadow of death" (Isaiah 9 : 2). The Israelites "wandered in the wilderness in a solitary way . . . hungry and thirsty, their soul fainted in them," and the Lord "brought them out of darkness and the shadow of death" (Psalms 107); "The terrors of the shadow of death" (Job 24 : 17).

In Job 38 the Lord speaks: "Who shut up the sea with doors [barriers], when it brake forth. . . . When I made the cloud the garment thereof, and thick darkness a swaddling band for it . . . and caused the dayspring to know his place; that it might take hold of the ends of the earth, that the wicked might be shaken out of it?"[2]

The low and slowly drifting clouds enshrouded the wanderers in the desert. These clouds dimly glowed at night; their upper portion reflected the sunlight. The glow being pale during the day and red after sunset, the Israelites were able to distinguish between day and night.[3] They were protected by the clouds from the sun during the wandering in the desert, and according to the Midrashic literature, they saw sun and moon for the first time only at the end of the wandering.[4]

The clouds that covered the desert during the wandering of the Twelve Tribes were called a "celestial garment" or "clouds of glory." "He spread a cloud for a covering; and fire to give light in the night." "And the cloud of the Lord was upon them by day."[5] For days or months the cloud tarried in one place, and the Israelites "journeyed not"; but when the cloud moved, the wanderers followed it, and revered it because of its celestial origin.[6]

[1] Numbers 11 : 3; 16 : 35.
[2] Cf. also Job 28 : 3 and 36 : 32.
[3] Baraita d'Melekhet ha-Mishkan 14; Ginzberg, *Legends*, V, 439. Cf. also Job 37 : 15.
[4] Ginzberg, *Legends*, VI, 114.
[5] Psalms 105 : 39; Numbers 10 : 34.
[6] Numbers 9 : 17-22; 10 : 11 ff. The names Bezalel and Rafael mean "in the shadow of God" and "the shade of God."

In Arabic sources, too, we read that the Amalekites, who left Hedjaz because of plagues, followed the cloud in their wandering through the desert.[1]

On their way to Palestine and Egypt they met the Israelites, and in the battles between them the screen of clouds played an important part.[2]

*Nihongi*, a chronicle of Japan from the earliest period, refers to a time when there was "continuous darkness" and "no difference of day and night." It describes in the name of the Emperor Kami Yamato an ancient time when "the world was given over to widespread desolation; it was an age of darkness and disorder. In this gloom Hiko-ho-no-ninigi-no-Mikoto fostered justice, and so governed this western border." [3]

In China the annals telling of the time of the Emperor Yahou refer to the Valley of Obscurity and to the Sombre Residence as places of astronomical observations.[4]

The name "shadow of death" expresses the influence of the sunless gloom upon the life processes. The Chinese annals of Wong-shi-Shing, in the chapter dealing with the Ten Stems (the ten stages of the earth's primeval history), relate that "at Wu, the sixth stem . . . darkness destroys the growth of all things." [5]

Buddhist scholars declare that with the beginning of the sixth world age or "sun," "the whole world becomes filled with smoke and saturated with greasiness of that smoke." There is "no distinction of day and night." The gloom is caused by a "cycle-destroying great cloud" of cosmic origin and dimensions.[6]

On the Samoan islands the aborigines narrate: "Then arose smell . . . the smell became smoke, which again became clouds. . . . The sea too arose, and in a stupendous catastrophe of nature the land sank into the sea. . . . The new earth (the Samoan islands) arose out of the womb of the last earth." [7] In the darkness that enveloped the world, the islands of Tonga,

[1] Kitab-Alaghaniyy (French transl. F. Fresnel), *Journal asiatique*, 1838. Cf. El-Macoudi (Mas'udi), *Les Prairies d'or*, III, Chap. 39. In *Ages in Chaos* these events will be synchronized with the Exodus.
[2] Sources in Ginzberg, *Legends*, VI, 24, n. 141.
[3] *Nihongi* (transl. W. G. Aston), pp. 46 and 110.
[4] *Les Mémoires historiques de Se-ma Ts'ien* (transl. Chavannes, 1895), I. 47.
[5] Donnelly, *Ragnorak*, p. 211.
[6] Warren, *Buddhism in Translations*, pp. 322-327.
[7] Williamson, *Religious and Cosmic Beliefs of Central Polynesia*, I, 8.

Samoa, Rotuma, Fiji, and Uvea (Wallis Island), and Fotuna rose from the bottom of the ocean.[1]

Ancient rhymes of the inhabitants of Hawaii refer to a prolonged darkness:

> The earth is dancing . . .
> let darkness cease. . . .
> The heavens are enclosing. . . .
> Finished is the world of Hawaii.[2]

The Quiché tribe migrated to Mexico, the Israelites roamed in the desert, the Amalekites migrated toward Palestine and Egypt—an uneasy movement took place in all corners of the ruined world. The migration in Central Polynesia, shrouded in gloom, is narrated in the traditions of the aborigines of this part of the world about a chief named Te-erui who "lived long in utter darkness in Avaiki," who migrated in a canoe named "Weary of Darkness" to find a land of light, and who, after many years of wandering, saw the sky clearing little by little and arrived at a region "where they could see each other clearly." [3]

In the *Kalevala*, the Finnish epos which "dates back to an enormous antiquity," [4] the time when the sun and moon disappeared from the sky, and dreaded shadows covered it, is described in these words:

> Even birds grew sick and perished,
> men and maidens, faint and famished,
> perished in the cold and darkness,
> from the absence of the sunshine,
> from the absence of the moonlight. . . .
> But the wise men of the Northland
> could not know the dawn of morning,
> for the moon shines not in season
> nor appears the sun at midday,
> from their stations in the sky-vault.[5]

An explanation which would rationalize this picture as the description of a seasonal long night in northern regions will

[1] Williamson, *Religious and Cosmic Beliefs of Central Polynesia*, I, 37.
[2] *Ibid.*, I, 30.
[3] *Ibid.*, I, 28-29.
[4] Crawford, in the Preface to the English translation of the *Kalevala*, refers the poem to a time when Hungarians and Finns were still united as one people, "in other words, to a time at least three thousand years ago."
[5] *The Kalevala*, Rune 49.

stumble over the second part of the passage: the seasons did not return in their wonted order. The dreaded shadow covered the earth when Ukko, the highest of the Finnish deities, relinquished the support of the heavens. Hailstones of iron rained down furiously, and then the world became shrouded in a generation-long darkness.

This "twilight of the gods" of the Nordic races is but the "shadow of death" of the Scriptures. The entire generation of those who left Egypt perished in the lightless desert. Vegetation died in the catastrophe. The Iranian book of *Bundahis* says: "Blight was diffused over the vegetation, and it withered away immediately."[1] When the sky was shattered, the day became dark, and the earth teemed with noxious creatures. For a long time there was no green thing seen; seeds would not germinate in a sunless world. It took many years before the earth again brought forth vegetation; this is told in the written and oral traditions of many peoples. According to American sources, the regeneration of the world and of humankind took place under the veil of the gloomy shadows, and the time is indicated as the end of the fifteenth year of the darkness, ten years before the end of the gloom.[2] In the scriptural narration it was probably the day when Aaron's dried twig budded for the first time.[3]

The eerie world, dark and groaning, was unpleasant to all the senses save the sense of smell: the world was fragrant. When the breeze blew, the clouds conveyed a sweet odour.

In the Papyrus Anastasi IV, written "in the year of misery," in which it is said that the months are reversed, the planet-god is described as arriving "with the sweet wind before him."[4]

In a similar text of the Hebrews we read that the times and seasons were confused, and "a fragrance perfumed all the world," and the perfume was brought by the pillar of smoke. The fragrance was like that of myrrh and frankincense. "Israel was surrounded by clouds," and as soon as the clouds were set in motion, the winds "breathed myrrh and frankincense."[5]

[1] *The Bundahis*, Chap. 3, Sec. 16.
[2] Gómara, *Conquista*, cxix.
[3] Numbers 17 : 8. The cover of clouds remained over the desert until after the death of Aaron. Cf. Ginzberg, *Legends*, VI, 114.
[4] Erman, *Egyptian Literature*, p. 309.
[5] Ginzberg, *Legends*, III, 158 and 235; VI, 71. According to Targum Yerushalmi, Exodus 35 : 28: "The clouds brought the perfumes from paradise and placed them in the wilderness for Israel."

The Vedas contain hymns to Agni which "glows from the sky." Its fragrance became the fragrance of the earth.

That fragrance of thine . . .
which the immortals of yore gathered up.[1]

The generation of those days, when the star conveyed its fragrance to men on the earth, is immortalized in the tradition of the Hindus. The Vedic hymn compares the fragrance of the star Agni to the scent of the lotus.

## AMBROSIA

In what way did this veil of gloom dissolve itself?

When the air is overcharged with vapour, dew, rain, hail, or snow falls. Most probably the atmosphere discharged its compounds, presumably of carbon and hydrogen, in a similar way.

Has any testimony been preserved that during the many years of gloom carbohydrates precipitated?

"When the dew fell upon the camp in the night, the manna fell upon it." It was like "the hoar frost on the ground." It had the shape of coriander seed, the yellowish colour of bdellium, and an oily taste like honeycomb. It was called "corn of heaven" and it was ground between stones and baked in pans.[2] The manna fell from the clouds.[3]

After the nightly cooling, the carbohydrates precipitated and fell with the morning dew. The grains dissolved in the heat and evaporated; but in a closed vessel the substance could be preserved for a long time.[4]

The exegetes have endeavoured to explain the phenomenon of manna and were helped by the naturalists who discovered that a tamarisk in the desert of Sinai sheds its seeds during certain months of the year.[5] But why should this seed be called

[1] *Hymns of the Atharva-Veda* (transl. M. Bloomfield, 1897), 201-202.
[2] Exodus 16 : 14-34; Numbers 11 : 7-9.
[3] Psalms 78 : 23-24.
[4] Exodus 16 : 21, 33-34.
[5] See A. P. Stanley, *Lectures on the History of the Jewish Church* (1863), Pt. I, p. 147: "The manna . . . according to the Jewish tradition of Josephus, and the belief of the Arab tribes, and of the Greek church at the present day, is still found in the dropping from the tamarisk bushes." However, Josephus, in his *Antiquities*, III, 26 ff., does not speak of tamarisks but of dew which looked like snow and still falls in the desert, being a "mainstay to dwellers in these parts."

An expedition of Jerusalem University in 1927 investigated the tamarisk in "corn of heaven," "bread of heaven," [1] or why should it be said it "will rain bread from heaven?" [2] It is also not easy to explain how a multitude of men and animals could have existed for many years in a wilderness on the scarce and seasonal seeds of some desert plant. Were such a thing possible, the desert would be preferable to tillable land that yields bread to the labourer only in the sweat of his brow.

The clouds brought the heavenly bread, it is also said in the Talmud.[3] But if the manna fell from the clouds that enveloped the entire world, it must have fallen not only in the Desert of Wanderings, but everywhere; and not only the Israelites, but other peoples, too, must have tasted it and spoken of it in their traditions.

There was a world fire, says the Icelandic tradition, followed by the Fimbul-winter, and only one human pair remained alive in the north. "This human pair lie hidden in the holt during the fire of Surt." Then came "the terrible Fimbul-winter at the end of the world [age]; meanwhile they feed on morning dew, and from them come the folk who people the renewed earth." [4]

Three elements are connected in the Icelandic tradition which are the same three we met in the Israelite tradition: the world fire, the dark winter that endured many years, and the morning dew that served as food during these years of gloom when nothing budded.

The Maoris of New Zealand tell of fiery winds and fierce clouds that lashed the waters into tidal waves that touched the sky and were accompanied by furious hailstorms. The ocean fled. The progeny of the storm and hail were "Mist, and Heavy-dew and Light-dew." After the catastrophe "but little of the dry land was left standing above the sea. Then clear light increased in the world, and the beings who had been

---

the Sinai Desert. See F. S. Bodenheimer and O. Theodor, *Ergebnisse der Sinai Expedition* (1929), Pt. III.

A German professor suggested also *Blattläuse*. "Blattläuse wie Blattsauger schwitzen zuweilen auch aus dem After einen honigartigen Saft in solcher Menge aus, dass die Pflanzen, besonders im Juli, damit gleichsam überfirnisst sind" (W. H. Roscher, *Nektar und Ambrosia* [1883], p. 14). But where are forests in a desert where lice would prepare on the leaves of the trees three meals a day for a myriad of migrants?

[1] Psalms 78 : 24 and 105 : 40.

[2] Exodus 16 : 4.

[3] Tractate Yoma 75a.

[4] J. A. MacCulloch, *Eddic Mythology* (1930), p. 168.

hidden between [sky and earth] before they were parted, now multiplied upon the earth." [1]

This tradition of the Maoris has substantially the same elements as the Israelite tradition. The destruction of the world was accompanied by hurricanes, hail (meteorites), and sky-high billows; the land submerged; a mist covered the earth for a long time; heavy dew fell to the ground together with light dew, as in the passage quoted from Numbers 11 : 9.

The writings of Buddhism relate that when a world cycle comes to an end with the world destroyed and the ocean dried up, there is no distinction of day and night and heavenly ambrosia serves as food.[2]

In the hymns of *Rig-Veda*,[3] it is said that honey (*madhu*) comes from the clouds. These clouds originated from the pillar of cloud. Among the hymns of the *Atharva-Veda* there is one to the honeylash: "From heaven, from earth, from the atmosphere, from the sea, from the fire, and from the wind, the honey-lash hath verily sprung. This, clothed in amrite (ambrosia), all the creatures revering, acclaim in their hearts." [4]

The Egyptian *Book of the Dead* speaks of "the divine clouds and the great dew" that bring the earth into contact with the heavens.[5]

The Greeks called the heavenly bread ambrosia. It is described by the Greek poets in identical terms with manna; it had the taste of honey and a fragrance. This heavenly bread has given classical scholars many headaches. Greek authors from Homer and Hesiod down through the ages continually referred to ambrosia as the heavenly food which in its fluid state is called nectar.[6] But it was used also as ointment [7] (it had the fragrance of a lily), and as food for the horses of Hera when she visited Zeus in the sky.[8] Hera (Earth) was veiled in it when she hurried from her brother Ares (Mars) to Zeus Jupiter). What could it be, this heavenly bread, which served also as a veil for a goddess-planet, and was used as an ointment, too? It was honey, said some scholars. But honey is a regular

[1] Tylor, *Primitive Culture*, I, 324.
[2] Warren, *Buddhism in Translations*, p. 322.
[3] Cf. Roscher, *Nektar und Ambrosia*, p. 19.
[4] *Hymns of the Atharva-Veda*, p. 229, *Rigveda* I, 112.
[5] E. W. Budge, *The Book of the Dead* (2nd ed., 1928), Chap 98; cf. G. A. Wainwright, *Journal of Egyptian Archaeology*, XVIII (1932), 167.
[6] Roscher, *Nektar und Ambrosia.*
[7] *Iliad* xiv. 170 ff.
[8] *Iliad* v. 368 ff.; see also *ibid.*, v. 775 ff.; xiii. 34 ff., and Ovid, *Metamorphoses*, ii. 119 ff.

food for mortals, whereas ambrosia was given only to the generation of heroes.

Then what was this substance that served as fodder on the ground for horses, as a veil for planets, bread from the sky for heroes, and that also turned fluid for their drink, and was oil and perfume for ointments?

It was the manna that was baked into bread, had an oily taste and also a honey taste, was found on the ground by man and beast, wrapped the earth and the heavenly bodies in a veil, was called "corn of heaven" and "bread of the mighty," [1] had a fragrant odour, and served the women in the wilderness as ointment.[2] Manna, like ambrosia, was compared with honey and with morning dew.

The belief of Aristotle and other writers[3] that honey falls from the atmosphere with the dew was based on the experience of those days when the world was veiled in the carbon clouds that precipitated honey-frost.

These clouds are described as "dreaded shades" in the *Kalevala.* From these "dreaded shades," says the epos, honey dropped. "And the clouds their fragrance sifted, sifted honey . . . from their home within the heavens." [4]

The Maoris in the Pacific, the Jews on the border of Asia and Africa, the Hindus, the Finns, the Icelanders, all describe the honey-food being dropped from the clouds, dreary shades of the shadow of death, that enveloped the earth after a cosmic catastrophe. All traditions agree also that the source of the heavenly bread falling from the clouds with the morning dew was a celestial body. The Sibyl says that the sweet heavenly bread came from the starry heavens.[5] The planet-god Ukko, or Jupiter, is said to have been the source of the honey that dropped from the clouds.[6] Athena covered other planet-goddesses with a "robe ambrosial," and provided nectar and ambrosia to the heroes.[7] Other traditions, too, see the origin of the honey-dew in a celestial body that enveloped the earth in clouds. For this reason ambrosia or manna is called "heavenly bread."

[1] Tractate Yoma 75a.
[2] Ginzberg, *Legends*, III, 49.
[3] Aristotle, *Historia Animalium* ("Generation of Animals"), v. 22. 32; Galen (ed. by C. G. Kühn, 1821-1823), VI, 739; Pliny, *Natural History*, xi. 30; Diodorus, *The Library of History*, xvii. 75.
[4] *The Kalevala* (transl. Crawford), p. xvi and Rune 9.
[5] Ginzberg, *Legends*, VI, 17.
[6] The *Kalevala*, Rune 15.
[7] *Iliad* xiv. 170 ff. Cf. Plutarch, *On the Face* (Die facie quae in orbe lunae apparet).

The honey-frost fell in enormous quantities. The haggadic literature says that the quantity which fell every day would have sufficed to nourish the people for two thousand years.[1] All the peoples of the East and West could see it.[2]

A few hours after the break of day, the heat under the cloud cover liquefied the grains and volatilized them.[3] The ground absorbed some of the liquefied mass, as it absorbs dew. The grains also fell upon the water, and the rivers became milky in appearance.

The Egyptians relate that the Nile flowed for a time blended with honey.[4] The strange appearance of the rivers of Palestine —in the desert the Israelites saw no river—caused the scouts who returned from a survey of the land to call it the land that "floweth with milk and honey" (Numbers 13 : 27). "The heavens rain oil, the wadis run with honey," says a text found in Ras-Shamra (Ugarit) in Syria.[5]

In the rabbinical literature it is said that "melting of manna formed streams that furnished drink to many deer and other animals." [6]

The *Atharva-Veda* hymns say that honey-lash came down from fire and wind; ambrosia fell, and streams of honey flowed upon the earth. "The broad earth shall milk for us precious honey . . . shall pour out milk for us in rich streams." [7] The Finnish tradition narrates that land and water were covered successively by black, red, and white milk. The first and second were the colours of the substances, ashes and "blood," that constituted the plagues (Exodus 7 and 9); the last one was the colour of ambrosia that turned into nectar on land and water.

A memory of a time when "streams of milk and streams of sweet nectar flowed" is also preserved in Ovid.[8]

[1] Midrash Tehillim to Psalm 23; Tosefta Sota 4, 3.
[2] Tractate Yoma 76a.
[3] Exodus 16 : 21.
[4] Manetho refers this phenomenon to the time of Pharaoh Nephercheres. See the volume of Manetho in the Loeb Classical Library, pp. 35, 37, 39.
[5] C. H. Gordon, *The Loves and Wars of Baal and Anat* (1943), p. 10.
[6] Midrash, Tannaim, 191; Targum Yerushalmi on Exodus 16 : 21; Tanhuma Beshalla 21, and other sources.
[7] "Hymn to Goddess Earth," *Hymns of the Atharva-Veda* (transl. Bloomfield), pp. 199 f.
[8] *Metamorphoses* (transl. F. J. Miller, 1916), i. 111-112.

# JERICHO

The earth's crust trembled and cracked again and again as its strata settled after the major displacement. Chasms opened up, springs disappeared, and new springs appeared.[1] When the Israelites approached the river Jordan, a slice of one bank fell, blocking the stream long enough for the tribes to cross over. "The waters which came down from above stood and rose up upon a heap very far from the city Adam, that is beside Zaretan: and those that came down toward the sea of the plain, even the salt sea, failed, and were cut off: and the people passed over right against Jericho." [2]

A similar occurrence took place on the eighth of December, 1267, when the Jordan was dammed for sixteen hours, and again following the earthquake of 1927, when a slice of one bank fell into the river not far from Adam and blocked the water for over twenty-one hours; at Damieh (Adam) the people crossed the river on its dry bed.[3]

The fall of the walls of Jericho at the blast of the trumpets is a well-known episode, but it is not well interpreted. The horns blown by the priests for seven days played no greater natural role than Moses' rod with which, in the legend, he opened a passage in the sea. "When the people heard the sound of the trumpet," it happened that "the wall fell down flat." [4] The great sound of the trumpet was produced by the earth; the Israelite tribes, believing in magic, thought that the sound of the earth came in response to the blowing of the rams' horns for seven days.

The great walls of Jericho—they were twelve feet wide—have been excavated.[5] They were found to have been destroyed by an earthquake. The archaeological evidences also prove that these walls collapsed at the beginning of the Hyksos period, or shortly after the close of the Middle Kingdom.[6] The earth had not yet recovered from the previous world catastrophe, and reacted with continuous tremors when the hour of a new

[1] Numbers 16 : 31-35; 20 : 11; Psalms 78 : 16; 107 : 33-35.
[2] Joshua 3 : 16. A correct translation requires: "very far *at* the city Adam."
[3] J. Garstang, *The Foundations of Bible History* (1931), p. 137.
[4] Joshua 6 : 20.
[5] E. Sellin and C. Watzinger, *Jericho: Die Ergebnisse der Ausgrabungen* (1913).
[6] J. Garstang and G. B. E. Garstang, *The Story of Jericho* (1940).

cosmic disaster approached: the event we described at the beginning of this book only to go back to the cataclysm of the Exodus—the upheaval of the days of Joshua, when the earth stood still on the day of the battle at Beth-horon.

CHAPTER VII

# STONES SUSPENDED IN THE AIR

"THE HOT HAILSTONES which, at Moses' intercession, had remained suspended in the air when they were about to fall upon the Egyptians, were now cast down upon the Canaanites."[1] These words mean that a part of the meteorites of the cometary train of the days of Exodus remained in the celestial sphere for about fifty years, falling in the days of Joshua, in the valley of Beth-horon, on the same forenoon when the sun and the moon stood still for the length of a full day.

The language of the Talmud and Midrash suggests that the same comet returned after some fifty years. Once more it passed very close to the earth. This time it did not reverse the poles of the earth, but kept the terrestrial axis tilted for a considerable length of time. Again the world was, in the language of the rabbis, "consumed in the whirlwind," "and all the kingdoms tottered," "the earth quaked and trembled from the noise of thunder"; terrified mankind was decimated once more, and carcasses were like rubbish in this Day of Anger.[2]

On the day when this took place on the earth, the sky was in confusion. Stones fell from the heavens, sun and moon stopped in their paths, and a comet must also have been seen. Habakkuk describes the portent in the sky on that memorable day when, in his words, "the sun and moon stood still in their habitation": it had the form of a man on a chariot drawn by horses and was regarded as God's angel.

In the King James version the passage read:

"His glory covered the heavens . . . and his brightness was as the light; he had horns coming out of his hand . . . burning coals went forth at his feet . . . [he] drove asunder the nations; and the everlasting mountains were scattered. . . . Was thine anger against the rivers? Was thy wrath against the sea, that thou didst ride upon thine horses and thy chariots of

[1] Ginzberg, *Legends,* IV, 10; the Babylonian Talmud, Tractate Berakhot 54b. See also Mishna of Rabbi Elieser or of 32 Midot.
[2] See the Section, "The Most Incredible Story."

salvation . . . ? Thou didst cleave the earth with rivers. The mountains saw thee, and they trembled: the overflowing of the water passed by: the deep uttered his voice. . . . The sun and moon stood still in their habitation: at the light of thy arrows they went, and at the shining of thy glittering spear. Thou didst march through the land in indignation, thou didst thresh the heathen in anger. . . . Thou didst walk through the sea with thine horses, through the heap of great waters."[1]

Since the texts of the Scriptures have, for some psychological reason rooted in the readers, the quality of being easily misread, misunderstood, or misinterpreted, I give also some of the passages of the third chapter of Habakkuk in another, modernized reading:

> His splendour over all the sky,
> his glory filling all the earth,
> his radiance is a lightning blaze,
> on either side flash rays. . . .
> At his step the earth is shaken,
> at his look nations are scattered,
> the ancient hills are shattered,
> mountains of old sink low. . . .
> Art thou wrathful at the sea,
> that thou art storming on the steeds,
> upon the chariots in triumph . . . ?
> The hills writhe at thy sight . . .
> the sun forgets to rise,
> the moon to move,
> before the flashes of thy darting arrows,
> before the sheen of thy lightning, thy lance.
> Thou trampest earth in fury,
> thou art threshing the peoples in thine anger.[2]

With the earth disturbed in its spinning on its axis, the mechanical friction of displaced strata and magma must have set the world on fire.

The world burned. The Greek story of Phaëthon will be introduced here because of the interpretation heard by Solon during his visit to Egypt.

[1] Habakkuk 3 : 3-15.
[2] *The Old Testament: A New Translation* (transl. James Moffat, 1924-1925).

## PHAËTHON

The Greeks as well as the Carians and other peoples on the shores of the Aegean Sea told of a time when the sun was driven off its course and disappeared for an entire day, and the earth was burned and drowned.

The Greek legend says that the young Phaëthon, who claimed parentage of the sun, on that fatal day tried to drive the chariot of the sun. Phaëthon was unable to make his way "against the whirling poles," and "their swift axis" swept him away. Phaëthon in Greek means "the blazing one."

Many authors have dealt with the story of Phaëthon; the best known version is a creation of the Latin poet Ovid. The chariot of the sun, driven by Phaëthon, moved "no longer in the same course as before." The horses "break loose from their course" and "rush aimlessly, knocking against the stars set deep in the sky and snatching the chariot along through uncharted ways." The constellations of the cold Bears tried to plunge into the forbidden sea, and the sun's chariot roamed through unknown regions of the air. It was "borne along just as a ship driven before the headlong blast, whose pilot has let the useless rudder go and abandoned the ship to the gods and prayers."[1]

"The earth bursts into flame, the highest parts first, and splits into deep cracks, and its moisture is all dried up. The meadows are burned to white ashes; the trees are consumed, green leaves and all, and the ripe grain furnishes fuel for its own destruction. . . . Great cities perish with their walls, and the vast conflagration reduces whole nations to ashes."

"The woods are ablaze with the mountains. . . . Aetna is blazing boundlessly . . . and twin-peaked Parnassus. . . . Nor does its chilling clime save Scythia; Caucasus burns . . . and the heaven-piercing Alps and cloud-capped Apennines."

The scorched clouds belched forth smoke. Phaëthon sees the earth aflame. "He can no longer bear the ashes and whirling sparks, and is completely shrouded in the dense, hot smoke. In this pitchy darkness he cannot tell where he is or whither he is going."

"It was then, as men think, that the peoples of Aethiopit

[1] Ovid, *Metamorphoses* (transl. F. J. Miller), Bk. II.

became black-skinned, since the blood was drawn to the surface of their bodies by the heat."

"Then also Libya became a desert, for the heat dried up her moisture. . . . The Don's waters steam; Babylonian Euphrates burns; the Ganges, Phasis, Danube, Alpheus boil; Spercheos' banks are aflame. The golden sands of Tagus melt in the intense heat, and the swans . . . are scorched. . . . The Nile fled in terror to the ends of the earth . . . the seven mouths lie empty, filled with dust; seven broad channels, all without a stream. The same mischance dries up the Thracian rivers, Hebrus and Strymon; also the rivers of the west, the Rhine, Rhone, Po and the Tiber. . . . Great cracks yawn everywhere. . . . Even the sea shrinks up, and what was but now a great watery expanse is a dry plain of sand. The mountains, which the deep sea had covered before, spring forth, and increase the numbers of the scattered Cyclades."

How could the poets have known that a change in the movement of the sun across the firmament must cause a world conflagration, blazing of volcanoes, boiling of rivers, disappearance of seas, birth of deserts, emergence of islands, if the sun never changed its harmonious journey from sunrise to sunset?

The disturbance in the movement of the sun was followed by a period as long as a day, when the sun did not appear at all. Ovid continues: "If we are to believe report, one whole day went without the sun.[1] But the burning world gave light."

A prolonged night in one part of the world must be accompanied by a prolonged day in another part; in Ovid we see the phenomenon related in the Book of Joshua, but from another longitude. This may stimulate surmise as to the geographical origin of the Indo-Iranian or Carian migrants to Greece.

The globe changed the inclination of its axis; latitudes changed, too. Ovid ends the description of the world catastrophe contained in the story of Phaëthon: "Causing all things to shake with her mighty trembling, she [the earth] sank back a little lower than her wonted place."

Plato recorded the story heard two generations before from Solon, the wise ruler of Athens.[2] Solon, on his visit to Egypt, questioned the priests, versed in the lore of antiquity, on early history. He discovered that "neither he himself nor any other

[1] "Si modo credimus, unum isse diem sine sole ferunt."
[2] Plato *Timaeus* (transl. R. G. Bury, 1929).

Greek knew any thing at all, one might say, about such matters." Solon unfolded before the priests the tale of the deluge, the only ancient tradition he was aware of. One of the priests, an old man,[1] said:

"There have been and there will be many and divers destructions of mankind, of which the greatest are by fire and water, and lesser ones by countless other means. For in truth the story that is told in your country as well as ours, how once upon a time Phaëthon, son of Helios, yoked his father's chariot, and, because he was unable to drive it along the course taken by his father, burnt up all that was upon the earth and himself perished by a thunderbolt—that story, as it is told, has the fashion of a legend, but the truth of it lies in the occurrence of a shifting of the bodies in the heavens which move around the earth, and a destruction of the things on the earth by fierce fire, which recurs at long intervals." [2]

The Egyptian priest explained to Solon that in these catastrophes the literary works of many peoples and their learned men perished; for that reason the Greeks were still childish, as they no longer knew the true horrors of the past.

These words of the priest were only an introduction to a revelation of his knowledge about lands that were erased when Greece also and the entire world were visited with heavenly wrath. He told the story of a mighty kingdom on a great island in the middle of the Atlantic Ocean that submerged and sank forever into its waters.

## ATLANTIS

The story narrated by Plato of the island of Atlantis that ruled Africa as far as the border of Egypt and Europe as far as Tuscany on the Apennine peninsula and that in one fatal night was shattered by earthquakes and sank, never ceased to occupy the imagination of the literati. Strabo and Pliny thought that the story of Atlantis was an illusion of the elderly Plato. But to this day the tradition, as revised by Plato, has not died. Poets and novelists have exploited the story freely; scientists have done so with caution. An incomplete catalogue of the

[1] According to Plutarch (*Isis and Osiris*) the name of the priest was Sonchis of Sais.
[2] Plato *Timaeus* 22 C-D.

literature on Atlantis in 1926 included 1,700 titles.[1] Although Plato said clearly that Atlantis was situated behind the Pillars of Hercules (Gibraltar), in the Atlantic Ocean, as is also indicated by the name of the island, travellers and other guessers have placed Atlantis in all parts of the world, even on dry land, as, for example, in Tunisia,[2] Palestine,[3] and South America. Ceylon, Newfoundland, and Spitzbergen have also been considered. This was due to the fact that traditions of inundations and submersion of islands exist in all parts of the world.

Plato set down what Solon had heard in Egypt from the learned priest. "The [Atlantic] ocean there was at that time navigable; for in front of the mouth which you Greeks call, as you say, 'the Pillars of Heracles' [Hercules], there lay an island which was larger than Libya and Asia [Asia Minor] together; and it was possible for the travellers of that time to cross from it to the other islands, and from the islands to the whole of the continent over against them which encompasses that veritable ocean. . . . Yonder is a real ocean, and the land surrounding it may most rightly be called, in the fullest and truest sense, a continent. Now in this island of Atlantis there existed a confederation of kings, a great and marvellous power, which held sway over all the island, and over many other islands also and parts of the continent; and, moreover, of the lands here within the Straits they ruled over Libya as far as Egypt and over Europe as far as Tuscany."[4]

In the nineteenth century ships sailed the Atlantic Ocean to explore its bed in search of Atlantis, and before the Second World War scientific societies existed for the sole purpose of exploring the problem of the sunken island.

Much speculation was offered, not only on the whereabouts of Atlantis, but also on the cultural achievements of its inhabitants. Plato, in another work of his (*Critias*), wrote a political treatise, and, as no real place in the world could have been the scene of his utopia, he chose for that purpose the sunken island. Modern scholars, finding some affinity between American, Egyptian, and Phoenician cultures, think that Atlantis may have been the intermediary link. There is much

[1] J. Gattefossé and C. Roux, *Bibliographie de l'Atlantide et des questions connexes* (1926).
[2] A. Herrmann, *Unsere Ahnen und Atlantis* (1934).
[3] F. C. Baer, *L'Atlantique des anciens* (1835).
[4] Plato, *Timaeus* 24 E-25 B.

probability in these speculations; if they are justified, Crete, a maritime base of Carian navigators, may disclose some information about Atlantis as soon as the Cretan scripts are satisfactorily deciphered.

One point in Plato's story about the submersion of Atlantis requires correction. Plato said that Solon told the story to Critias the elder, and that young Critias, Plato's friend, heard it from his grandfather when he was a ten-year-old boy. Critias the younger remembered having been told that the catastrophe which befell Atlantis happened 9,000 years before. There is one zero too many here. We do not know of any vestiges of human culture, aside from that of the Neolithic age, nor of any navigating nation, 9,000 years before Solon. Numbers we hear in childhood easily grow in our memory, as do dimensions. When revisiting our childhood home, we are surprised at the smallness of the rooms—we had remembered them as much larger. Whatever the source of the error, the most probable date of the sinking of Atlantis would be in the middle of the second millennium, 900 years before Solon, when the earth twice suffered great catastrophes as a result of "the shifting of the heavenly bodies." These words of Plato received the least attention, though they deserved the greatest.

The destruction of Atlantis is described by Plato as he heard it from his source: "At a later time there occurred portentous earthquakes and floods, and one grievous day and night befell them, when the whole body of your [Greek] warriors was swallowed up by the earth, and the island of Atlantis in like manner was swallowed up by the sea and vanished; wherefore also the ocean at that spot has now become impassable and unsearchable, being blocked up by the shoal mud which the island created as it settled down."[1]

At the time when Atlantis perished in the ocean, the people of Greece were destroyed: the catastrophe was ubiquitous.

As if recalling what happened, the Psalmist wrote: "Destructions are come to a perpetual end: and thou hast destroyed cities, their memorial is perished with them."[2] He prayed also: "God is our refuge and strength . . . therefore will not we fear, though the earth be removed and though the mountains be carried into the midst of the sea; though the waters thereof roar and be troubled."[3]

[1] Plato *Timaeus* 25 C-D. [2] Psalms 9 : 6. [3] Psalms 46 : 1-3.

## THE FLOODS OF DEUCALION AND OGYGES

The history of Greece knows two great natural catastrophes: the floods of Deucalion and of Ogyges. One of them, usually that of Deucalion, is described by Greek authors as having been simultaneous with the conflagration of Phaëthon. The floods of Deucalion and Ogyges brought overwhelming destruction to the mainland of Greece and to the islands around and caused changes in the geographical profile of the area. That of Deucalion was most devastating: water covered the land and annihilated the population. According to the legend, only two persons—Deucalion and his wife—remained alive. This last detail must not be taken more literally than similar statements found in descriptions of great catastrophes all around the world; for example, two daughters of Lot, who hid with him in a cave after the catastrophe of Sodom and Gomorrah, believed that they and their father were the only survivors in the land.[1]

The chronologists among the Fathers of the Church found material for assuming that one of the two catastrophes, the flood of Deucalion or that of Ogyges, had been contemporaneous with the Exodus.

Julius Africanus wrote: "We affirm that Ogygus [Ogyges] from whom the first flood [in Attica] derived its name, and who was saved when many perished, lived at the time of the Exodus of the people from Egypt along with Moses."[2] He further expressed his belief in the coincidence of the catastrophe of Ogyges and the one that occurred in Egypt in the days of the Exodus in the following words:

"The Passover and the Exodus of the Hebrews from Egypt took place, and also in Attica the flood of Ogygus. And that is according to reason. For when the Egyptians were being smitten in the anger of God with hail and storms, it was only to be expected that certain parts of the earth should suffer with them."[3]

Eusebius placed the Flood of Deucalion and the conflagration of Phaëthon in the fifty-second year of Moses' life.[4] Augustine

[1] Genesis 19 : 31.
[2] Julius Africanus in *The Ante-Nicene Fathers*, ed. A. Roberts and J. Donaldson (1896), VI, 132.
[3] *Ibid.*, p. 134.
[4] Eusebius, *Werke*, Vol. V, *Die Chronik*, "Chronikon-Kanon."

also synchronized the Flood of Deucalion with the time of Moses;[1] he assumed that the Flood of Ogyges took place earlier.

A chronologist of the seventh century (Isidore, bishop of Seville)[2] dated the Flood of Deucalion in the time of Moses; chronologists of the seventeenth century likewise calculated that the Flood of Deucalion took place in the time of Moses, close to but not simultaneous with the Exodus.[3]

It would seem to be more probable that, if the catastrophes occurred one shortly after the other, the catastrophe of Ogyges took place after that of Deucalion which practically destroyed the land, depopulated it, and erased every memory of what had happened up to that time. In the words of Plato, who quoted the Egyptian priest speaking to Solon, the catastrophes must have escaped the notice of the future generations because, as a result of the devastation, "for many generations the survivors died with no power to express themselves in writing." The memory of the catastrophe of Ogyges would have vanished in the catastrophe of Deucalion if Ogyges had preceded Deucalion.[4]

Apparently, the truth is with those who placed the catastrophe of Deucalion in the days of Exodus; but those who reckoned that Ogyges was a contemporary of Moses were also correct, except that Moses did not live until the Flood of Ogyges—it took place in the days of Joshua.

In commemoration of the Deucalion flood, the people of Athens observed a feast in the month of Anthesterion, which is a spring month; the feast was called Anthesteria. On the thirteenth of the month, the main day of the feast, honey and flour were poured into a fissure in the earth as a sacrifice.[5]

The date of this ceremony—the thirteenth day of Anthesterion in the spring—is revealing if we remember what was said in the section entitled "13." It was on the thirteenth day

[1] *The City of God*, Bk. XVIII, Chaps. 10, 11.

[2] See J. G. Frazer, *Folklore in the Old Testament* (1918), I, 159.

[3] Seth Calvisius, in *Opus chronologicum* (1629), assigns the year 2429 *anno mundi* or 1519 before the present era to Phaëthon's conflagration, and 2432 (-1516) to the Flood of Deucalion, and 2453 (-1495) to the Exodus.

Christopher Helvicus (1581-1617), in *Theatrum historicum* (1662), assigns 2437 *anno mundi* to the Flood of Deucalion and Phaëthon's conflagration, and 2453 (or 797 *a Diluvio universali*) to the Exodus from Egypt.

[4] But cf. Frazer, "Ancient Stories of a Great Flood," *Journal of the Royal Anthropological Institute*, XLVI (1916). However, Eusebius placed Deucalion before Ogyges.

[5] Cf. Pausanias, *Description of Greece*, I, xviii, 7. Pauly-Wissowa, *Real-Encyclopädie*, *s. v.* "Anthesterion"; also Andree, *Die Flutsagen*, p. 41.

of the spring month (Aviv) that the great planetary contact occurred which preceded by a few hours the Exodus of the Israelites from Egypt.

The offering of honey and flour as the main ceremony of the feast is also revealing if we recollect that manna, or heavenly corn, tasting like honey, fell on the earth after the contact of the earth with a celestial body.

As to the provenance of the name Deucalion, scholars admit that it is not known.[1] For the name and the person of Ogyges we have some concrete information. Although Ogyges was a king, the Greek annalists who wrote of the "flood of Ogyges" as one of the outstanding events of the past of their country, at the same time did not know anything about a king of that name in Greece.[2] Who was Ogyges?

We can solve this problem. When the Israelites under Moses approached the border of Moab, Balaam in his blessing of Israel used these words: "His king shall be higher than Agag [Agog]."[3] Agog must have been the most important king of that time in the area around the eastern Mediterranean.

In my reconstruction of ancient history, I shall put forward proofs that the Amalekite king, Agog I, was identical with the Hyksos king whose name the Egyptologists tentatively read Apop I, and who, a few decades after the invasion of Egypt by the Amu (Hyksos), laid the foundation of Thebes, the future capital of the New Kingdom in Egypt.

In conformity with this assertion, I can point to the fact that Greek tradition, which does not know of any activities of King Ogyges in Attica, occasionally places the domicile of Ogyges in Egyptian Thebes, and Aeschylus calls Thebes of Egypt "the Ogygian Thebes," to differentiate it from the Greek Thebes in Boeotia. Ogyges is also credited with founding Thebes in Egypt.[4]

Agog was a contemporary of the ageing Moses; he was a

[1] "While the meaning of the legend is clear, the meaning of the name Deucalion is enigmatic." Roscher, "Deukalion," *Lexikon d. griech. und römisch. Mythologie.* According to Homer, Deucalion was a son of Minos, king of Crete, and a grandson of Zeus and Europa (*The Iliad*, xiv, 321 ff; xiii, 450 f.). According to Apollodorus (*The Library*, I, vii), Deucalion was a son of Prometheus.

[2] Julius Africanus wrote: "After Ogygus [Ogyges], by reason of the vast destruction caused by the flood, the present land of Attica remained without a king up to Cecrops, a period of 189 years." Fragment of the *Chronography* in *The Ante-Nicene Fathers*, VI.

[3] Numbers 24 : 7. Cf. the vowels in the name in the Hebrew text of I Samuel 15.

[4] Aeschylus, *The Persians*, l. 37. See also *Scholium to Aristides.* Cf. Roscher, "Ogyges, als König des ägyptischen Thebes," *Lexikon d. griech. und römisch. Mythologie*, Vol. 3 [1], Col. 689.

ruler who, in his time, had no equal in the region bordering the eastern Mediterranean;[1] the catastrophe in the time of Joshua, successor to Moses, was called by his, Agog's, name.

The assertion of Solinus, the author of *Polyhistor*, that the flood of Ogyges was followed by a night of nine months' duration does not necessarily signify a confusion with the darkness that ensued after the cataclysm of the Exodus; as the causes were similar, similar results must have followed. The eruption of thousands of volcanoes would suffice to produce this darkness, of a shorter duration than that which followed the cataclysm of the Exodus.[2]

Thus, the Greek traditions of the floods of Ogyges and Deucalion contain elements which, though interchanged, can be traced to two great upheavals in the middle of the second millennium before the present era.[3]

[1] The rabbinical sources say that Amalek went to conquer "the entire world." Seals of the Hyksos kings were found on Crete, in Palestine, in Mesopotamia, and in other places outside Egypt.

[2] Cf. *Polyhistor*, translated by A. Golding (London, 1587), Chap. xvi, and the translation by Agnant (Paris, 1847), Chap. xi.

[3] It seems that the legend of Deucalion contains also elements of the story of the universal Deluge (of Noah).

## CHAPTER VIII

# THE FIFTY-TWO-YEAR PERIOD

THE WORKS OF Fernando de Alva Ixtlilxochitl, the early Mexican scholar (circa 1568—1648) who was able to read old Mexican texts, preserve the ancient tradition according to which the multiple of fifty-two-year periods played an important role in the recurrence of world catastrophes.[1] He asserts also that only fifty-two years elapsed between two great catastrophes, each of which terminated a world age.

As I have already pointed out, the Israelite tradition counts forty years of wandering in the desert; between the time when the Israelites left the desert and started the difficult task of the conquest, and the time of the battle at Beth-horon twelve years may well have passed. The conquest of Canaan took fourteen years, and the entire duration of Joshua's leadership amounted to twenty-eight years.[2]

Now there exists a remarkable fact: the natives of pre-Columbian Mexico expected a new catastrophe at the end of every period of fifty-two years and congregated to await the event. "When the night of this ceremony arrived, all the people were seized with fear and waited in anxiety for what might take place." They were afraid that "it will be the end of the human race and that the darkness of the night may become permanent: the sun may not rise anymore." [3] They watched for the appearance of the planet Venus, and when, on the feared day, no catastrophe occurred, the people of Maya rejoiced. They brought human sacrifices and offered the hearts

[1] Ixtlilxochitl, *Obras históricas* (ed. 1891-1892 in 2 vols.). French translation of his annals is *Histoire des Chichimèques* (1840).

In the *Codex Vaticanus* the world ages are reckoned in multiples of fifty-two years with a changing number of years as an addition to these figures. A. Humboldt (*Researches*, II, 28) contraposed the lengths of the world ages in the Vatican manuscript (No. 3738) and their lengths in the system of the tradition preserved by Ixtlilxochitl.

Four ages of 105 years are referred to by Censorinus (*Liber de die natali*) as having taken place, according to the belief of the Etruscans, between world catastrophes presaged by celestial portents.

[2] Seder Olam 12. Augustine speaks of 27 years of Joshua's leadership (*The City of God*, Bk. XVIII, Chap. II).

[3] B. de Sahagun, *Historia general de la cosas de Nueva España* (French transl. by D. Jourdanet and R. Simeon, 1880), Bk. VII, Chaps. X-XIII.

of prisoners whose chests they opened with knives of flint. On that night, when the fifty-two-year period ended, a great bonfire announced to the fearful crowds that a new period of grace had been granted and a new Venus cycle started.[1]

The period of fifty-two years, regarded by the ancient Mexicans as the interval between two world catastrophes, was definitely related by them to the planet Venus; and this period of Venus was observed by both the Mayas and the Aztecs.[2]

The old Mexican custom of sacrificing to the Morning Star survived in human sacrifices by the Skidi Pawnee of Nebraska in years when the Morning Star "appeared especially bright, or in years when there was a comet in the sky." [3]

What had Venus to do with the catastrophes that brought the world to the brink of destruction? Here is a question that will carry us very far, indeed.

## JUBILEE

I shall postpone only a little giving the answer to the question just posed. First, I should like to find an explanation for the institution of the jubilee year of the Israelites.

Every seventh year, according to the law, was a sabbatical year during which the land had to be left fallow and Jewish slaves set free. The fiftieth year was a jubilee year, when the land not only had to be left fallow, but had to be returned to its original proprietors. According to the law, one could not convey his land for ever; the deed of sale was but a lease for whatever number of years remained until the jubilee year. The year was proclaimed by the blowing of horns on the Day of Atonement. "In the Day of Atonement shall ye make the trumpet sound throughout all your land. And ye shall hallow the fiftieth year, and proclaim liberty throughout all the land unto all the inhabitants thereof: it shall be a jubilee unto you, and ye shall return every man unto his possession, and ye shall return every man unto his family."[4]

[1] Cf. Seler, *Gesammelte Abhandlungen*, I, 618 ff.
[2] W. Gates in De Landa, *Yucatan*, note to p. 60.
[3] This ceremony was described by G. A. Dorsey. See *infra*, the Section, "Venus in the Folklore of the Indians."
[4] Leviticus 25 : 9 ff.

Ever since, exegetes have laboured over the biblical statement that the jubilee year was to be observed every fiftieth year. The seventh sabbatical year is the forty-ninth year: "And the space of the seven sabbaths of years shall be unto thee forty and nine years. . . . And ye shall hallow the fiftieth year." [1] To leave the land fallow for two consecutive years was too great a demand and cannot be explained by the need of the soil under cultivation for rest. The festival of the jubilee, with the return of land to its original owners and the release of slaves, bears the character of an atonement, and its proclamation on the Day of Atonement emphasizes this still further. Was there any special reason why fear returned every fifty years? The jubilee of the Mayas must have had a genesis similar to that of the jubilee of the Israelites. The difference lies in the human character of the festival of the Jews and its inhuman character among the Mayas; but with both peoples it was a year of atonement, repeating itself every fiftieth year in the one case and every fifty-second year in the other.

Comets do not return at exact periods because of perturbations caused by larger planets.[2] The Mayas expected the return of a catastrophe every fifty-second year because that was the interval between two cataclysms that had taken place. It may be that the comet was actually seen again at such intervals. The Jews fasted and prepared themselves for the Day of Judgment on the earliest possible date of its return; the Mayas had their festival when the dreaded time had passed without harm.

On the Day of Atonement the Israelites used to send a scapegoat to "Azazel" in the desert.[3] It was a ceremony of propitiation of Satan. In Egypt the goat was an animal dedicated to Seth-Typhon.[4] Azazel was a fallen star or Lucifer. It was also called Azzael, Azza, or Uzza.[5] According to the rabbinical legend, Uzza was the star angel of Egypt: it was thrown into the Red Sea when the Israelites made their passage.[6] The Arab name of the planet Venus is al-Uzza.[7] The

[1] Leviticus 25 : 8-10.

[2] Halley's comet has an average period of 77 years, with single periods as short as 74½ years or as long as 79½ years.

[3] Leviticus 16 : 8-26. The priests used to cast lots for two goats: one goat for the Lord and the other as the scapegoat for Azazel.

[4] Plutarch, *Isis and Osiris*, 73; cf. Herodotus ii. 46, Diodorus i. 84.4, and Strabo xvii. 1.19.

[5] Ginzberg, *Legends*, V, 152, 170.

[6] *Ibid.*, VI, 293. According to another legend, the fallen angel Uzza is chained to the Mountains of Darkness (*ibid.*, V, 170), the Caucasus.

[7] See "al-Uzza." *Encyclopaedia of Islam* (1913-1934), Vol. IV.

Arabs used to bring human sacrifices to al-Uzza; Mohammed, too, in his early days, worshipped it, and even today the Arabs seek its help.[1]

On the day on which the jubilee year was proclaimed, the Israelites dispatched a placating offering of a scapegoat to Lucifer. But what had Venus to do with the jubilee and the atonement?

## THE BIRTH OF VENUS

A planet turns and revolves on a quite circular orbit around a greater body, the sun; it makes contact with another body, a comet, that travels on a stretched out ellipse. The planet slips from its axis, runs in disorder off its orbit, wanders rather erratically, and in the end is freed from the embrace of the comet.

The body on the long ellipse experiences similar disturbances. Drawn off its path, it glides to some new orbit; its long train of gaseous substances and stones is torn away by the sun or by the planet, or runs away and revolves as a smaller comet along its own ellipse; a part of the tail is retained by the parent comet on its new orbit.

Ancient Mexican records give the order of the occurrences. The sun was attacked by Quetzal-cohuatl; after the disappearance of this serpent-shaped heavenly body, the sun refused to shine, and during four days the world was deprived of its light; a great many people died at that time. Thereafter, the snakelike body transformed itself into a great star. The star retained the name of Quetzal-cohuatl (Quetzal-coatl). This great and brilliant star appeared for the first time in the east.[2] Quetzal-cohuatl is the well-known name of the planet Venus.[3]

Thus we read that "the sun refused to show itself and during four days the world was deprived of light. Then a great star . . . appeared; it was given the name Quetzal-cohuatl . . . the sky, to show its anger . . . caused to perish a great number of people who died of famine and pestilence." [4] The sequence

[1] J. Wellhausen, *Reste arabischen Heidentums* (2nd ed., 1897), pp. 40-44; C. M. Doughty, *Travels in Arabia Deserta* (new ed., 1921), II, 516; P. K. Hitti, *History of the Arabs* (1937), pp. 98 ff.

[2] Brasseur, *Histoire des nations civilisées du Mexico*, I, 181.

[3] Seler, *Gesammelte Abhandlungen*, I, 625.

[4] Brasseur, *Histoire des nations civilisées du Mexique*, I, 311.

of seasons and the duration of days and nights became disarranged. "It was then . . . that the people [of Mexico] regulated anew the reckoning of days, nights, and hours, according to the difference in time." [1]

"It is a remarkable thing, moreover, that time is measured from the moment of its [Morning Star's] appearance. . . . Tlahuizcal-panteuctli or the Morning Star appeared for the first time following the convulsions of the earth overwhelmed by a deluge." It looked like a monstrous serpent. "This serpent is adorned with feathers: that is why it is called Quetzal-cohuatl, Gukumatz or Kukulcan. Just as the world is about to emerge from the chaos of the great catastrophe, it is seen to appear." [2] The feather arrangement of Quetzal-cohuatl "represented flames of fire." [3]

Again, the old texts speak "of the change that took place, at the moment of the great catastrophe of the deluge, in the condition of many constellations, principal among them being precisely Tlahuiz-calpanteuctli or the star of Venus." [4]

The cataclysm, accompanied by a prolonged darkness, appears to have been that of the days of the Exodus, when a tempest of cinders darkened the world disturbed in its rotation. Some of the references may allude to the subsequent catastrophe of the time of the conquest by Joshua, when the sun remained for more than a day in the sky of the old world. Since it was the same comet that on both occasions made contact with the earth, and at each of the contacts the comet changed its own orbit, the relevant question is not, "On which occasion did the comet change its orbit?" but first of all, "Which comet changed to a planet?" or "Which planet was a comet in historical times?" The transformation of the comet into a planet began on contact with the earth in the middle of the second millennium before the present era and was carried a step further one jubilee period later.

After the dramatic events of the time of Exodus, the earth was shrouded in dense clouds for decades, and observation of stars was not possible; after the second contact, Venus, the new and splendid member of the solar family, was seen moving along its orbit. It was in the days of Joshua, a time designation meaningful to the reader of the sixth book of the Scriptures;

[1] Brasseur, *Histoire des nations civilisées du Mexique,* I, 120.
[2] Brasseur, *Sources de l'histoire primitive du Mexique,* p. 82.
[3] Sahagun, *A History of Ancient Mexico* (transl. F. R. Bandelier, 1932), p. 26.
[4] Brasseur, *Sources de l'histoire primitive du Mexique,* p. 48.

but for the ancients it was "the time of Agog." As I explained above, he was the king by whose name the cataclysm (the Deluge of Ogyges) was known, and who, according to Greek tradition, laid the foundations of Thebes in Egypt.

In *The City of God* by Augustine it is written:

"From the book of Marcus Varro, entitled *Of the Race of the Roman People,* I cite word for word the following instance: 'There occurred a remarkable celestial portent; for Castor records that in the brilliant star Venus, called Vesperugo by Plautus, and the lovely Hesperus by Homer, there occurred so strange a prodigy, that it changed its colour, size, form, course, which never happened before nor since. Adrastus of Cyzicus, and Dion of Naples, famous mathematicians, said that this occurred in the reign of Ogyges.' "[1]

The Fathers of the Church considered Ogyges a contemporary of Moses. Agog, mentioned in the blessing of Balaam, was the king Ogyges. The upheaval that took place in the days of Joshua and Agog, the deluge that occurred in the days of Ogyges, the metamorphosis of Venus in the days of Ogyges, the star Venus which appeared in the sky of Mexico after a protracted night and a great catastrophe—all these occurrences are related.

Augustine went on to make a curious comment on the transformation of Venus: "Certainly that phenomenon disturbed the canons of the astronomers . . . so as to take upon them to affirm that this which happened to the Morning Star (Venus) never happened before nor since. But we read in the divine books that even the sun itself stood still when a holy man, Joshua, the son of Nun, begged this from God."

Augustine had no inkling that Castor, as quoted by Varro, and the Book of Jasher, as quoted in the Book of Joshua, refer to the same occurrence.

Are Hebrew sources silent on the birth of a new star in the days of Joshua? They are not. It is written in a Samaritan chronicle that during the invasion of Palestine by the Israelites under Joshua, a new star was born in the east: "A star arose out of the east against which all magic is vain."[2]

Chinese chronicles record that "a brilliant star appeared in the days of Yahu [Yahou]."[3]

[1] Bk. XXI, Chap. 8 (transl. M. Dods).
[2] Ginzberg, *Legends*, VI, 179.
[3] Legge, *The Chinese Classics* (Hong Kong ed., 1865), III, Pt. 1, 112, note.

Plato, citing the Egyptian priest, said that the world conflagration associated with Phaëthon was caused by a shifting of bodies in the sky which move around the earth. As we have reason to assume that it was the comet Venus that, after two contacts with the earth, eventually became a planet, we shall do well to inquire: Did Phaëthon turn into the Morning Star?

Phaëthon, which means "the blazing star,"[1] became the Morning Star. The earliest writer who refers to the transformation of Phaëthon into a planet is Hesiod.[2] This transformation is related by Hyginus in his *Astronomy*, where he tells how Phaëthon, that caused the conflagration of the world, was struck by a thunderbolt of Jupiter and was placed by the sun among the stars (planets).[3] It was the general belief that Phaëthon changed into the Morning Star.[4]

On the island of Crete, Atymnios was the name of the unlucky driver of the sun's chariot; he was worshipped as the Evening Star, which is the same as the Morning Star.[5]

The birth of the Morning Star, or the transformation of a legendary person (Istehar, Phaëthon, Quetzal-cohuatl) into the Morning Star was a widespread motif in the folklore of the oriental[6] and occidental[7] peoples. The Tahitian tradition of the birth of the Morning Star is narrated on the Society Island in the Pacific;[8] the Mangaian legend says that with the birth of a new star, the earth was battered by countless fragments.[9] The Buriats, Kirghiz, and Yakuts of Siberia, and the Eskimos of North America also tell of the birth of the planet Venus.[10]

A blazing star disrupted the visible movement of the sun, caused a world conflagration, and became the Morning-Evening Star. This may be found not only in the legends and traditions, but also in astronomical books of the ancient peoples of both hemispheres.

[1] Cf. Cicero, *De natura deorum* (transl. H. Rackham), ii. 52.
[2] *Theogony*, ll. 989 ff.
[3] Hyginus, *Astronomy*, ii. 42.
[4] See Roscher, "Phaëthon" in Roscher's *Lexikon d. griech. und röm. Mythologie*, Col. 2182.
[5] Nonnos *Dionysiaca* xi. 130 f.; xii, 217; xix. 182; Solinus, *Polyhistor* xi.
[6] Ginzberg, *Legends*, V, 170.
[7] Brasseur, *Histoire des nations civilisées du Mexique*, I, 311-312.
[8] Williamson, *Religious and Cosmic Beliefs of Central Polynesia*, I, 120.
[9] *Ibid.*, p. 43.
[10] Holmberg, *Siberian Mythology*, p. 432; Alexander, *North American Mythology*, p. 9.

## THE FOUR-PLANET SYSTEM

By asserting that the planet Venus was born in the first half of the second millennium, I assume also that in the third millennium only four planets could have been seen, and that in astronomical charts of this early period the planet Venus cannot be found.

In an ancient Hindu table of planets, attributed to the year —3102, Venus alone among the visible planets is absent.[1] The Brahmans of the early period did not know the five-planet system,[2] and only in a later ("middle") period did the Brahmans speak of five planets.

Babylonian astronomy, too, had a four-planet system. In ancient prayers the planets Saturn, Jupiter, Mars and Mercury are invoked; the planet Venus is missing; and one speaks of "the four-planet system of the ancient astronomers of Babylonia." [3] These four-planet systems and the inability of the ancient Hindus and Babylonians to see Venus in the sky, even though it is more conspicuous than the other planets, are puzzling unless Venus was not among the planets.

On a later date "the planet Venus receives the appellative: 'The great star that joins the great stars.' The great stars are, of course, the four planets Mercury, Mars, Jupiter and Saturn . . . and Venus joins them as the fifth planet." [4]

Apollonius Rhodius refers to a time "when not all the orbs were yet in the heavens." [5]

[1] J. B. J. Delambre, *Histoire de l'astronomie ancienne* (1817), I, 407: "Venus alone is not found there."

[2] "It is often denied that the Veda-Hindus knew of the existence of the five planets." "The striking fact that the Brahmans . . . never mention five planets." G. Thibaut, "Astronomie, Astrologie und Mathematik" in *Grundriss der indo-arischen Philol. und Altertumskunde*, III (1899).

[3] E. F. Weidner, *Handbuch der babylonischen Astronomie* (1915), p. 61, writes of a star list found in Boghaz Keui in Asia Minor: "That the planet Venus is missing will not startle anybody who knows the eminent importance of the four-planet system in the Babylonian astronomy." Weidner supposes that Venus is missing in the list of planets because "she belongs to a triad with the moon and the sun." On Ishtar in early inscriptions cf. infra, p. 175.

[4] *Ibid.*, p. 83.

[5] Apollonius Rhodius, *The Argonautica*, Bk. iv, ll. 257 ff.

# ONE OF THE PLANETS IS A COMET

Democritus (circa —460 to circa —370), a contemporary of Plato and one of the great scholars of antiquity, is accused by the moderns of not having understood the planetary character of Venus.[1] Plutarch quotes him as speaking of Venus as if it were not one of the planets. But apparently the author of the treatises on geometry, optics, and astronomy, no longer extant, knew more about Venus than his critics think. From quotations which have survived in other authors, we know that Democritus built a theory of the creation and destruction of worlds which sounds like the modern planetesimal theory without its shortcomings. He wrote: "The worlds are unequally distributed in space; here there are more, there fewer; some are waxing, some are in their prime, some waning: coming into being in one part of the universe, ceasing in another part. The cause of their perishing is collision with one another." [2] He knew that " the planets are at unequal distances from us" and that there are more planets than we are able to discover with our eyes.[3] Aristotle quoted the opinion of Democritus: "Stars have been seen when comets dissolve." [4]

Among the early Greek scholars, Pythagoras of the sixth century is generally credited with having had access to some secret science. His pupils, and their pupils, the so-called Pythagoreans, were cautious not to disclose their science to anyone who did not belong to their circle. Aristotle wrote of their interpretation of the nature of comets: "Some of the Italians called Pythagoreans say that the comet is one of the planets, but that it appears at great intervals of time and only rises a little above the horizon. This is the case with Mercury too; because it only rises a little above the horizon it often fails to be seen and consequently appears at great intervals of time." [5]

[1] "Democritus [says] that the fixed stars are in the highest place; after those the planets; after which the sun, Venus, and the moon, in their order." Plutarch, *Morals* (transl. "by several hands," revised by W. W. Goodwin), Vol. III, Chap. XV. Cf. Roscher's *Lexikon der Griech. u. Röm. Myth.*, col. 2182.

[2] Hippolytus, *The Refutation of All Heresies*, I, Chap. XI. Plato, who was a contemporary of Democritus, similarly described the destruction of the earth and its future rebirth in a far-away region of the universe (*Timaeus* 56 D).

[3] Seneca *Naturales quaestiones* vii. iii. 2.

[4] Aristotle *Meteorologica* i. 6.

[5] *Ibid.*

This is a confused presentation of a theory; but it is possible to trace the truth in the Pythagorean teaching, which was not understood by Aristotle. A comet is a planet which returns at long intervals. One of the planets, which rises only a little above the horizon, was still regarded by the Pythagoreans of the fourth century as a comet. With the knowledge obtained from other sources, it is easy to guess that by "one of the planets" is meant Venus; only Mercury and Venus rise a little above the horizon.

Aristotle disagreed with the Pythagorean scholars who considered one of the five planets to be a comet.

"These views involve impossibilities. . . . This is the case, first, with those who say that the comet is one of the planets . . . more comets than one have often appeared simultaneously . . . as a matter of fact, no planet has been observed besides the five. And all of them are often visible above the horizon together at the same time. Further, comets are often found to appear, as well when all the planets are visible as when some are not." [1]

With these words, Aristotle, who did not learn the secrets of the Pythagoreans directly, tried to refute their teaching by arguing that all five planets are in their places when a comet appears, as if the Pythagoreans thought that all comets were one and the same planet leaving its usual path at certain times. But the Pythagoreans did not think that one planet represents all comets. According to Plutarch,[2] they taught that each of the comets has its own orbit and period of revolution. Hence the Pythagoreans apparently knew that the comet which is "one of the planets" is Venus.

## THE COMET VENUS

During the centuries when Venus was a comet, it had a tail. The early traditions of the peoples of Mexico, written down in pre-Columbian days, relate that Venus smoked. "The star that smoked *la estrelle que humeava*, was Sitlae choloha, which the Spaniards call Venus." [3]

[1] Aristotle *Meteorologica* i. 6.

[2] Plutarch, "Les Opinions des philosophes," in *Œuvres de Plutarque* (transl. Amyot), Vol. XXI, Chap. III, Sec. 2.

[3] Humboldt, *Researches*, II, 174; see E. T. Hammy, *Codex Telleriano-Remensis* (1899).

"Now, I ask," said Alexander Humboldt, "what optical illusion could give Venus the appearance of a star throwing out smoke?" [1]

Sahagun, the sixteenth century Spanish authority on Mexico, wrote that the Mexicans called a comet "a star that smoked." [2] It may thus be concluded that since the Mexicans called Venus "a star that smoked," they considered it a comet.

It is also said in the *Vedas* that the star Venus looks like fire with smoke.[3] Apparently, the star had a tail, dark in the daytime and luminous at night. In very concrete form this luminous tail, which Venus had in earlier centuries, is mentioned in the Talmud, in the Tractate Shabbat: "Fire is hanging down from the planet Venus." [4]

This phenomenon was described by the Chaldeans. The planet Venus "was said to have a beard." [5] This same technical expression ("beard") is used in modern astronomy in the description of comets.

These parallels in observations made in the valley of the Ganges, on the shores of the Euphrates, and on the coast of the Mexican Gulf prove their objectivity. The question must then be put, not in the form, What was the illusion of the ancient Toltecs and Mayas? but, What was the phenomenon and what was its cause? A train, large enough to be visible from the earth and giving the impression of smoke and fire, hung from the planet Venus.

Venus, with its glowing train, was a very brilliant body; it is therefore not strange that the Chaldeans described it as a "bright torch of heaven," [6] also as a "diamond that illuminates like the sun," and compared its light with the light of the rising sun.[7] At present, the light of Venus is less than one millionth of the light of the sun. "A stupendous prodigy in the sky," the Chaldeans called it.[8]

[1] Humboldt, *Researches*, II, 174.
[2] Sahagun, *Historia general de las cosas de Nueva España*, Bk. VII, Chap. 4.
[3] J. Scheftelowitz, *Die Zeit als Schicksalsgottheit in der iranischen Religion* (1929), p. 4; Venus: "aussieht wie ein mit Rauch versehenes Feuer" ("looks like a fire accompanied by smoke"). Cf. *Atharva-Veda* vi. 3, 15.
[4] Babylonian Talmud, Tractate Shabbat 156a.
[5] M. Jastrow, *Religious Belief in Babylonia and Assyria* (1911), p. 221; cf. J. Schaumberger, "Der Bart der Venus" in F. X. Kugler, *Sternkunde und Sterndienst in Babel* (3rd supp., 1935), p. 303.
[6] "A Prayer of the Raising of the Hand to Ishtar," in *Seven Tablets of Creation*, ed. L. W. King.
[7] Schaumberger in Kugler, *Sternkunde in Sterndienst in Babel*, 3rd supp., p. 291.
[8] *Ibid.*

The Hebrews similarly described the planet: "The brilliant light of Venus blazes from one end of the cosmòs to the other end."[1]

The Chinese astronomical text from Soochow refers to the past when "Venus was visible in full daylight and, while moving across the sky, rivalled the sun in brightness."[2]

As late as the seventh century, Assurbanipal wrote about Venus (Ishtar) "who is clothed with fire and bears aloft a crown of awful splendour."[3] The Egyptians under Seti thus described Venus (Sekhmet): "A circling star which scatters its flame in fire . . . a flame of fire in her tempest."[4]

Possessing a tail and moving on a not yet circular orbit, Venus was more of a comet than a planet, and was called a "smoking star" or a comet by the Mexicans. They also called it by the name of Tzontemocque, or "the mane."[5] The Arabs called Ishtar (Venus) by the name Zebbaj or "one with hair," as did the Babylonians.[6]

"Sometimes there are hairs attached to the planets," wrote Pliny;[7] an old description of Venus must have served as a basis for his assertion. But hair or *coma* is a characteristic of comets, and in fact "comet" is derived from the Greek word for "hair." The Peruvian name "Chaska" (wavy-haired)[8] is still the name for Venus, though at present the Morning Star is definitely a planet and has no tail attached to it.

The coma of Venus changed its form with the position of the planet. When the planet Venus approaches the earth now, it is only partly illuminated, a portion of the disc being in shadow; it has phases like the moon. At this time, being closer to the earth, it is most brilliant. When Venus had a coma, the horns of its crescent must have been extended by the illuminated portions of the coma. It thus had two long appendages and looked like a bull's head.

[1] Midrash Rabba, Numeri 21, 245a: "Noga shezivo mavhik me'sof haolam ad sofo." Cf. "Mazal" and "Noga" in J. Levy, *Wörterbuch über die Talmudim und Midrashim* (2nd ed., 1924).

[2] W. C. Rufus and Hsing-chih tien, *The Soochow Astronomical Chart* (1945).

[3] D. D. Luckenbill, *Ancient Records of Assyria* (1926-1927), II, Sec. 829.

[4] Breasted, *Records of Egypt*, III, Sec. 117.

[5] Brasseur, *Sources de l'histoire primitive du Mexique*, p. 48, note.

[6] H. Winckler, *Himmels- und Weltenbild der Babylonier* (1901), p. 43.

[7] Pliny, *Natural History*, ii. 23.

[8] "The Peruvians call the planet Venus by the name Chaska, the wavy-haired." H. Kunike, "Sternmythologie auf ethnologischer Grundlage" in *Welt und Mensch*, IX-X. E. Nordenskiöld, *The Secret of the Peruvian Quipus* (1925), pp. 533 ff.

Sanchoniathon says that Astarte (Venus) had a bull's head.[1] The planet was even called Ashteroth-Karnaim, or Astarte of the Horns, a name given to a city in Canaan in honour of this deity.[2] The golden calf worshipped by Aaron and the people at the foot of Sinai was the image of the star. Rabbinical authorities say that "the devotion of Israel to this worship of the bull is in part explained by the circumstance that, while passing through the Red Sea, they beheld the celestial Throne, and most distinctly of the four creatures about the Throne, they saw the ox." [3] The likeness of a calf was placed by Jeroboam in Dan, the great temple of the Northern Kingdom.[4]

Tistrya of the *Zend-Avesta*, the star that attacks the planets, "the bright and glorious Tistrya mingles his shape with light moving in the shape of a golden-horned bull." [5]

The Egyptians similarly pictured the planet and worshipped it in the effigy of a bull.[6] The cult of a bull sprang up also in Mycenaean Greece. A golden cow head with a star on its brow was found in Mycenae, on the Greek mainland.[7]

The people of far away Samoa, primitive tribes that depend on oral tradition as they have no art of writing, repeat to this day: "The planet Venus became wild and horns grew out of her head." [8]

Examples and references could be multiplied *ad libitum*.

The astronomical texts of the Babylonians describe the horns of the planet Venus. Sometimes one of the two horns became more prominent. Because the astronomical works of antiquity have so much to say about the horns of Venus, modern scholars have asked themselves whether the Babylonians could have seen the phases of Venus, which cannot now be distinguished with the naked eye; [9] Galileo saw them for the first time in modern history when he used his telescope.

[1] Cf. L. Thorndike, *A History of Magic and Experimental Science* (1923-1941), I, Chap. X.
[2] Genesis 14 : 5. See also I Maccabee v. 26, 43, and II Maccabee xii. 21-26; G. Rawlinson, *The History of Herodotus* (1858), II, 543.
[3] Ginzberg, *Legends*, III, 123. [4] I Kings 12 : 28.
[5] *The Zend-Avesta* (transl. James Darmesteter, 1883), Pt. II, p. 93.
[6] Cf. E. Otto, *Beiträge zur Geschichte der Stierkulte in Ägypten* (1938).
[7] H. Schliemann, *Mycenae* (1870), p. 217.
[8] Williamson, *Religious and Cosmic Beliefs of Central Polynesia*, I, 128.
[9] "It is well known that not a few passages in the cuneiform texts on astrology speak of the right or the left horn of Venus. It was deduced that the phases of Venus were observed already by the Babylonians and that Galileo, in the sixteenth century, was not the first to see them." Schaumberger, "Die Hörner der Venus" in Kugler, *Sternkunde*, 3rd Supp., pp. 302 ff.

The long horns of Venus could have been seen without the aid of a telescopic lens. These horns were the illuminated portions of the coma of Venus, which stretched toward the earth. These horns could also have extended toward the sun as Venus approached the solar orb, since comets were repeatedly observed with projections in the direction of the sun, while the tails of the comets are regularly directed away from the sun.

When Venus approached close to one of the planets, its horns grew longer: this is the phenomenon the astrologers of Babylon observed and described when Venus neared Mars.[1]

[1] Schaumberger, "Die Hörner der Venus", Kugler, *Sternkunde*, 3rd Supp., pp. 302 ff.

CHAPTER IX

# PALLAS ATHENE

IN EVERY COUNTRY of the ancient world we can trace cosmological myths of the birth of the planet Venus. If we look for the god or goddess who represents the planet Venus, we must inquire which among the gods or goddesses did not exits from the beginning, but was born into the family. The mythologies of all peoples concern themselves with the birth only of Venus, not with that of Jupiter, Mars, or Saturn. Jupiter is described as heir to Saturn, but his birth is not a mythological subject. Horus of the Egyptians and Vishnu, born of Shiva, of the Hindus, were such newborn deities. Horus battled in the sky with the monster-serpent Seth; so did Vishnu. In Greece the goddess who suddenly appeared in the sky was Pallas Athene. She sprang from the head of Zeus-Jupiter. In another legend she was the daughter of a monster, Pallas-Typhon, who attacked her and whom she battled and killed.

The slaying of the monster by a planet-god is the way in which the peoples perceived the convulsion of the pillar of smoke when the earth and the comet Venus disturbed each other in their orbits, and the head of the comet and its tail leaped against each other in violent electrical discharges.

The birth of the planet Athene is sung in the Homeric hymn dedicated to her, "the glorious goddess, virgin, Tritogeneia." When she was born, the vault of the sky—the great Olympus—"began to reel horribly," "earth round about cried fearfully," "the sea was moved and tossed with dark waves, while foam burst forth suddenly," and the sun stopped for "a long while."[1] The Greek text speaks of "purple waves" [2] and of "the sea [that] rises up like a wall," and the sun stopping on its course.[3]

Aristocles said that Zeus hid the unborn Athene in a cloud and then split it open with lightning,[4] which is the mythological

[1] "The Homeric Hymns to Athena" (transl. Evelyn-White) in Hesiod's volume in the Loeb Classical Library.
[2] The correct translation requires "purple waves"; see "The Homeric Hymn to Minerva" (transl. A. Buckley) in *The Odyssey of Homer with the Hymns* (1878).
[3] L. R. Farnell, *The Cults of the Greek States* (1896), I, 281.
[4] *Ibid.*

way to describe the appearance of a celestial body from the pillar of cloud.

Athene, or Latin Minerva, is called Tritogeneia (or Tritonia) after the lake Triton.[1] This lake disappeared in a catastrophe in Africa when it broke into the ocean, leaving the desert of Sahara behind it, a catastrophe connected with the birth of Athene.

Diodorus,[2] referring to undisclosed older authorities, says that Lake Triton in Africa "disappeared from sight in the course of an earthquake, when those parts of it which lay toward the ocean were torn assunder." This account implies that a great lake or marsh in Africa, separated from the Atlantic Ocean by a mountainous barrier, disappeared when the barrier was broken or lowered in a catastrophe. Ovid says that Libya became a desert in consequence of Phaëthon's conflagration.

In the *Iliad* it is said that Pallas Athene "darted down to earth a gleaming star" with sparks springing from it; it darted as a star "sent by Jupiter to be a portent for seamen or for a wide host of warriors, a gleaming star," [3] Athene's counterpart in the Assyro-Babylonian pantheon is Astarte (Ishtar) who shatters mountains, "bright torch of heaven" at whose appearance "heaven and earth quake," who causes darkness and appears in a hurricane.[4] Like Astarte (Ashteroth-Karnaim), Athene was pictured with horns. "Athena, daughter of Zeus . . . upon her head she set the helmet with two horns," said Homer.[5] Pallas Athene is identified with Astarte (Ishtar) or the planet Venus of the Babylonians.[6] Anaitis of the Iranians, too, is identified as Pallas Athene and as the planet Venus.[7]

Plutarch identified Minerva of the Romans or Athene of the Greeks with Isis of the Egyptians, and Pliny identified the planet Venus with Isis.[8]

It is necessary to recall this here because it is generally supposed that the Greeks had no deity of importance who

[1] "Minerva . . . is reported to have appeared in virgin age in the times of Ogyges at the lake called Triton, from which she is also styled Tritonia." Augustine, *The City of God*, Bk. XVIII, Chap. 8.

[2] Diodorus of Sicily iii. 55 (transl. C. H. Oldfather).

[3] *Iliad* iv. 75 f.

[4] "A Prayer . . . to Ishtar" in *Seven Tablets of Creation* (transl. King); Farnell, *The Cults of the Greek States*, I, 258 ff.

[5] *Iliad* v. 735.

[6] S. Langdon, *Tammuz and Ishtar* (1914), p. 97.

[7] F. Cumont, *Les Mystères de Mithra* (3rd ed., 1913), p. 111.

[8] Plutarch, *Isis and Osiris*, Chap. 62: "They often call Isis by the name of Athena." See G. Rawlinson, *The History of Herodotus*, II, 542; Pliny, *Natural History*, ii, 37.

personified the planet Venus[1] and that, on the other hand, they "did not find even a star in which to place" Athene.[2] Modern books on the mythology of the Greeks repeat today what Cicero wrote: "Venus, called in Greek Phosphorus and in Latin Lucifer when it preceded the sun, but when it follows it Hesperos."[3] Phosphorus does not play any role on Olympus. But following Cicero in his description of the planets, we read also of "the planet called Saturn's, the Greek name of which is Phaenon," though we know a more common name, Cronus, by which the Greeks called the planet Saturn. Cicero gives the Greek names of other planets which are not the common ones. It is therefore entirely wrong to think that Phosphorus and Hesperos are the chief or only names of the planet Venus in Greek. Athene, in whose honour the city of Athens was named, was the planet Venus. Next to Zeus she was the most honoured deity of the Greeks. The name Athene in Greek, according to Manetho, "is indication of self-originated movement." He wrote of the name Athene as meaning, "I came from myself."[4] Cicero, speaking of Venus, explained the origin of the name thus: "Venus was so named by our countrymen as the goddess who 'comes' [*venire*] to all things."[5] The name Vishnu signifies "pervader," from the Sanskrit *vish*, to "enter" or "pervade."

The birth of Athene was assigned to the middle of the second millennium. Augustine wrote: "Minerva [Athene] is reported to have appeared . . . in the times of Ogyges." This statement is found in *The City of God*,[6] the book containing the quotation from Varro that the planet Venus changed its course and form in the time of Ogyges. Augustine also synchronized Joshua with the time of Minerva's activities.[7]

[1] The name Venus or Aphrodite belonged to the moon.

[2] Augustine, *The City of God*, Bk. VII, Chap. 16. Farnell, *The Cults of the Greek States*, I, 263, discusses the various hypotheses of the physical nature of Athene and, unable to agree with any, asks: "Is there any proof that Athene, as a goddess of the Hellenic religion, ever was a personification of some part of the physical world?"

Cicero *De natura deorum* i. 41, referred to a treatise by the Stoic Diogenes Babylonius, *De Minerva*, in which its author gave a natural explanation of the birth of Athene. The work is not extant.

[3] Cicero *De natura deorum* ii. 53.

[4] "The usage of the Egyptians is also similar: they often call Isis by the name of Athena, which expresses some such meaning as 'I came from myself,' and is indication of self-originated movement." Manetho, cited by Plutarch, *Isis and Osiris* (transl. Waddell), Chap. 62. But cf. Farnell, *The Cults of the Greek States*, I, 258: "The meaning of the name remains unknown."

[5] Cicero *De natura deorum* ii. 69.

[6] *The City of God*, Bk. XVIII, Chap. 8.

[7] *Ibid.*, Bk. XVIII, Chap. 12.

The cover of carbonigenous clouds in which the earth was enveloped by the comet is the "robe ambrosial" wrought by Athene for Hera (Earth).[1] The source of ambrosia was closely connected with Athene.[2] The origin of Athene as a comet is implied in her epithet Pallas which, as is commonly known, is synonymous with Typhon; Typhon, as Pliny said, was a comet.

The bull and the cow, the goat and the serpent, were animals dedicated to Athene. "The goat being usually tabooed but chosen as an exceptional victim for her," the animal was annually sacrificed on the Acropolis of Athens.[3] With the Israelites the goat was the victim for Azazel, or Lucifer.

In the Babylonian calendar "the nineteenth day of all months is marked 'day of wrath' of goddess Gula (Ishtar). No work was done. Weeping and lamentation filled the land. . . . Any explanation of *dies irae* of Babylonia must be sought in some myth concerning the nineteenth of the first month. Why should the nineteenth day after the moon of the spring equinox be a day of wrath? . . . It corresponds to the quinquatrus of the Roman farmer's calendar, the nineteenth of March, five days after the full moon. Ovid says that Minerva was born on that day, she being the Pallas Athene of the Greeks."[4] The nineteenth of March was Minerva's day.

The first appearance of Athene-Minerva took place on the day the Israelites crossed the Red Sea. The night between the thirteenth and the fourteenth days of the first month after the vernal equinox was the night of the great earthshock; six days later, on the last day of Passover week, according to the Hebrew tradition, the waters were heaped up like mountains and the fugitives crossed on the dry bed of the sea.

The birth of Pallas Athene or her first visit to earth was the cause of a cosmic disturbance, and the memory of that catastrophe was "a day of wrath in all the calendars of ancient Chaldea."

[1] *Iliad* xiv. 170 ff. In the Babylonian mythology Marduk cuts Tiamat in two and makes from one part a cover or veil for the sky.

[2] T. Bergk, "Die Geburt der Athene" in Fleckeisen's *Jahrbücher für classische Philologie* (1860), Chap. VI, refers to the relation of Athene to the "Quellen der Ambrosia" ("the sources of ambrosia"). Apollodorus (*The Library*) says that Athene "slayed Pallas and used his skin," which appears to refer to the envelope of Venus that previously formed the tail of the comet.

[3] Farnell, *The Cults of the Greek States*, I, 290.

[4] Langdon, *Babylonian Menologies and the Semitic Calendars* (1935), pp. 86-87.

## ZEUS AND ATHENE

If there was a problem in this research which caused prolonged deliberation on the part of the author, it was the question: Was it the planet Jupiter or Venus that caused the catastrophe of the time of Exodus? Some ancient mythological sources point to Venus, other sources point to Jupiter. In one group of legends Jupiter (Zeus) is the protagonist of the drama: he leaves his place in the sky, rushes to battle Typhon, and strikes him with thunderbolts. But other legends and historical sources, too, which I have quoted on previous pages, indicate that it was the planet Venus, or Pallas Athene of the Greeks. Athene killed her father, Typhon-Pallas, the celestial monster, and the description of this battle is not different from that of the battle in which Zeus killed Typhon.

Under the weight of many arguments, I came to the conclusion—about which I no longer have any doubt—that it was the planet Venus, at the time still a comet, that caused the catastrophe of the days of Exodus. Then why do a part of the legends tie up this event with Jupiter?

The cause of this duality in the mythological handling of an historical event lies in the fact that the ancients themselves did not know for certain which of the planets had caused the destruction. Some saw the pillar of cloud—Typhon defeated by Jupiter, the ball of fire that emerged from the pillar and battled with it. Others interpreted the globe as a body different from Jupiter.

The Greek authors described the birth of Athene (planet Venus), saying she sprang from the head of Jupiter. "And mighty Olympus trembled fearfully . . . and the earth around shrieked fearfully, and the sea was stirred, troubled with its purple waves."[1] One or two authors thought that Athene was born of Cronus. But the consensus of ancient authors makes Athene-Venus the offspring of Jupiter: she sprang from his head, and this birth was accompanied by great disturbances in the celestial and terrestrial spheres. The comet rushed toward the earth, and it could not be very well distinguished whether the planet Jupiter or its offspring was approaching. I may divulge here something that belongs to the second book of this

[1] "The Homeric Hymn to Minerva" (transl. Buckley) in *The Odyssey of Homer with the Hymns*. Cf. the translation on p. 168.

work; namely, that at an earlier time, Jupiter had already caused havoc in the planetary family, the earth included, and it was therefore only natural to see in the approaching body the planet Jupiter.

I referred in the introductory part of this work to the modern theory which ascribes the birth of the terrestrial planets to the process of expulsion by larger ones. This appears to be true in the case of Venus. The other modern theory, which ascribes the origin of comets of short period to expulsion by large planets, is also correct: Venus was expelled as a comet and then changed to a planet after contact with a number of members of the solar system.

Venus, being an offspring of Jupiter, bore all the characteristics known to men from early cataclysmic encounters. When a ball of fire tore the pillar of cloud and pelted the pillar with thunderbolts, the imagination of the people saw in this the planet-god Jupiter-Marduk rushing to save the earth by killing the serpent-monster Typhon-Tiamat.

It is not strange, therefore, that, in places as remote from Greece as the islands of Polynesia, it is related that "the planet Jupiter suppressed the tail of the great storm." [1] But we are told that in the same places, notably on the Harvey Islands, "Jupiter was often mistaken for the Morning Star." [2] On other islands of Polynesia, "the planets Venus and Jupiter seem to have been confused with each other." Explorers found "that the name Fauma or Paupiti was given to Venus . . . and that the same names were given to Jupiter." [3]

Early astronomy shared Ptolemy's opinion that "Venus has the same powers" and also the nature of Jupiter,[4] an opinion reflected also in the astrological belief that "Venus, when she becomes sole ruler of the event, in general brings about results similar to those of Jupiter." [5]

In one local cult in Egypt the name of Isis, as I shall show in the next volume, originally belonged to Jupiter, Osiris being Saturn. In another local cult Amon was the name for Jupiter. Horus originally was also Jupiter.[6] But when a new planet

[1] Williamson, *Religious and Cosmic Beliefs of Central Polynesia*, I, 123.
[2] *Ibid.*, p. 132. See also W. W. Gill, *Myths and Songs from the South Pacific* (1876), p. 44, and his *Historical Sketches of Savage Life in Polynesia* (1880), p. 38.
[3] Williamson, I, 122, See also J. A. Moerenhut, *Voyages aux isles du Grand Océan* (1837), II, p. 181.
[4] Ptolemy, *Tetrabyblos* (transl. F. E. Robbins, 1940), I, 4.
[5] *Ibid.*, II, 8.
[6] S. A. B. Mercer, *Horus, Royal God of Egypt* (1942).

was born of Jupiter and became supreme in the sky, the onlookers could not readily recognize the exact nature of this change. They gave the name of Isis to the planet Venus, and sometimes the name of Horus. This must have caused confusion. "One is confused by the various relations which exist between mother and son (Isis and Horus). Now he is her consort, now her brother; now a youth . . . now an infant fed at her breast."[1] "A noteworthy representation shows her [Isis] in association with Horus as the Morning Star, and thus in a strange relation . . . which we cannot yet explain from the texts."[2]

Also Ishtar of Assyria-Babylonia was in early times the name of the planet Jupiter; later it was transferred to Venus, Jupiter retaining the name of Marduk.

Baal, still another name for Jupiter, was an earlier name for Saturn, and later on became the name of Venus, sometimes the feminine form Baalath or Belith being used.[3] Ishtar, also, was at first a male planet, subsequently becoming a female planet.[4]

## WORSHIP OF THE MORNING STAR

Now that it has been shown it was Venus which, at an interval of fifty-two years, caused two cosmic catastrophes in the fifteenth century before the present era, we understand also the different historical connections between Venus and these catastrophes.

In numerous biblical and rabbinical passages it is said that when the Israelites went from Mount Sinai into the desert, they were covered by clouds. These clouds were illuminated by the pillar of fire, so that they gave a pale light.[5] With this should be connected a statement of Isaiah: "The people that walked in darkness have seen a great light; they that dwell in the land of the shadow of death, the light of Noga was upon them."[6] Noga is Venus; it is, in fact, the usual name of this

[1] Langdon, *Tammuz and Ishtar*, p. 24.
[2] W. M. Müller, *Egyptian Mythology*, p. 56.
[3] J. Bidez and F. Cumont, *Les Mages hellénisés* (1938), II, 116.
[4] C. Bezold in F. Ball, *Sternglaube und Sterndeutung* (1926), p. 9.
[5] See the Section, "The Shadow of Death."
[6] Isaiah 9 : 2.

planet in Hebrew,[1] and it is therefore an omission not to translate it so.

Amos says that during the forty years in the wilderness the Israelites did not sacrifice to the Lord, but carried "the star of your god, which you made to yourselves."[2] St. Jerome interprets this "star of your god" as Lucifer (the Morning Star).[3]

What image of the star was carried in the wilderness? Was it the bull (calf) of Aaron or the brazen serpent of Moses? "And Moses made a serpent of brass, and put it upon a pole."[4] Of this serpent it is said that it was made with the purpose of providing a cure for those bitten by snakes.[5] Seven and a half centuries later this brazen serpent of Moses was broken by King Hezekiah, guided in his monotheistic zeal by the prophet Isaiah, "for unto those days the children of Israel did burn incense to it."[6]

The brazen serpent was most probably the image of the pillar of cloud and fire which appeared as a moving serpent to all peoples of the world. St. Jerome apparently had this image in view when he interpreted the star mentioned by Amos as Lucifer. Or was it the "star of David," the six-pointed star?

The Egyptian Venus-Isis, the Babylonian Venus-Ishtar, the Greek Venus-Athene were goddesses pictured with serpents, and sometimes represented as dragons. "Ishtar, the fearful dragon," wrote Assurbanipal.[7]

[1] Tractate Shabbat 156a; Midrash Rabba, Numbers 21, 245a; J. Levy, *Wörterbuch über die Talmudim und Midraschim* (2nd ed. 1924), s.v. In the Hindu pantheon Naga or snake gods are apparently the comets. Cf. J. Hewitt, "Notes on the Early History of Northern India," *Journal of the Royal Asiatic Society* (1827), p. 325.

[2] Amos 5 : 26.

[3] Cf. Vulgate (Latin) version of the Prophet Amos and Jerome's Commentary on the Prophets.

[4] Numbers 21 : 9.

[5] Those who were bitten by serpents looked at the brazen serpent for cure. Can a psychosomatic relationship go such a long way? The practices of the snake worshippers lend some credence to the physiological background of Numbers 21 : 9. But it is outside the scope of the present research to go into these details.

The fact that Moses made an image—in violation of the second commandment of the Decalogue—is not necessarily inconsistent with his being monotheist: there are many churches today where symbolic and even human figures are deified by people who profess to be monotheists. But as time passed, the presence of the serpent of Moses in the Temple of Jerusalem became so objectionable to the spirit of the prophets that in the days of Isaiah the serpent was broken into pieces. Even though its original purpose may have been curative, it being the image of the angel who was sent in the pillar of fire and cloud to save the people of Israel from slavery, the brazen serpent with the lapse of time became an object of worship.

[6] II Kings 18 : 4. An astrological opinion is found in the rabbinical literature that the brazen serpent was a magic image, which obtained its power from the star under the protection of which Moses made it.

[7] Langdon, *Tammuz and Ishtar*, p. 67.

The Morning Star of the Toltecs, Quetzal-cohuatl (Quetzalcoatl), also is represented as a great dragon or serpent: "cohuatl" in Nahuatl is "serpent", and the name means "a feathered serpent." [1] The Morning Star of the Indians of the Chichimec tribe in Mexico is called "Serpent cloud," [2] a remarkable name because of its relation to the pillar of cloud and the clouds that covered the globe after the contact of the earth with Venus.

When Quetzal-cohuatl, the lawgiver of the Toltecs, disappeared on the approach of a great catastrophe and the Morning Star that bore the same name rose for the first time in the sky, the Toltecs, "regulated the reckoning of the days, the nights, and the hours according to the difference in the time." [3]

The people of Ugarit (Ras-Shamra) in Syria addressed Anat, their planet Venus: "You reverse the position of the dawn in the sky." [4] In the Mexican *Codex Borgia*, the Evening Star is represented with the solar disc on its back.[5]

In the Babylonian psalms Ishtar says: [6]

By causing the heavens to tremble and the earth to quake,
By the gleam which lightens in the sky,
By the blazing fire which rains upon the hostile land,
I am Ishtar.
Ishtar am I by the light that arises in heaven,
Ishtar the queen of heaven am I by the light that arises in heaven.

I am Ishtar; on high I journey . . .
The heavens I cause to quake, the earth I cause to shake,
That is my fame. . . .
She that lightens in the horizon of heaven,
Whose name is honoured in the habitations of men,
That is my fame.
"Queen of heaven above and beneath" let be spoken,
That is my fame.
The mountains I overwhelm altogether,
That is my fame.

[1] Brasseur, *Sources de l'histoire primitive du Mexique*, pp. 81, 87.
[2] Alexander, *Latin American Mythology*, p. 87.
[3] Brasseur, *Histoire des nations civilisées du Mexique*, I, 120.
[4] Virolleaud, "La déesse Anat," *Mission de Ras Shamra*, IV.
[5] Seler, *Wandmalereien von Mitla* (1895), p. 45.
[6] Langdon, *Sumerian and Babylonian Psalms* (1909), pp. 188, 194.

The Morning-Evening Star Ishtar was called also "the star of lamentation."[1]

The Persian Mithra, the same as Tistrya, descended from the heavens and "let a stream of fire flow toward the earth," "signifying that a blazing star, becoming in some way present here below, filled our world with its devouring heat."[2]

In Aphaca in Syria fire fell from the sky, and it was asserted that it fell from Venus: "by which one would think of fire that had fallen from the planet Venus."[3] The place became holy and was visited each year by pilgrims.

The festivals of the planet Venus were held in the spring. "Our ancestors dedicated the month of April to Venus," wrote Macrobius.[4]

Baal of the Canaanites and of the Northern Kingdom of Israel was worshipped in Dan, the city of the cult of the calf, and throngs visited there during the week of Passover. The cult of Venus spread to Judea also. According to II Kings (23 : 5), King Josiah in the seventh century "put down the idolatrous priests, whom the kings of Judah had ordained to burn incense in the high places in the cities of Judah, and in the places round about Jerusalem; them also that burned incense unto Baal, to the sun, and to the moon, and to the planets, and to all the host of heaven." Baal, the sun, the moon, and the planets, is the division used also by Democritus: Venus, the sun, the moon, and the planets.

In Babylonia the planet Venus was distinguished from other planets and worshipped as a member of a trinity: Venus, Moon, and Sun.[5] This triad became the Babylonian holy trinity in the fourteenth century before the present era.[6]

In the *Vedas* the planet Venus is compared to a bull: "As a bull thou hurlest thy fire upon earth and heaven."[7] The Morning Star of the Phoenicians and Syrians was Ashteroth-Karnaim, Astarte of the Horns. Belith of Sidon was likewise Venus, and Izebel, wife of Ahab, made her the chief deity of

[1] Langdon, *Tammuz and Ishtar*, p. 86.
[2] F. Cumont, "La Fin du monde selon les mages occidentaux," *Revue de l'histoire des religions* (1931), p. 41.
[3] F. K. Movers, *Die Phönizier* (1841-1856), I, 640. Sources: Sozomen, *The Ecclesiastical History* ii. 5; Zosimus i. 58.
[4] Macrobe, *Oeuvres* (ed. Panckoncke, 1845), I, 253.
[5] H. Winckler, *Die babylonische Geisteskultur* (1919), p. 71.
[6] C. Bezold in F. Boll, *Sternglaube und Sterndeutung* (1926), p. 12.
[7] *Hymns of the Atharva-Veda* (transl. Bloomfield), Hymn ix.

the Northern Kingdom.[1] The "queen of heaven," referred to repeatedly by Jeremiah, was Venus. The women of Jerusalem made cakes for the queen of heaven and worshipped her from the roofs of their houses.[2]

On Cyprus it was neither Jupiter nor any other god but "Kypris Queen whom they with holy gifts were wont to appease . . . pouring libations out upon the ground of yellow honey." [3] Such libation, as already mentioned, was made in Athens in commemoration of the Flood of Deucalion.

Not long ago, in Polynesia, human sacrifices were offered to the Morning Star, Venus.[4] To the Arabian Morning Star, queen of the heaven—al-Uzza—boys and girls were sacrificed down to modern times.[5] Likewise, human sacrifices were brought to the Morning Star in Mexico; this was described by early Spanish authors,[6] and was still practised by Indians only a generation ago.[7] Quetzal-cohuatl "was called the god of winds" and of "flames of fire"; [8] the Greek Athene, too, was not only the planet, but also the goddess of storm and fire. The planet Venus was Lux Divina, the Divine Light, in the worship of the Roman imperial colonies.[9]

In Babylonia, Venus was pictured as a six-pointed star—which is also the shape of David's shield—or as a pentagram—a five-pointed star (seal of Solomon)—and sometimes as a cross; as a cross it was pictured in Mexico, too.

The attributes and deeds of the Morning Star was not invented by the peoples of the world: this star shattered mountains, shook the globe with such a violence, that it looked as if the heavens were shaking, was a storm, a cloud, a fire, a heavenly dragon, a torch, and a blazing star, and it rained naphtha on the earth.

Assurbanipal speaks of Ishtar-Venus, "who is clothed with fire and bears aloft a crown of awful splendour, [and who] rained fire over Arabia." [10] It has been shown previously that

[1] I Kings 18; Josephus, *Jewish Antiquities*, VIII, xiii, 1; Philo of Byblos, Fragment 2. 25; D. Chwolson, *Die Ssabier und der Ssabismus* (1856), II, 660.

[2] Jeremiah 7 : 18; 44 : 17-25. Wellhausen, *Reste arabischen Heidentums*, p. 41.

[3] *The Fragments of Empedocles* (transl. W. E. Leonard, 1908), Fragment 128, p. 59.

[4] Williamson, *Religious and Cosmic Beliefs of Central Polynesia*, II, 242.

[5] Wellhausen, *Reste arabischen Heidentums*, pp. 40-44, 115.

[6] *Manuscrit Ramírez*.

[7] G. A. Dorsey, *The Sacrifice to the Morning Star by the Skidi Pawnee*. This ceremony is described later in the present book.

[8] De Sahagun, *Historia general de las cosas de Nueva España*, I, Chap. V.

[9] Movers, *Die Phönizier*, II, 652.

[10] Luckenbill, *Records of Assyria*, II, Sec. 829.

the comet of the days of the Exodus rained naphtha over Arabia.

In the attributes and in the deeds ascribed to the planet Venus—Isis, Ishtar, Athene—we recognize the attributes and deeds of the comet described in the earlier sections of this book.

## THE SACRED COW

The comet Venus, of which it is said that "horns grew out of her head," or Astarte of the horns, Venus *cornuta*, looked like the head of a horned animal; and since it moved the earth out of its place, like a bull with its horns, the planet Venus was pictured as a bull.

The worship of a bullock was introduced by Aaron at the foot of Mount Sinai. The cult of Apis originated in Egypt in the days of the Hyksos, after the end of the Middle Kingdom,[1] shortly after the Exodus. Apis, or the sacred bull, was very much venerated in Egypt; when a sacred bull died, its body was mummified and placed in a sarcophagus with royal honours, and memorial services were held. "All the coffins and everything excellent and profitable for this august god (the bull Apis)" were prepared by the Pharaoh,[2] when "this god was conducted in peace to the necropolis, to let him assume his place in his temple."

The worship of a cow or bull was widespread in Minoan Crete and in Mycenaean Greece, for golden images of this animal with large horns were found in excavations.

Isis, the planet Venus,[3] was represented as a human figure with two horns, like Astarte (Ishtar) of the horns; and sometimes it was fashioned in the likeness of a cow. In time, Ishtar changed from male to female, and in many places worship of the bull changed to worship of the cow. The main reason for this seems to have been the fall of manna which turned the rivers into streams of honey and milk. A horned planet that produced milk most closely resembled a cow. In the *Hymns of the Atharva-Veda*, in which the ambrosia that falls from the sky is glorified, the god is exalted as the "great cow" which "drips

[1] "The Book of Sothis," in *Manetho* (transl. W. G. Waddell, Loeb Classical Library, 1940) says that in the days of the Hyksos king Aseth, "the bull-calf was deified and called Apis."

[2] The Apis inscription of Necho-Wahibre in Breasted, *Records of Egypt*, IV, 976 ff.

[3] Pliny, *Natural History*, ii. 37.

with streams of milk" and as "a bull" that "hurlest thy fire upon earth and heaven."[1] A passage of the *Ramayana* about the "celestial cow" says: "Honey she gave, and roasted grain . . . and curdled milk, and soup in lakes with sugared milk,"[2] which is the Hindu version of "rivers of milk and honey."

The "celestial cow" or "the heavenly Surabhi" ("the fragrant") was the daughter of the Creator: she "sprung from his mouth"; at the same time nectar and "excellent perfume" were spread, according to the Indian epic.[3] This description of the birth of the daughter from the mouth of the Creator is a Hindu parallel of Athene springing from the head of Zeus. Fragrance and nectar are mentioned in connection with the birth of the celestial cow, a combination that can be understood if we recall what we learned in the Sections "Ambrosia" and "Birth of the Planet Venus."

Down to the present day, the Brahmans worship the cow. Cows are regarded as daughters of the "heavenly cow." In India, as in other places, the worship of cows began in some period of recorded history. "We find in early Hindu literature sufficient information to establish the thesis that cows were once victimised at sacrifices and used at times as articles of food."[4] Then came the change. Cows became sacred animals, and ever since the religious law has forbidden the use of their meat for food. The *Atharva-Veda* repeatedly deprecates cow-killing as "the most heinous of crimes." "All that kill, eat or permit the slaughter of cows rot in hell for as many years as there are hairs on the body of the cow slain."[5] Capital punishment was prescribed for those who either stole, hurt, or killed a cow. "Whoever hurts or causes another to hurt, or steals or causes another to steal, a cow, should be slain." Even cows' urine and dung are sacred to the Brahmans. "All its excreta are hallowed. Not a particle ought to be thrown away as impure. On the contrary, the water it ejects ought to be preserved as the best of holy waters. . . . Any spot which a cow has condescended to honour with the sacred deposit of her excrement is forever afterwards consecrated ground."[6] Sprinkled on a sinner, it "converts him into a saint."

[1] Hymn to the honey-lash in *Hymns of the Atharva-Veda*, IX.
[2] L. L. Sundara Ram, *Cow-protection in India* (1927), p. 56.
[3] *Mahabharata*, XIII.
[4] Ram, *Cow-Protection in India*, p. 43.
[5] "Visistha Dharmasastra." See Ram, *Cow-Protection in India*, p. 40.
[6] M. Monier-Williams, *Brahmanism and Hinduism* (1891), pp. 317-319.

The bull is sacred to Shiva, "the god of destruction in the Hindu Trinity." "The consecration of the bulls and letting them loose as privileged beings to roam at their will and draw respect from all people is to be noted with particular interest. . . . The freedom and privileges of the Brahman bull are inviolate." Even when it is destructive, the bull must not be restrained.[1]

These quotations show the Apis cult preserved until our times. The "celestial cow" that gored the earth with its horns and turned rivers and lakes into honey and milk is still revered in the common cow and bull by hundreds of millions of the people of India.

## BAAL ZEVUV (BEELZEBUB)

The beautiful Morning Star was related to Ahriman, Seth, Lucifer, name equivalents of Satan. It was also Baal of the Canaanites and of the Northern Kingdom of the Ten Tribes, the god hated by the biblical prophets, also Beelzebub or Baal Zevuv, or Baal of the fly.

In the Pahlavi text of the Iranian book, the *Bundahis*, describing the catastrophes caused by celestial bodies, it is written that at the close of one of the world ages "the evil spirit [Ahriman] went toward the luminaries." "He stood upon one-third of the inside of the sky, and he sprang, like a snake, out of the sky down to the earth." It was the day of the vernal equinox. "He rushed in at noon," and "the sky was shattered and frightened." "Like a fly, he rushed out upon the whole creation, and he injured the world and made it dark at midday as though it were in dark night. And noxious creatures were diffused by him over the earth, biting and venomous, such as the snake, scorpion, frog, and lizard, so that not so much as the point of a needle remained free from noxious creatures."[2]

Then the *Bundahis* proceeds: "The planets, with many demons [comets], dashed against the celestial sphere, and they mixed the constellations; and the whole creation was as disfigured as though fire disfigured every place and smoke arose over it."

A similar plague of vermin is described in the Scriptures, in

[1] Ram, *Cow-Protection in India*, p. 58.
[2] *Bundahis* (in the *Pahlavi Texts*, transl. West), Chap III.

Exodus, Chapters 8 to 10, and also in Psalm 78 where it is told that there were sent "divers sorts of flies among them [the people of Egypt], which devoured them; and frogs, which destroyed them." Their labour was given to the caterpillar and the locust. "The dust of the land became lice throughout all the land of Egypt." [1] "And there came a grievous swarm of flies . . . into all the land of Egypt." [2] The second, third, fourth, and eighth plagues were caused by vermin. The plague of *eruv*, "swarms of flies" of the King James Version, is translated in the *Septuagint*, "a stinging fly," and Philo calls it "the dog-fly," a ferocious insect; [3] it is also called "gnat" by the rabbis. Psalm 105 narrates that darkness was sent upon the country and "locusts came, and caterpillars, and that without number, and did eat up all the herbs." "Their land brought forth frogs in abundance, in the chambers of their kings," and "there came divers sorts of flies, and lice in all their coasts."

The Amalekites left Arabia because of "ants of the smallest kind" and wandered toward Canaan and Egypt at the same time that the Israelites went from Egypt toward the desert and Canaan.

In the Chinese annals describing the time of Yahou, from which I quoted previously, it said that when the sun did not set for ten days and the forests of China were destroyed by fire, multitudes of loathsome vermin were bred in the entire land.

During their wanderings in the desert, the Israelites were plagued by serpents.[4] A generation later, hornets preceded the Israelites under Joshua, plaguing the land of Canaan and driving entire nations from their domiciles.[5]

The inhabitants of the islands in the South Seas relate that when the clouds lay only a few feet from the ground and "the sky was so close to the earth that men could not walk," "myriads of dragonflies with their wings severed the clouds confining the heavens to the earth." [6]

At the close of the Middle Kingdom, the Egyptian standard bore the emblem of a fly.

When Venus sprang out of Jupiter as a comet and flew very close to the earth, it became entangled in the embrace of the earth. The internal heat developed by the earth and the

[1] Exodus 8 : 17. [2] Exodus 8 : 24. [3] Philo *Vita Mosis* i. 23.
[4] Numbers 21 : 6, 7; Deuteronomy 8 : 15.
[5] Exodus 23 : 28; Deuteronomy 7 : 20.
[6] Williamson, *Religious and Cosmic Beliefs of Central Polynesia*, I, 45.

scorching gases of the comet were in themselves sufficient to make the vermin of the earth propagate at a very feverish rate. Some of the plagues, like the plague of the frogs ("the land brought forth frogs") or of the locusts, must be ascribed to such causes. Anyone who has experienced a khamsin (sirocco), an electrically charged wind blowing from the desert, knows how, during the few days that the wind blows, the ground around the villages begins to teem with vermin.[1]

The question arises here whether or not the comet Venus infested the earth with vermin which it may have carried in its trailing atmosphere in the form of larvae together with stones and gases. It is significant that all around the world peoples have associated the planet Venus with flies.

In Ekron, in the land of the Philistines, there was erected a magnificent temple to Baal Zevuv, the god of the fly. In the ninth century King Ahaziah of Jezreel, after he was injured in an accident, sent his emissaries to ask advice of this god at Ekron and not of the oracle at Jerusalem.[2] This Baal Zevuv is Beelzebub of the Gospels.[3]

Ahriman, the god of darkness who battled with Ormuzd, the god of light, is compared in the *Bundahis* to a fly. Of the flies that filled the earth buried in gloom it is said: "His multitudes of flies scatter themselves over the world that is poisoned through and through."[4]

Ares (Mars) in the *Iliad* calls Athene "dog-fly." "The gods clashed with a mighty din, and the wide earth rang, and round about great heaven pealed as with a trumpet." And Ares spoke to Athene: "Wherefore now again, thou dog-fly, art making gods to clash with gods in strife?"[5]

The people of Bororo in central Brazil call the planet Venus "the sand fly,"[6] an appellation similar to that which Homer used for Athene. The Bantu tribes of central Africa relate that the "sand fly brought fire from the sky,"[7] which appears to be a reference to the Promethean role of Beelzebub, the planet Venus.

[1] A change in atmospheric conditions can cause galloping germination among insects.
[2] II Kings 1 : 2 ff.
[3] Matthew 10 : 25; 12 : 24, 27; Mark 7 : 22; Luke 11 : 15 ff.
[4] *Bundahis*, Chap. III, Sec. 12. Cf. H. S. Nyberg, "Die Religionen des alten Iran," *Mitteil. d. Vorderasiat.-ägypt. Ges.*, Vol. 43 (1938), pp. 28 ff.
[5] *Iliad* xxi. 385 ff. In Greek mythology, Metis, pregnant with Pallas, took the shape of a fly.
[6] See Kunike, "Sternmythologie," *Welt und Mensch*, IX-X.
[7] A. Werner, *African Mythology* (1925), p. 135.

The *Zend-Avesta*, describing the battle of Tistrya, "the leader of the stars against the planets" (Darmesteter), refers to worm-stars that "fly between the earth and heaven," and that supposedly signify the meteorites.[1] Possibly it is a reference to their infesting property.

This idea of contaminating planets is found in a belief of the Mexicans described by Sahagun: "The Mexicans called the comet *citalin popoca* which means a smoking star. . . . These natives called the tail of such a star *citlalin tlamina*, exhalation of the comet; or, literally, 'the star shoots a dart.' They believed that when such a dart fell on a living organism, a hare, a rabbit, or any other animal, worms suddenly formed in the wound and made the animal unfit to serve as food. It was for this reason that they took great care to cover themselves during the night so as to protect themselves from this inflaming emanation." [2]

The Mexicans thus thought that larvae from the emanation of the comet fell on all living things. As I have already mentioned, they called Venus a "smoking star." Sahagun says also that at the rising of the Morning Star, the Mexicans used to shut the chimneys and other apertures in order to prevent mishap from penetrating into the house together with the light of the star.[3]

The persistence with which the planet Venus is associated with a fly in the traditions of the peoples of both hemispheres, also the emblems carried by the Egyptian priests and the temple services conducted in honour of the planet-god "of the fly," create the impression that the flies in the tail of Venus were not merely the earthly brood, swarming in heat like other vermin, but guests from another planet.

The old question, whether there is life on other planets, has been debated time and again without much progress.[4] Atmospheric and thermal conditions are so different on other planets that it seems incredible that the same forms of life exist there as on the earth; on the other hand, it is wrong to conclude that there is no life on them at all.

Modern biologists toy with the idea that micro-organisms arrive on the earth from interstellar spaces, carried by the pressure of light. Hence, the idea of the arrival of living organisms from interplanetary spaces is not new. Whereas there

[1] *Zend-Avesta*, Pt. II, p. 95.
[2] Sahagun, *Historia general de las cosas de la Nueva España*, Bk. VIII, Chap. 3.
[3] *Ibid.*
[4] See H. Spencer Jones, *Life on Other Worlds* (1940) and Sir James Jeans, "Is There Life on Other Worlds?" *Science*, June 12, 1942.

is truth in this supposition of larval contamination of the earth is anyone's guess. The ability of many small insects and their larvae to endure great cold and heat and to live in an atmosphere devoid of oxygen renders not entirely improbable the hypothesis that Venus (and also Jupiter, from which Venus sprang) may be populated by vermin.

## VENUS IN THE FOLKLORE OF THE INDIANS

Primitive peoples often are bound by inflexible customs and beliefs that date back hundreds of generations. The traditions of many primitive peoples speak of a "lower sky" in the past, a "larger sun," a swifter movement of the sun across the firmament, a shorter day that became longer after the sun was arrested on its path.

World conflagration is a frequent motif in folklore. According to the Indians of the Pacific coast of North America the "shooting star" and the "fire drill" set the world aflame. In the burning world one "could see nothing but waves of flames; rocks were burning, the ground was burning, everything was burning. Great rolls and piles of smoke were rising; fire blew up toward the sky in flames, in great sparks and brands. . . . The great fire was blazing, roaring all over the earth, burning rocks, earth, trees, people, burning everything. . . . Water rushed in . . . it rushed in like a crowd of rivers, covered the earth, and put out the fire as it rolled on toward the south. . . . Water rose mountain high." A celestial monster flew with "a whistle in his mouth; as he moved forward he blew it with all his might, and made a terrible noise. . . . He came flowing and blowing; he looked like an enormous bat with wings spread . . . [his] feathers waved up and down, [and] grew till they could touch the sky on both sides."[1]

The shooting star that made the earth into a sea of flames, the terrible noise, the water that rose mountain high, and the appearance of a monster in the sky, like Typhon or a dragon, all these elements were not brought together in this Indian narrative by sheer invention; they belong together.

The Wichita, an Indian tribe of Oklahoma, tell the following story of "The Deluge and the Repeopling of the Earth":[2]

[1] Alexander, *North American Mythology*, p. 223.
[2] G. A. Dorsey, *The Mythology of the Wichita* (1904).

"There came to the people some signs, which showed that there was something in the north that looked like clouds; and the fowl of the air came, and the animals of the plains were seen. All of this indicated that something was to happen. The clouds that were seen in the north were a deluge. The deluge was all over the face of the earth."

The water monsters succumbed. Only four giants remained, but they fell, too, each on his face. "The one in the south as he was falling said that the direction he fell should be called south." The other giant said that "the direction in which he was falling should be called west—Where-the-sun-goes." The third fell and named the direction of his fall north; the last called his direction "east—Where-the-sun-rises."

Only a few men survived. The wind also survived on the face of the earth; everything else was destroyed. A child was born to a woman (from the wind), a Dream-girl. The girl grew rapidly. A boy child was born to her. "He told his people that he would go in the direction of the east, and he was to become the Morning Star."

This tale sounds like an incoherent story, but let us note its various elements: "Something in the north that looked like clouds" which made people and animals huddle together in apprehension of an approaching catastrophe; wild beasts emerging from the forests and coming to human abodes; an engulfing tide that destroyed everything, even the monster animals; the determination of the new four quarters of the horizon; a generation later the birth of the Morning Star.

This combination of elements cannot be accidental; all these events, and in the same sequence, were found to have occurred in the middle of the second millennium before the present era.

The Indians of the Chewkee tribe on the Gulf Coast tell: "It was too hot. The sun was put 'a handbreadth' higher in the air, but it was still too hot. Seven times the sun was lifted higher and higher under the sky arch, until it became cooler." [1]

In eastern Africa we can trace the same tradition. "In very old times the sky was very close to the earth." [2]

The Kaska tribe in the interior of British Columbia relate: "Once a long time ago the sky was very close to the earth." [3] The sky was pushed up and the weather changed.

[1] Alexander, *North American Mythology*, p. 60.
[2] L. Frobenius, *Dichten und Denken in Sudan* (1925).
[3] J. A. Teit, "Kaska Tales," *Journal of American Folk-Lore*, XXX (1917).

The sun, after being stopped on its way across the firmament, "became small, and small it has remained since then." [1]

Here is a story, told to Shelton by the Snohomish tribe on Puget Sound, about the origin of the exclamation "Yahu,"[2] to which I have already referred briefly. "A long time ago, when all the animals were still human beings, the sky was very low. It was so low that the people could not stand erect. . . . They called a meeting together and discussed how they could raise the sky. But they were at a loss to know how to do so. No one was strong enough to lift the sky. Finally the idea occurred to them that possibly the sky might be moved by the combined efforts of the people, if all of them pushed against it at the same time. But then the question arose of how it would be possible to make all the people exert their efforts at exactly the same moment. For the different peoples would be far away from one another, some would be in this part of the world, others in another part. What signal could be given that all people would lift at precisely the same time? Finally, the word 'Yahu!' was invented for this purpose. It was decided that all the people should shout 'Yahu!' together, and then exert their whole strength in lifting the sky. In accordance with this, the people equipped themselves with poles, braced them against the sky, and then all shouted 'Yahu!' in unison. Under their combined efforts the sky rose a little. Again the people shouted 'Yahu!' and lifted the heavy weight. They repeated this until the sky was sufficiently high." Shelton says that the word "Yahu" is used today when some heavy object like a large canoe is being lifted.

It is easy to recognize the origin of this legend. Clouds of dust and gases enveloped the earth for a long time; it seemed that the sky had descended low. The earth groaned repeatedly because of the severe twisting and dislocation it had experienced. Only slowly and gradually did the clouds lift themselves from the ground.

The clouds that enveloped the Israelites in the desert, the trumpet-like sounds that they heard at Mount Sinai, and the gradual lifting of the clouds in the years of the Shadow of Death are the same elements that we find in this Indian legend.

[1] Frobenius, *Das Zeitalter des Sonnengottes*, pp. 205 ff.
[2] Shelton, "Mythology of Puget Sound," *Journal of American Folk-Lore*, XXXVII (1924).

Because the same elements can be recognized in very different settings, we can affirm that there was no borrowing from one people by another. A common experience created the stories, so dissimilar at first, and so much alike on second thought.

The story of the end of the world, as related by the Pawnee Indians, has an important content. It was written down [1] from the mouth of an old Indian :

"We are told by the old people that the Morning Star ruled over all the minor gods in the heavens. . . . The old people told us that the Morning Star said that when the time came for the world to end, the Moon would turn red . . . that when the Moon should turn red, the people would know that the world was coming to an end.

"The Morning Star said further that in the beginning of all things they placed the North Star in the north, so that it should not move. . . . The Morning Star also said that in the beginning of all things they gave power to the South Star for it to move up close, once in a while, to look at the North Star to see if it were still standing in the north. If it were still standing there, it was to move back to its place. . . . When the time approached for the world to end, the South Star would come higher. . . . The North Star would then disappear and move away and the South Star would take possession of the earth and of the people. . . . The old people knew also that when the world was to come to an end, there were to be many signs. Among the stars would be many signs. Meteors would fly through the sky. The Moon would change its colour once in a while. The Sun would also show different colours.

"My grandchild, some of the signs have come to pass. The stars have fallen among the people, but the Morning Star is still good to us, for we continue to live. . . . The command for the ending of all things will be given by the North Star, and the South Star will carry out the command. . . . When the time comes for the ending of the world, the stars will again fall to the earth."

In this narrative of the Pawnee Indians, elements are brought together which, as we know now, actually belong together. The planet Venus established the present order on the earth and placed the north and south polar stars in their places. The Pawnee delieve that the future destruction of the world

[1] Dorsey, ed., *The Pawnee Mythology* (1906), Pt. I, p. 35.

depends on the planet Venus. When the end of the world will come, the North and South poles will change places. In the past the South Star left its place a few times and came up higher, bringing about a shifting of the poles, but on these occasions the polar stars did not reverse their positions.

The change in the colour of the sun and the moon was conditioned by the presence of cometary gases between the earth and these bodies; it is referred to in the Prophets of the Scriptures. Stones falling from the sky belong to the same complex of phenomena.

The Pawnee Indians are not versed in astronomy. For one hundred and twenty generations father has transmitted to son and grandfather to grandchild the story of the past and the signs of future destruction.

The belief that the world is endangered by the planet Venus plays an important role in the ritual of the Skidi Pawnee Indians of Nebraska.

Next in rank to Tirawa (Jupiter) stands the Morning Star. "Tirawa gave most of his power to the Morning Star."[1] "Through her four assistants, Wind, Cloud, Lightning, and Thunder, she transmitted the mandates of Tirawa to the people upon earth." Next in rank to the Morning Star "were the gods of four world-quarters, who stood in the northeast, southeast, southwest, and northwest and supported the heavens. Next in rank was the North Star. Below these in turn were the Sun and Moon." "The greater part of the heavenly gods were identified with stars. The sacred bundle of each village was believed to have been given to its ancestors by one of these heavenly beings."

The ceremony of sacrifice to the Morning Star is the main ritual of the Pawnee Indians. It is a "dramatization of the acts performed by the Morning Star." A human offering was sacrificed when Venus "appeared especially bright or in years when there was a comet in the sky." The act of appeasing Venus when a comet was seen in the sky takes on clearer meaning in the light of the present research.[2]

The sacrificial procedure took the following form. A captive girl was turned over by her captor to a man who would howl

[1] This and the following quotations are from *The Thunder Ceremony of the Pawnee* and *The Sacrifice to the Morning Star*, compiled by R. Linton from unpublished notes of G. A. Dorsey, Field Museum of Natural History, Department of Anthropology, Chicago (1922).

[2] See the Section, "The Fifty-two Year Period."

like a wolf. She was kept by the guardian until the day of the sacrifice. "Her guardian then painted her whole body red and dressed her in a black skirt and robe. His face and hair were painted red, and a fan-shaped headdress of twelve eagle feathers was attached to his hair." "This was the costume in which the Morning Star usually appeared in visions."

The scaffold was erected between four poles that pointed to the four quarters (northeast, southeast, southwest, northwest). A few words were pronounced about the darkness that threatened to endure forever, and in the name of the Morning Star a command was addressed to the poles to keep upright "so that you will always hold up the heavens."

The chief priest then "painted the right half of her body red and the left half black. A headdress of twelve black-tipped eagle feathers, arranged like a fan, was fastened on her head."

"At the moment the Morning Star appeared, two men came forward bearing firebrands." The breast of the girl was cut open and the heart taken out, and "the guardian thrust his hand into the thoracic cavity and painted his face with the blood." The people around shot arrows into the body of the victim. "Boys too young to draw a bow were helped by their fathers or mothers." Four bundles were laid northeast, northwest, southeast, and southwest of the scaffold and were ignited.

"There seem to have been astronomical beliefs connected with the sacrifices."

These human sacrifices, as described by Dorsey, were executed by the Indians only a few decades ago. They recall the Mexican sacrifices to the Morning Star described by the authors of the sixteenth century.

The meaning of these ceremonies and their relation to the planet Venus, especially in the years of a comet, the references to the cardinal points and to prolonged darkness, the anxiety that the sky should not fall, and even such details as the black and red colours so important in the ceremonies, become understandable now that we know the role Venus played in world upheavals.

## CHAPTER X

# THE SYNODICAL YEAR OF VENUS

THE PLANET VENUS, at the present time, revolves around the sun in 224.7 days, which is the sidereal year of the planet. However, seen from the earth, which revolves around the sun on a larger orbit and at a lower speed, Venus returns to the same position with respect to the earth after 584 days, which is its synodical year. It rises before the sun, earlier every day for seventy-one days, until it reaches the western elongation or its westernmost point away from the rising sun. Each morning thereafter the Morning Star rises lower and lower and for 221 days approaches the superior conjunction. About a month before the end of this period, it is eclipsed by the rays of the sun, and for over sixty days it is not seen because of the sun's rays: it is behind the sun or in superior conjunction. Then it appears for a moment after the setting sun, being now the Evening Star and east of the western sun. For 221 nights it retreats from the middle point of the superior conjunction, and beginning with the evening on which it first appears as an Evening Star, each night it appears farther from the setting sun until it reaches the eastern elongation. Then for seventy-one nights it approaches the sun. Finally it enters the inferior conjunction, when it is between the earth and the sun. It is usually invisible for one or two days, and thereafter appears west of the rising sun and is again the Morning Star.

These movements of Venus and their exact duration have been known to the people of the Orient and the Occident for over two thousand years. Actually a "Venus year," which follows the synodical revolution of Venus, was employed in calendars of the Old and New World alike. Five synodical years of Venus equal 2919.6 days, whereas eight years of 365 days equal 2920 days, and eight Julian years of 365¼ days equal 2922 days. In other words, in four years there is a difference of approximately one day between the Venus and the Julian calendars.

As I shall show in more detail in my reconstruction of ancient history, the Egyptians of the second part of the first pre-Christian millennium observed the Venus year. A decree published in Egyptian and in Greek by the conclave of priests which took place in Canopus in the reign of Ptolemy III (Euergetes) in –239 was intended to reform the calendar "according to the present arrangement of the world" and "an amendment of the faults of the heaven," replacing the year regulated by the rising of the star Isis—and Pliny says that Isis is the planet Venus [1]—with a year regulated by the rising of the fixed star Sothis (Sirius); this would make a difference of one day in four years, so that, as the decree says, "the festivals of the winter should not arrive in the summer because of the change of a day every four years in the rising of the star Isis." [2]

The reform intended by the Canopus Decree did not take root because the people and the conservatives among the priests kept faith with Venus and observed the New Year and other festivals on the days regulated by it. As a matter of fact, we know that the Ptolemaic pharaohs were obliged to swear in the temple of Isis (Venus) that they would not reform the calendar, nor add a day every four years. Julius Caesar actually followed the Canopus Decree by fixing a calendar of 365¼ days. In –26 Augustus introduced the Julian year in Alexandria, but the Egyptians outside Alexandria still continued to observe the Venus year of 365 days, and Claudius Ptolemy, the Alexandrian astronomer of the second Christian century, wrote in his *Almagest:* "Eight Egyptian years without a sensible error equals five circlings of Venus." [3]

As this period of eight years can be divided in two, each part being equal to two and a half synodical periods, the dividing point being alternately at a heliacal (simultaneous with the sun) rising or setting of Venus, the Egyptians of the second half of the last millennium before the present era observed a four-year cycle. This is the meaning of Horapollo's information that the Egyptian year is equal to four years.[4] In like manner the Greeks counted by four-year cycles dedicated to Athene: the Olympic Games took place every fourth year

[1] Pliny, *Natural History,* ii. 37.
[2] S. Scharpe, *The Decree of Canopus in Hieroglyphics and Greek* (1870).
[3] Bk. X. Chap. iv.
[4] A. T. Cory, *The Hieroglyphics of Horapollo Nilous* (1840), II, lxxxix. See also Wilkinson in G. Rawlinson, *The History of Herodotus,* II, 285.

(in the beginning, every eighth year[1]), and time was reckoned by the Olympiads. The Olympic games were started in the eighth century. At the Parthenon in Athens every fourth year was celebrated by the Panathenaic processions in honour of Athene.

The Incas of Peru in South America and the Mayas and Toltecs in Central America observed the synodical revolution of Venus and the Venus year in addition to the solar year.[2] They also calculated by groups of five Venus years equal to eight years of 365 days. Like the Egyptians and the Greeks the Mayas observed the four-year cycles,[3] from the inferior to the superior and from the superior to the inferior conjunctions of Venus. The Incas correctly marked the Venus calendar by tying knots in their quipus,[4] and the Mayas, in the *Dresden Codex*, correctly gave the length of the Venus synodical cycle as 584 days.[5] The astronomical observations of the Mayas were so precise that in computing the solar year, they arrived at figures not only more accurate than the Julian year, but also more accurate than the Gregorian year, introduced in Europe in 1582, ninety years after the discovery of America, which is our calendar year today.[6]

All this proves that the Venus calendar preserved its religious significance for a long time, down to the end of the Middle Ages and the discovery of America, and even thereafter, but that already in the eighth century before the present era an eight or double four-year cycle of Venus was observed in time reckoning and therefore must have been established in the celestial sphere.

A few decades after the discovery of America, the Augustinian friar Ramón y Zamora wrote that the Mexican tribes held the Morning Star in great veneration and kept a precise record of its appearance: "So exact was the book-record of the day when it appeared and when it concealed itself, that they never made mistakes."[7]

[1] E. N. Gardiner, *Olympia* (1925), p. 71; Farnell, *The Cults of the Greek States,* IV, 293; Frazer, *The Dying God* (1911), p. 78.
[2] Brasseur, *Sources de l'histoire primitive du Mexique*, p. 27.
[3] J. E. Thompson, "A Correlation of the Mayan and European Calendars," *Field Museum of Natural History Anthropological Series,* Vol. XVII.
[4] Nordenskiöld, *The Secret of the Peruvian Quipus*, II, 35.
[5] W. Gates, *The Dresden Codex*, Maya Society Publication No. 2 (1932).
[6] Gates in De Landa, *Yucatan*, p. 60.
[7] Seler, *Gesammelte Abhandlungen*, I, 624.

This was a very old custom originating in a past when Venus moved on an elongated orbit.

The movements of Venus were carefully watched by the ancient astronomers of Mexico, India, Iran, and Babylonia. Temple observatories for the cult of the planets were built in both hemispheres. The *bamot* or "high places" so often mentioned in the Scriptures were observatories as well as places for offerings to the planet-gods, chiefly Venus (Baal). On these high places idolatrous priests, ordained by the erring kings of Judah, burned incense to Baal, to the sun, and the moon, and to the planets.[1]

In the second half of the second millennium and in the beginning of the first millennium, Venus was still a comet; and though a comet can have a circular orbit—there is such a comet in the solar system[2]—Venus was not then moving on a circular orbit as it does now; its orbit crossed the orbit of the earth and endangered it every fifty years. Since, by the second half of the eighth century before the present era, Venus' cycle was similar to what it is today, it follows that some time before then Venus must have changed its orbit and achieved its present circular path between Mercury and the earth and became the Morning and Evening Star.

The irregularities in the movements of Venus must have been observed by the ancients; the data in the ancient records must differ very much from the figures on Venus' movements given at the head of this section.

## VENUS MOVES IRREGULARLY

In the library of Assurbanipal in Nineveh were stored astronomical books of his and of previous ages; in the ruins of this library Sir Henry Layard found the Venus tablets.[3]

There arose the question: From what period do the observations of these tablets date? Schiaparelli investigated this problem and "as an example of method his work is excellent." [4]

[1] II Kings 23 : 5.

[2] The Schwassmann-Wachmann comet, the orbit of which is between the orbits of Jupiter and Saturn.

[3] Published by H. C. Rawlinson and G. Smith, *Table of the Movements of the Planet Venus and Their Influences.* Sayce's translation was printed in the *Transactions of the Society of Biblical Archaeology*, 1874; a more recent translation by S. Langdon and J. K. Fotheringham was published as *The Venus Tablets of Ammizaduga* (1928).

[4] Fotheringham in Langdon and Fotheringham, *The Venus Tablets of Ammizaduga*,

He decided that "the inquiry could be limited to the seventh and eighth centuries."

The year-formula of an early king, Ammizaduga, was discovered on one of the tablets, and since then the tablets are usually ascribed to the first Babylonian dynasty; however, a scholar has offered evidence to the effect that the year-formula of Ammizaduga was inserted by a scribe in the seventh century.[1] (If the tablets originated in the beginning of the second millennium, they would prove only that Venus was even then an errant comet.)

Following are a few excerpts from the Venus tablets:

"On the 11th of Sivan, Venus disappeared in the west, remaining absent in the sky for 9 months and 4 days, and on the 15th of Adar she was seen in the east."

The next year, "on the 10th of Arahsamna, Venus disappeared in the east, remaining absent 2 months and 6 days in the sky, and was seen on the 16th of Tebit in the west."

The following year Venus disappeared in the west on the 26th of Ulul (Elul), remaining absent from the sky for eleven days, and was seen on the 7th of intercalary Ulul in the east.

The year thereafter Venus disappeared in the east on the 9th of Nisan, remaining absent for 5 months and 16 days, and was seen on the 25th of Ulul in the west.

In the fifth year of the observations, Venus disappeared in the west on the 5th of Ayar (Ijar), remaining absent from the sky for seven days, and reappeared in the east on the 12th of Ayar; the same year it disappeared on the 20th of Tebit in the east, remaining absent from the sky one month, and on the 21st day of Sabat (Shevat) it appeared in the west, and so on.

How explain these observations of the ancient astronomers, modern astronomers and historians have asked. Were they written in a conditional form ("If Venus disappeared on the 11th of Sivan . . .")? No, they were expressed categorically.

The observations were "inaccurately" registered, decided some authors. However, inaccuracy may account for a few days' difference but not for a difference of months.

---

p. 32. See Schiaparelli, "Venusbeobachtungen und Berechnungen der Babylonier," *Das Weltall*, Vols. VI, VII.

[1] Kugler ascribed the Venus tablets to the first Babylonian Dynasty, because he read a year-formula of Ammizaduga in one of them. In 1920, F. Hommel (*Assyriologische Bibliothek,* XXV, 197-199) declared that the year-formula of Ammizaduga was inserted into the Venus tablets by a scribe in the reign of Assurbanipal, in the seventh century.

"The invisibility of Venus at superior conjunction is given as 5 months 16 days instead of the correct difference of 2 months 6 days," noted the translators of the text, wonderingly.[1]

"The period between the heliacal setting of Venus and its rise is 72 days. But in the Babylonian-Assyrian astrological texts, the period varies from one month to five months—too long and too short: the observations were defective," wrote another scholar.[2]

"The impossible interval shows that the data are not trustworthy." "Obviously, the days of the month have been mixed up. As the impossible intervals show, the months are also wrong," wrote still another author.[3]

It is difficult to imagine how such obvious errors could have been committed. The dates are written in a contemporary document; they are not a poetical composition but a dry record, and each item in the record is stated in dates as well as in the number of days between the dates.

Similar difficulties are encountered by the scholars who try to understand the Hindu tables of the movements of the planets. The only explanation proposed is: "All the manuscripts are completely corrupted. . . . The details referring to Venus . . . are very difficult to unriddle."[4] "No attention at all was paid to the actual movements in the sky."[5]

The Babylonians did not note these irregular movements merely as matters of factual interest; they were dismayed by them. In their prayers they expressed this dismay.

> O Ishtar, queen of all peoples . . .
> Thou art the light of heaven and earth. . . .
> At the thought of thy name the heaven and the earth quake . . .
> And the spirits of the earth falter.
> Mankind payeth homage unto thy mighty name,
> for thou art great, and thou art exalted.
> All mankind, the whole human race,
> boweth down before thy power. . . .
> How long wilt thou tarry, O lady of heaven and earth . . . ?
> How long wilt thou tarry, O lady of all fights and of the battle?

[1] Langdon-Fotheringham, *The Venus Tablets*, p. 106.
[2] M. Jastrow, *Religious Belief in Babylonia and Assyria*, p. 220.
[3] A. Ungnad "Die Venustafeln und das neunte Jahr Samsuilunas," *Mitteilungen der altorientalischen Gesellschaft* (1940), p. 12.
[4] Thibaut, "Astronomie, Astrologie und Mathematik," Vol. 3, Pt. 9 (1899) of *Grundriss der indo-arisch. Philol. und Altertumskunde*, p. 27.
[5] *Ibid.*, p. 15.

O thou glorious one, that . . . art raised on high, that art firmly established,
O valiant Ishtar, great in thy might!
Bright torch of heaven and earth, light of all dwellings,
Terrible in the fight, one who cannot be opposed, strong in the battle!
O whirlwind, that roarest against the foe and cuttest off the mighty!
O furious Ishtar, summoner of armies![1]

As long as Venus returned at regular intervals, fear of the planet was kept in bounds; when the star passed without causing harm, as it had already done for a few centuries, the peoples were calmed and felt themselves out of danger for another period. But when Venus, for some reason, began to move irregularly, fear grew intense.

The priests of Iran prayed:[2]

We sacrifice to Tistrya, the bright and glorious star,
for whom long flocks and herds and men,
looking forward for him and deceived in their hope:
When shall we see him rise up, the bright and glorious star Tistrya?

The *Zend-Avesta* answered for the star:

If men would worship me with a sacrifice
in which I were invoked by my own name . . .
then I should come to the faithful
at the appointed time.

The priests responded:

The next ten nights, O Spitama Zarathustra!
the bright and glorious Tistrya mingles his shape with light,
moving in the shape of a golden-horned bull.

They glorified the star that made "all the shores of the ocean boiling over, all the middle of it boiling over." They heaped up sacrifices to the star, imploring it not to change its course.

[1] A "Prayer of the Raising of the Hand" to Ishtar (transl. L. W. King) in *The Seven Tablets of Creation.*

[2] *Zend-Avesta* (transl. Darmesteter), Pt. II, pp. 94 ff. The belief sometimes expressed, that Tistrya is Sirius, is an obvious error: Sirius does not travel in a winding course. The star in the shape of a golden-horned bull was Venus. Also, inaccurate movements of Sirius could not occur without similar irregularity on the part of all the stars.

We sacrifice unto Tistrya, the bright and glorious star
who from the shining east moves along his long winding course,
along the path made by the gods. . . .
We sacrifice unto Tistrya the bright and glorious star,
whose rising is watched by the chiefs of deep understanding.

The star of Venus did not appear in the prescribed seasons. In the Book of Job the Lord asks him: "Canst thou bring forth Mazzaroth in his season . . . ? Knowest thou the changes of heaven?" [1]

There exists an extensive exegetic literature on this Mazzaroth,[2] from which it can be concluded only that "the meaning of Mazzaroth is uncertain."[3] But the Vulgate (Latin) translation of the Bible has Lucifer for Mazzaroh. The (Greek) translation of the Seventy (Septuagint) reads: "Canst thou bring forth Mazzaroth in his season and guide the Evening Star by his long hair?" The words of the Septuagint seem very strange. I have already mentioned that the Greek word *komet* means "the long-haired one," or a star with hair, a comet. In Latin, *coma* is "hair."

Mazzaroth means a comet, wrote an exegete, and therefore, he argued, it cannot mean Venus.[4] But in any case it is said that the Evening Star has hair. Actually, Mazzaroth means Venus and a hairy star.

Venus ceased to appear in its seasons. What had happened?

## VENUS BECOMES THE MORNING STAR

Since the latter part of the eighth century before the present era, Venus has followed an orbit between Mercury and earth, which it has maintained ever since. It became the Morning and Evening Star. Seen from the earth, it is never removed more than 48 degrees (when at its eastern and western elongation) or three hours and a few minutes east or west of the sun. The dreaded comet became a tame planet. It has the most nearly circular orbit among the planets.

The end of the terror which Venus kept alive for eight centuries after the days of the Exodus was the inspiration for Isaiah when he said:[5]

[1] Job 38 : 32-33. The King James translation has, "Knowest thou the ordinances of heaven?" The Septuagint has "the changes of heaven."
[2] See Schiaparelli, *Astronomy in the Old Testament*, p. 74.
[3] Cambridge Bible, Book of Job, by A. B. Davidson and H. C. Lanchester.
[4] J. S. Suschken, *Unvorgreifliche Kometen-Gedanken: Ob der Kometen in der heiligen Schrift gedacht werde?* (1744).
[5] Isaiah 14 : 12-13. See also *infra*, p. 259.

"How art thou fallen from heaven, O Lucifer, son of the morning! how art thou cut down to the ground, which didst weaken the nations! For thou has said in thine heart, I will ascend to heaven, I will exalt my throne above the stars of God."

Septuagint and Vulgate both translate Morning Star or Lucifer. What does it mean, that the Morning Star was assailing the heavens and rising high, and that it was cut down low to the horizon, and would weaken no more the nations?

More than a hundred generations of commentators have occupied themselves with this passage, but have met with failure.

Why, it is also asked, should the beautiful Morning Star, called Lucifer, the Light Bearer, live in the imaginations of peoples as an evil power, a fallen star? What is in this lovely planet that makes her name an equivalent of Satan, or Seth of the Egyptians, the dark power? In his confusion, Origen wrote this question to the quoted verses of Isaiah: "Most evidently by these words is he shown to have fallen from heaven, who formerly was Lucifer, and who used to arise in the morning. For if, as some think, he was a nature of darkness, how is Lucifer said to have existed before? Or how could he arise in the morning, who had in himself nothing of the light?"[1]

Lucifer was a feared prodigy in the sky, and its origin, as illuminated in this book, explains how it became to be regarded as a dark power and a fallen star.

After a great struggle, Venus achieved a circular orbit and a permanent place in the family of planets. During the perturbations which brought about this metamorphosis, Venus also lost its cometary tail.

In the valley of the Euphrates, "Venus then gives up her position as a great stellar divinity, equal with sun and moon, and joins the ranks of the other planets."[2]

A comet became a planet.

Venus was born as a comet in the second millennium before the present era. In the middle of that millennium it twice made contact with the earth and changed its cometary orbit. In the tenth to eighth centuries of the first millennium, it was still a comet. What caused such further changes in the motion of Venus in the first millennium that it became a planet on a circular orbit?

[1] *The Writings of Origen*, "De principiis" (transl. F. Crombie, 1869), p. 51.
[2] A. Jeremias, *The Old Testament in the Light of the Ancient East* (1911), I, 18.

# PART II

## *Mars*

## CHAPTER I

# AMOS

ABOUT SEVEN HUNDRED fifty years passed after the great catastrophe of the days of the Exodus, or seven centuries after the cosmic disturbance in the days of Joshua. During all this time the world was afraid of the recurrence of the catastrophe at the end of every jubilee period. Then, starting about the middle of the eighth century before the present era, a new series of cosmic upheavals took place at intervals of short duration.

It was the time of the Hebrew prophets whose books are preserved in writing, of Assyrian kings whose annals are excavated and deciphered, and of Egyptian pharaohs of the Libyan and Ethiopian dynasties; in short, the catastrophes which we are now about to describe did not take place in a mist-shrouded past: the period is part of the well authenticated history of the lands of the eastern Mediterranean. The eighth century also saw the beginning of the nations of Greece and Rome.

The seers who prophesied in Judea were versed in the lore of heavenly motion; they observed the ways of the planetary and cometary bodies and, like the stargazers of Assyria and Babylonia, they were aware of future changes.

In the eighth century, in the days of Uzziah, king of Jerusalem, there occurred a devastating catastrophe called *raash* or "commotion."[1] Amos, who lived at the time of Uzziah, began to predict a cosmic upheaval before the *raash* took place, and after the catastrophe, Isaiah, Joel, Hosea, and Micah insisted unanimously and with great emphasis on the inevitability of another encounter of the earth with some cosmic body.

The prophecy of Amos was made two years before the *raash* (1 : 1). He declared that fire sent by the Lord would devour Syria, Edom, Moab, Ammon, and Philistia, as well as the

[1] *Raash* is translated "earthquake," which is incorrect here; cf. Jeremiah 10 : 22: "a great commotion [*raash*] out of the north." "Earthquake" is rendered in the Scriptures by words derived from the roots *raad, hul, regoz, hared, palez, ruf,* and *raash* (commotion).

far-off countries, "with a tempest in the day of the whirlwind" (1 : 14). The land of Israel would not be exempted; "great tumult" would be on its mountains, and "great houses shall have an end" (3 : 15). "He will smite the great house with breaches, and the little house with clefts" (6 : 11).[1]

Amos warned those who invited the day of the Lord and waited for it: "Woe unto you that desire the day of the Lord! To what end is it for you? The day of the Lord is darkness, and not light . . . even very dark, and no brightness in it" (5 : 18-20).

Amos, the earliest among the prophets of Judah and Israel whose speeches are preserved in writing,[2] reveals the concept of Yahweh in that remote period of history. Yahweh orders the planets. "He who maketh [ordains] Khima and Khesil,[3] and turneth the shadow of death into the morning, and maketh the day dark with night, and calleth for the waters of the sea, and poureth them out upon the face of the earth, the Lord [Yahweh] is his name: He strengtheneth the spoiled against the strong" (5 : 8-9).

Amos prophesied: The land "shall rise up wholly as a flood; and it shall be cast out and drowned, as by the flood of Egypt. And it shall come to pass in that day, saith the Lord God, that I will cause the sun to go down at noon, and I will darken the earth in the clear day" (8 : 8-9).

The "flood of Egypt" mentioned by Amos may be a reference to the catastrophe of the day of the Passage of the Sea; but more probably it refers to an event within the memory of the generation to which Amos spoke.

In the reign of Osorkon II of the Libyan Dynasty in Egypt, in the third year, the first month of the second season, on the twelfth day, according to a damaged inscription, "the flood came on, in this whole land . . . this land was in its power like the sea; there was no dyke of the people to withstand its fury. All the people were like birds upon it . . . the tempest . . . suspended . . . like the heavens. All the temples of Thebes were like marshes."[4]

[1] *Rsisim*, translated as "breaches," is not strong enough; it would be better to say, "smite great houses into pieces." Hebrew words translated as "breach" in the King James Version are *bedek, bkia, peretz, shever*.

[2] Some rabbinical authorities regard Hosea as the oldest among the prophets of the time (Hosea, Amos, Isaiah).

[3] The material for the identification of Khima as Saturn and Khesil as Mars will be presented in a subsequent part of this work.

[4] Breasted, *Records of Egypt*, IV, Sec. 743. Cf. J. Vandier, *La Famine dans l'Egypte ancienne* (1936), p. 123. "The water reduced the land to the same state as when it was still covered with the primeval water of creation."

That it was not a seasonal inundation of the Nile is clear from the date. "This calendar date for the high level of inundation does not at all correspond to the place of the calendar in the seasons." [1]

On the day of the approaching catastrophe, Amos says, there will be no place of escape, not even on Mount Carmel, rich in caves. "Though they climb up to heaven, thence will I bring them down. And though they hide themselves in the top of Carmel, I will search and take them out thence" (9 : 2-3).

Earth will melt and the sea will be heaped up and thrown upon inhabited land. "And the Lord God of hosts is he that toucheth the land, and it shall melt. . . . He that calleth for the waters of the sea, and poureth them out upon the face of the earth" (9 : 5-6).

Amos was persecuted and killed. The catastrophe did not fail to come at the appointed time. In anticipation and in fear of it, King Uzziah went to the Temple to burn incense.[2] The priests opposed his appropriating their functions: "Suddenly the earth started to quake so violently that a great breach was torn in the Temple. On the west side of Jerusalem, half of a mountain was split off and hurled to the east." [3] Flaming seraphim leaped in the air.[4]

Earthquakes act suddenly, and the population has no means of knowing about them in advance in order to flee. But before the *raash* of Uzziah the population escaped from the cities and fled into caves and clefts between the rocks. Many generations later, in the post-Exilic period, it was remembered how the population "fled from before the *raash* in the days of Uzziah king of Judah." [5]

## THE YEAR −747

If the commotion of the days of Uzziah was of global character and was brought about by an extraterrestrial agent, it must have caused some disturbance in the motion of the earth on its axis and along its orbit. Such a disturbance would have made the old calendar obsolete and would have required the introduction of a new calendar.

In −747 a new calendar was introduced in the Middle East, and that year is known as "the beginning of the era of

[1] Breasted, *Records of Egypt*, IV, Secs. 742-743.
[2] II Chronicles 26 : 16 ff.
[3] Ginzberg, *Legends*, IV, 262.
[4] *Ibid.*, VI, 358.
[5] Zechariah 14 : 5.

Nabonassar." It is asserted that some astronomical event gave birth to this new calendar, but the nature of the event is not known. The beginning of the era of Nabonassar, otherwise an obscure Babylonian king, was an astronomical date used as late as the second Christian century by the great mathematician and astronomer of the Alexandrian school, Ptolemy, and also by other scholars. It was employed as a point of departure of ancient astronomical tables.

"This was not a political or religious era. . . . Farther back there was no certainty in regard to the calculation of time. It is from that moment that the records of eclipses begin which Ptolemy used."[1] What was the astronomical event that closed the previous era and gave birth to a new era?

According to retrospective calculations, there was no eclipse of the sun in the region of Assyro-Babylonia between the years –762 and –701,[2] if the earth has revolved and rotated uniformly since then, which is taken for granted.

Uzziah reigned from about –789 to about –740.[3] The last few years of his reign, beginning with the day of the "commotion," he spent in seclusion, having been pronounced a leper. It was apparently the upheaval in the days of Uzziah that separated the two ages. Time was counted "from the commotion in the days of Uzziah."[4]

If this conclusion is correct, the upheaval took place in –747. The computation, according to which the era started on the twenty-sixth day of February, must be re-examined in the light of the fact that further cosmic disturbances occurred during the decades that followed –747. It is worth noting, however, that the ancient inhabitants of Mexico celebrated their New Year on the day which corresponds, in the Julian calendar, to the same date: "The first day of their yeere was the sixe and twentie day of February."[5]

The chronographer and Byzantine monk, Georgius Syncellus, one of the chief sources of ancient chronology, synchronized the forty-eighth year of Uzziah and the first year of the first Olympiad.[6] But according to modern calculations, the first year

[1] F. Cumont, *Astrology and Religion among the Greeks and Romans* (1912), pp. 8-9. To be correct, the earliest eclipse Ptolemy calculated is dated March 21, –721.
[2] T. von. Oppolzer, *Canon der Finsternisse* (1887).
[3] K. Marti, "Chronology," *Encyclopaedia Biblica*, ed. by Cheyne and Black.
[4] Cf. Amos 1 : 1; Zechariah 14 : 5.
[5] J. de Acosta, *The Natural and Moral History of the Indies* (transl. E. Grimston, 1604; re-edited, 1880).
[6] Georgius Syncellus (ed. G. Dindorf, 1829), II, 203.

of the first Olympiad was −776.[1] The Olympiads most probably were inaugurated by some cosmic event. The text of the ancient Chinese book of Shiking refers to some celestial phenomenon in the days of the king Yen-Yang, in −776: the sun was obscured.[2] If the occurrence of −776 was of the same nature as that of −747, then Amos' prophecy was a prognostication based on an earlier experience.

## ISAIAH

According to Hebrew sources,[3] Isaiah began to prophesy immediately after the "commotion" of the days of Uzziah, even on the same day. The destruction in the land was very great. "Your country is desolate, your cities are burned with fire. . . . Except the Lord of hosts had left unto us a very small remnant, we should have been as Sodom, and we should have been like unto Gomorrah" (1 : 7 ff.). The very horizon of Jerusalem was disfigured by the splitting of the mountain on the west; and the cities were filled with debris and mutilated bodies. "The hills did tremble, and . . . carcasses were torn in the midst of the streets" (5 : 25).

This was the event that kindled in Isaiah the prophetic spirit. During his long life—he prophesied in "the days of Uzziah, Jotham, Ahaz, and Hezekiah, kings of Judah"—he did not cease to foretell the return of the catastrophes. Isaiah was skilled in the observation of the stars, and he apparently knew that at periodic intervals—every fifteen years—a catastrophe occurred, caused, he believed, by the messenger of God. "His anger is not turned away, but his hand [sign [4]] is stretched out still. And he will lift up an ensign to the nations from afar" (5 : 25-26).

Isaiah drew an apocalyptic picture of swiftly moving hostile troops. Was he prophesying a cruel and mighty people of warriors, or a host of missiles hurled from afar when he spoke of the army that would come swiftly from the end of the world, called by the Lord? Their horses' hoofs would be like flint, and their wheels like a whirlwind. "If one look unto the land,

[1] S. Newcomb, *The American Nautical Almanac*, 1891 (1890).
[2] A. Gaubil, *Traité de l'astronomie chinoise*, Vol. III of *Observations mathématiques, astronomiques, géographiques, chronologiques, et physiques . . . aux Indes et à la Chine*, ed. E. Souciet (1729-1732); J. B. du Halde, *A Description of the Empire of China* (1741), II, 128-129.
[3] Seder Olam 20.
[4] *Yad* is "hand" as well as "sign."

behold darkness and sorrow; and the light is darkened in the heavens thereof" (5 : 30).

It is not the Assyrians on horses and in chariots that are compared to the flint and the whirlwind, but the flint and the whirlwind that are likened to warriors.[1] The darkness at the end of the picture discloses that which is the object of comparison and that to which it is compared.

The catastrophe of the days of Uzziah was only a prelude: the day of wrath will return and will destroy the population "until the cities be wasted without inhabitant" (6 : 11). "Enter into the rock, and hide thee in the dust" (2 : 10)—all over the world caves in the rocks were regarded as the best places of refuge. "And they shall go into the holes of the rocks, and into the caves of the earth, for fear of the Lord, and for the glory of his majesty, when he ariseth to shake terribly the earth" (2 : 19).

Isaiah appeared before King Ahaz and offered him a sign, on the earth or "in the height above." Ahaz refused: "I will not ask, neither will I tempt the Lord" (7 : 12).

Then Isaiah faced the people. "And they shall look unto the earth; and behold trouble and darkness, dimness of anguish" (8 : 22). Nevertheless, he said, the dimness will not be as great as on two former occasions when "at the first he lightly afflicted the land of Zebulun and the land of Naphtali, and afterward did more grievously afflict her by the way of the sea, beyond Jordan, in Galilee of the nations" (9 : 11). He calculated that the next catastrophe would cause less harm than had been caused on previous occasions. But soon thereafter he changed his prognostication and became utterly pessimistic.

"Through the wrath of the Lord of hosts is the land darkened, and the people shall be as the fuel of the fire" (9 : 19). His rod will lift the sea up "after the manner of Egypt," as on the day of the crossing of the Red Sea (10 : 26). "And the Lord shall utterly destroy the tongue of the Egyptian sea; and with his mighty wind shall he shake his hand [sign] over the river, and shall smite it in the seven streams" (11 : 15). Nor will Palestine be spared. "He shall shake his hand [sign] against . . . the hill of Jerusalem" (10 : 32).

Thus, a war of the heavenly host, commanded by the Lord, was proclaimed against the nations of the earth. And the nations of the earth were aroused by the expectation of Dooms-

[1] See *infra* the Section, "The Terrible Ones."

day. "The noise of a multitude in the mountains, like as of a great people; a tumultuous noise of the kingdoms of nations gathered together: the Lord of hosts mustereth the host of the battle" (13 : 4). This multitude comes "from the end of heaven, even the Lord, and the weapons of his indignation, to destroy the whole land" (13 : 5).

The world will be darkened. "The stars of heaven and the constellations thereof shall not give their light: the sun shall be darkened in his going forth [in the forenoon], and the moon shall not cause her light to shine" (13 : 10).

The world will be thrown off its axis: the heavenly host "will shake the heavens, and the earth shall remove out of her place, in the wrath of the Lord of hosts, and in the day of his fierce anger" (13 : 13).

The nations "shall flee far off, and shall be chased as the chaff of the mountains before the wind, and like a rolling thing before the whirlwind" (17 : 13).

Isaiah, on his vigils, watched the firmament, and in "appointed times" expected "from the north a smoke" (14 : 31).

"All ye inhabitants of the world . . . see ye, when he lifteth up an ensign on the mountains; and when he bloweth a trumpet, hear ye" (18 : 3). The eyes of all "dwellers of the earth" were directed toward the sky, and they listened to the bowels of the earth.

Inquiries were sent to Jerusalem from Seir in Arabia: "Watchman, what of the night?" From his watchtower ("Prepare the table, watch in the watchtower") Isaiah gave his forecasts to inquirers (21 : 5; 21 : 11).

Nervous tension grew with the approach of the "appointed time," and a rumour sufficed to drive the population of the cities to the housetops. "What aileth thee now, that thou art wholly gone up to the housetops?" (22 : 1.)

Much of the city of David was damaged and many structures had fissures from almost continuous earth tremors (22 : 9). The seer frightened the population with his constant warnings of "a day of trouble . . . and of perplexity by the Lord God of hosts," with "breaking down the walls, and of crying to the mountains" (22 : 5). But many among the population took the attitude of those who before Doomsday say: "Let us eat and drink; for tomorrow we shall die" (22 : 13).

Joel, who prophesied at the same time, also spoke of "wonders in the heavens and in the earth, blood, and fire, and pillars of smoke. The sun shall be turned into darkness, and the moon into blood, before the great and the terrible day of the Lord come" (Joel 2 : 30-31).

Micah, another seer "in the days of Jotham, Ahaz, and Hezekiah, kings of Judah," warned that the day was close when "the mountains shall be molten . . . and the valleys shall be cleft, as wax before the fire", (Micah 1 : 4). "Marvellous things" will be shown, as in the days when Israel left Egypt: "The nations shall see and be confounded at all their might . . . their ears shall be deaf . . . they shall move out of their holes like worms of the earth" (7 : 15-17).

Joel, Micah, and Amos warned in similar terms of "a day of thick darkness" and "the day dark with night." Astronomers, who thought that all this refers to a common eclipse of the sun, wondered: "From –763 down to the destruction of the First Temple in –586 no total eclipse of the sun was visible in Palestine." [1] They took it for granted that the earth revolves along exactly the same orbit and on a slowly rotating axis, and so they questioned: Why did the prophets speak of eclipses when there were none? However, other descriptions of the world catastrophe in these prophets do not accord with the effects of an ordinary eclipse, either.

The word *shaog*, used by Amos and Joel, is explained by the Talmud [2] as an earthshock, the field of action of which is the entire world, whereas a regular earthquake is of local character. Such a shaking of the earth, disturbed in its rotation, is vizualised also as a "shaking of the sky," an expression found in the Prophets, in Babylonian texts, and in other literary sources.

Then the prophecy was fulfilled. Amid the catastrophe Isaiah raised his voice: "Fear, and the pit, and the snare [pitch [3]] are upon thee, O inhabitant of the earth . . . for the windows from on high are open, and the foundations of the earth do shake. The earth is utterly broken down, the earth is clean dissolved, the earth is moved exceedingly" (24 : 17-19).

The catastrophe came on the day on which King Ahaz was buried. There was a "commotion": the terrestrial axis shifted or was tilted, and the sunset was hastened by several hours.

[1] Schiaparelli, *Astronomy in the Old Testament*, p. 43. Oppolzer and Ginzel arranged canons of the solar eclipses in antiquity on the premise that there was no change in the movement of the earth or the moon.

[2] The Jerusalem Talmud, Tractate Berakhot 13b.

[3] *Pah* in Hebrew originally meant "bitumen" or "pitch," as can be inferred from Psalms 11 : 6.

This cosmic disturbance is described in the Talmud, in the Midrashim, and referred to by the Fathers of the Church.[1] It is related also in the records and told in the traditions of many peoples. It appears that a heavenly body passed very close to the earth, moving, as it seems, in the same direction as the earth on its nocturnal side.

"Behold, the Lord maketh the earth empty, and maketh it waste and turneth it upside down. . . . The inhabitants of the earth are burned, and few men left" (Isaiah 24 : 1, 6).

## THE ARGIVE TYRANTS

In *Ages in Chaos* I shall present proof that the large, raw stone structures of Mycenae and Tiryns on the Argive plain in Greece are the ruins of the palaces of the Argive tyrants, well remembered by the Greeks of subsequent centuries, and date from the eighth century before the present era. If the material remains of the palaces of Mycenae and Tiryns are ascribed to the second millennium, then nothing has been found on the Argive plain that can be ascribed to the Argive tyrants, although they are known to have built spacious palaces.

Thyestes and his brother Atreus were of these Argive tyrants. Living in the eighth century, they must have witnessed the cosmic catastrophes of the days of Isaiah. Greek tradition persists that a cosmic catastrophe occurred in the time of these tyrants: the sun changed its course and the night arrived before its proper time.

Men should be prepared for everything and not wonder at anything, wrote Archilochus, since the day that Zeus "turned midday into night, hiding the light of the dazzling sun; and sore fear came upon men."[2]

Many classical authors referred to the occurrence. I give here Seneca's description. In his drama, *Thyestes*, the chorus asks the sun:

"Whither, O father of the lands and skies, before whose rising thick night with all her glories flees, whither dost turn thy course and why dost blot out the day in mid-Olympus [midday]? Not yet does Vesper, twilight's messenger, summon

[1] Tractate Sanhedrin 96a; Pirkei Rabbi Elieser 52; Hippolytus on Isaiah. Cf. Ginzberg, *Legends*, VI, 367, n. 81.
[2] Archilochus, Fragment 74.

the fires of night; not yet does thy wheel, turning its western goal, bid free thy steeds from their completed task; not yet as day fades into night has the third trump sounded; the ploughman with oxen yet unwearied stands amazed at his supper hour's quick coming. What has driven thee from thy heavenly course? . . . Has Typhoeus [Typhon] thrown off the mountainous mass and set his body free?" [1]

This picture reminds us of the description of the day of Ahaz' burial.

Seneca relates the fear of world destruction experienced by those who lived at the time of Atreus and Thyestes, the tyrants of the Argive plain. The hearts of men were oppressed with terror at the sight of the untimely sunset. "The shadows arise, though the night is not yet ready. No stars come out; the heavens gleam not with any fires: no moon dispels the darkness' heavy pall. . . . Trembling, trembling are our hearts, sore smit with fear, lest all things fall shattered in fatal ruin and once more gods and men be overwhelmed by formless chaos; lest the lands, the encircling sea, and the stars that wander in the spangled sky, nature blot out once more."

Will the seasons be ended and the moon be carried away? "No more" shall the stars "mark off the summer and the winter times; no more shall Luna, reflecting Phoebus' rays, dispel night's terrors."

After the catastrophe of the days of Atreus and Thyestes, the luminaries crossed their former paths obliquely; the poles were shifted; the year lengthened—the orbit of the earth became wider. "The Zodiac, which, making passage through the sacred stars, crosses the zones obliquely, guide and sign-bearer for the slow moving years, falling itself, shall see the fallen constellations."

Seneca describes the change in position of each constellation —the Ram, the Bull, the Twins, the Lion, the Virgin, the Scales, the Scorpion, the Goat, and the Wain (the Great Bear). "And the Wain, which has never bathed in the sea, shall be plunged beneath the all-engulfing waves." A commentator who wondered about this description of the position of the Great Bear wrote: "There was no mythological reason why the Wain—otherwise known as the Great Bear—should not be bathed in the Ocean." [2] But Seneca said precisely this strange

[1] Translated by F. J. Miller (1917).
[2] A note by F. J. Miller to his translation of *Thyestes.*

thing: the Great Bear—or one of its stars—never set beneath the horizon, and thus the polar star was among its stars during the age that came to its end in the time of the Argive tyrants.

Seneca also says explicitly that the poles were torn up in this cataclysm. The polar axis now is turned toward one of the stars, the North Star, of the Little Bear.

In the face of the cataclysm, when humanity was overwhelmed with awe, the heartbroken Thyestes, longing for death, called upon the universe to go down in utter confusion. The picture was not invented by Seneca: it was familiar because of what had happened in earlier ages.

"O thou, exalted ruler of the sky, who sittest in majesty upon the throne of heaven, enwrap the whole universe in awful clouds, set the winds warring on every hand, and from every quarter of the sky let the loud thunder roll; not with what hand thou seekest houses and undeserving homes, using thy lesser bolts, but with that hand by which the threefold mass of mountains fell . . . these arms let loose and hurl thy fires."

## AGAIN ISAIAH

Time passed after the death of Ahaz, and the fourteenth year of King Hezekiah approached. Again the frightened world anxiously anticipated a catastrophe. On its two previous approaches, the celestial missile had come very close, indeed. This time the end of the world was feared. After the cataclysms of the days of Uzziah and of the funeral day of Ahaz, one did not have to be a prophet to foretell a new cosmic catastrophe. The earth will move out of its place, a scorching flame will devour the air, hot stones will fall from the sky, and the waters of the sea will mount and descend upon the continents.

"Behold, the Lord hath a mighty and strong one, which as a tempest [cataract] of hail and a destroying storm, as a flood of mighty waters overflowing, shall cast down to the earth with the hand" (Isaiah 28 : 2).

"The mighty and strong one" was a heavenly body, a missile of the Lord. Once more it was destined to scourge the earth. "The overflowing scourge shall pass through" (28 : 18), was Isaiah's new prognostication. Although the people of Jerusalem hoped that "when the overflowing scourge shall pass through, it shall not come unto us" (28 : 15), Jerusalem had no covenant with death.[1]

There will be no safe place of refuge. "The waters shall overflow the hiding place" (28 : 17). "A consumption even determined upon the whole earth" (28 : 22).

"For the Lord . . . shall be wroth as in the valley of Gibeon, that he may do his work, his strange work; and bring to pass his act, his strange act" (28 : 21).

What was the "strange act" in the valley of Gibeon? In that valley the host of Joshua witnessed a rain of bolides and saw the sun and the moon disturbed in their movement across the firmament.

"At an instant suddenly" the land will be invaded with "small dust" and with "the multitude of terrible ones," and it will be visited "with thunder, and with earthquake, and great noise, with storm and tempest, and the flame of devouring fire" (29: 5-6).

"A devouring fire" and "an overflowing stream" shall "sift the nations" with "tempest and hailstones" (30 : 27-30).

The prophet, reading the signs of the sky, took upon himself the role of sentinel of the universe, and from his watchtower in Jerusalem he spread the alarm:

"Let the earth hear. . . . For the indignation of the Lord is upon all nations. . . . He hath delivered them to the slaughter" (34 : 1 ff.).

Then follows the desolate picture of the destroyed earth and dissolved sky (34 : 4 ff.) :

> And all the host of heaven shall be dissolved,
> and the heavens shall be rolled together as a scroll:
> and all their host shall fall down. . . .
> For my sword shall be bathed in heaven. . . .
> And the streams . . . shall be turned into pitch,
> and the dust into brimstone,
> and the land shall become burning pitch.
> It shall not be quenched night nor day;
> the smoke shall go up for ever.

Isaiah referred his readers to the "Book of the Lord": "Seek ye out of the book of the Lord, and read: no one of these shall fail" (34 : 16). This book probably belonged to the same series

[1] Cf. Psalms 46 : 5: "God is in the midst of her [Jerusalem]; she shall not be moved: God shall help her."

as the Book of Jasher, in which the records of the days of Joshua at Gibeon were preserved; old traditions and astronomical observations must have been written down in the Book of the Lord, no longer extant.

## MAIMONIDES AND SPINOZA, THE EXEGETES

> Ego sum Dominus, faciens omnia, extendens caelos solus, stabiliens terram, et nullus mecum. Irrita faciens signa divinorum, et ariolos in furorem vertens. Convertens sapientes retrorsum: et scientiam eorum stultam faciens.
>
> —Prophetiae Isaiae 44 : 24-25 (Vulgate)

Here, before I go on to the description of the day on which the prophecies of Isaiah, pronounced after the death of Ahaz, were fulfilled, I should like to present the common view of generations of commentators. The books of the Mayas have come into the hands of only a few scholars; likewise the papyri of Egypt and the clay tablets of the Assyrians. But the Book of Isaiah and other books of the Scriptures have been read by millions during the centuries in hundreds of languages. Is the way in which Isaiah expressed himself obscure? It is a kind of collective psychological blind spot which prevents the understanding of the clearly revealed and scores-of-times-repeated description of astronomical, geological, and meteorological phenomena. The description was thought to be a peculiar kind of poetic metaphor, a flowery manner of expression.

Even a modest attempt to review the various commentaries on Isaiah would burst the frame of a book larger than this one. Therefore it should satisfy the orthodox and the liberal reader alike if the opinions presented by two great authorities in the world of thought are given here, and the thousands of commentators not quoted at all.

Moses ben Maimon, called Rambam, also Maimonides (1135–1204), in his *The Guide for the Perplexed*,[1] expressed the opinion that a belief in the Creation is a fundamental principle of Jewish religion, "but we do not consider it a principle of our faith that the Universe will again be reduced to nothing";

[1] English Translation by M. Friedlander (1928).

"it depends on His will," and "it is therefore possible that He will preserve the Universe for ever"; "the belief in the destruction is not necessarily implied in the belief in the Creation." "We agree with Aristotle in one half of his theory. . . The opinion of Aristotle is that the Universe being permanent and indestructible, is also eternal and without beginning."

With this theophilosophic approach to the problem at large, Maimonides was averse to finding any word or sentence in the Prophets or elsewhere in the Bible that would suggest a destruction of the world or even a change in its order.[1] Each and every such expression he explained as a poetical substitute for an exposition of political ideas and acts.

Maimonides says: " 'The stars have fallen,' 'The heavens are overthrown,' 'The sun is darkened,' 'The earth is waste and trembles,' and similar metaphors" are "frequently employed by Isaiah, and less frequently by other prophets, when they describe the ruin of a kingdom." In these phrases the term "mankind" is used occasionally; this is also a metaphor, says Maimonides. "Sometimes the prophets use the term 'mankind' instead of 'the people of a certain place,' whose destruction they predict; e.g., Isaiah, speaking of the destruction of Israel, says: 'And the Lord will remove man far away' (6 : 12). So also Zephaniah (1 : 3-4), 'And I will cut off man from off the earth.' "

He maintains that Isaiah and other seers of Israel, when examined by the realistic method of Aristotelianism, were persons inclined to exaggerated forms of speech, and instead of saying, "Babylon will fall," or "fell," they spoke in terms of some fantastic perturbation in the cosmos above and beneath.

"When Isaiah received the divine mission to prophesy the destruction of the Babylonian empire, the death of Sennacherib and that of Nebuchadnezzar, who rose after the overthrow of Sennacherib,[2] he commences in the following manner to describe their fall . . . : 'For the stars of heaven and the constellations thereof, shall not give their light' (13 : 10); again, 'Therefore I will shake the heavens, and the earth shall remove out of her place, in the wrath of the Lord of hosts, and in the

[1] Maimonides apparently follows Philo, the Greek-writing Jewish philosopher of the first century, who in his *The Eternity of the World* was of the opinion that the world was created but that it is indestructible; however, Philo admitted changes in nature caused by periodic floods and conflagrations on a large scale and of cosmic origin.

[2] Nebuchadnezzar lived a century after Sennacherib.

day of his fierce anger' (13 : 13). I do not think that any person is so foolish and blind, and so much in favour of the literal sense of figurative and oratorical phrases, as to assume that at the fall of the Babylonian kingdom a change took place in the nature of the stars of heaven, or in the light of the sun and moon, or that the earth moved away from its centre. For all this is merely the description of a country that has been defeated; the inhabitants undoubtedly find all light dark and all sweet things bitter: the whole earth appears too narrow for them, and the heavens are changed in their eyes."

"He speaks in a similar manner when he describes . . . the loss of the entire land of Israel when it came into the possession of Sennacherib. He says (24 : 18-20): '. . . for the windows from on high are open, and the foundations of the earth do shake. The earth is utterly broken down, the earth is clean dissolved, the earth is moved exceedingly. The earth shall reel to and fro like a drunkard.' "

The subjugation of Judah by Assyria was joyless, but what was so bad, from Isaiah's point of view, in the destruction of Babylon that the stars should not give their light?

A reading of the literature indicates that no exegete has ever been "so foolish and blind" as to read sky for sky, stars for stars, brimstone for brimstone, fire for fire, blast for blast.[1] Referring to the quoted verses—Isaiah 34 : 4-5—Maimonides writes: "Will any person who has eyes to see find in these verses any expression that is obscure, or that might lead him to think that they contain an account of what will befall the heavens? . . . The prophet means to say that the individuals, who were like stars as regards their permanent, high, and undisturbed position, will quickly come down."

Maimonides quotes Ezekiel, Joel, Amos, Micah, Haggai, Habakkuk, and Psalms, and in verses similar to those cited from Isaiah, he finds incidentally a description of "a multitude of locusts," or a speech appropriate for the destruction of Samaria or the "destruction of Medes and Persians," spoken "in metaphors which are intelligible to those who understand the context."

[1] But for what they were taken may be illustrated by the exegesis of Augustine. He writes: "Hail and coals of fire (Psalms 18): Reproofs are figured, whereby as by hail, the hard hearts are bruised." To the words, "And He sent out His arrows, and scattered them" (Psalms 15), Augustine writes: "And He sent out Evangelists traversing straight paths on the wings of strength." St. Augustine, *Expositions on the Book of Psalms*, ed. Ph. Schaff (1905).

In a settled world nothing alters the given order. To sustain this doctrine, the prophecies were translated into metaphors, for, in the opinion of Maimonides, if the world does not change its regimented harmony, true prophets would not declare that it does. "Our opinion, in support of which we have quoted these passages," writes Maimonides, "is clearly established, namely, that no prophet or sage has ever announced the destruction of the Universe, or a change of its present condition, or a permanent change of any of its properties." This standpoint of Maimonides, as far as a change of conditions in the Universe is concerned, is a deduction, not from the texts he interprets, but from a philosophical a priori approach. Prophets might err in their prophecies, but it could hardly be that in saying "stars" they meant "persons." The reading of subsequent chapters in Isaiah(36-39) and parallel chapters in Kings and Chronicles, as well as the Talmudic and Midrashic fragments (concerning the time of Sennacherib's invasion), makes it apparent that this time the prophets did not err, and that a change in harmonious conditions did occur in the lifetime of these very prophets, in the days of Hezekiah.

Maimonides asserts that Joel's prophecies referred to Sennacherib, but he is puzzled: "You may perhaps object—how can the day of the fall of Sennacherib, according to our explanation, be called 'the great and the terrible day of the Lord'? "

In the following pages it will be shown that on the very day which preceded the night when Sennacherib's army was destroyed, the order of nature was upset. The speeches of the seers must be interpreted not apart from, but in the light of, the description of these changes as they are preserved in the Scriptures and in the Talmud. There was keener insight during the times prior to Maimonides, and to these more ancient interpreters he referred when he wrote:

"The Universe [ever] since continues its regular course. This is my opinion; this should be our belief. Our Sages, however, said very strange things as regards miracles; they are found in Bereshith Rabba, and in Midrash Koheleth, namely, that the miracles are to some extent also natural."

Baruch Spinoza proceeds from the premise that "Nature always observes laws and rules . . . although they may not all be known to us, and therefore she keeps a fixed and immutable order." "Miracles" merely mean events of which the natural cause cannot be explained. "In so far as a miracle is supposed to destroy or interrupt the order of Nature or her laws, it not only gives us no knowledge of God, but, contrariwise . . . makes us doubt of God and everything else." "What is meant in Scripture by a miracle can only be a work of Nature."[1]

All these premises are philosophically true, and no objection can be raised against them. Of course, they are true only as long as the philosopher does not insist that the laws of nature as known to him are the real and only laws.

Discussing instances in the Scriptures to which the quoted principles should be applied, Spinoza insists that the subjective apperception and the peculiar manner of expression of the ancient Hebrews are the only reasons for the accounts of unnatural events.

"I will content myself with one instance from Scripture, and leave the reader to judge of the rest. In the time of Joshua the Hebrews held the ordinary opinion that the sun moves with a daily motion, and that the earth remains at rest; to this preconceived opinion they adapted the miracle which occurred during their battle with the five kings. They did not simply relate that the day was longer than usual, but asserted that the sun and moon stood still, or ceased from their motion."

The deduction made is: "Partly through religious motives, partly through preconceived opinions, they conceived of and related the occurrence as something quite different from what really happened." "It is necessary to know the opinions of those who first related them . . . and to distinguish such opinions from the actual impression made upon our senses, otherwise we shall confound opinions and judgments with the actual miracle as it really occurred: nay, further, we shall confound actual events with symbolical and imaginary ones."

The Book of Isaiah is offered by Spinoza as another example, and the chapter on Babylon's doomed destruction is quoted: "The stars of heaven . . . shall not give their light; the sun shall be darkened in his going forth, and the moon shall not cause her light to shine." The philosopher writes: "Now I suppose no one imagines that at the destruction of Babylon these phenomena actually occurred any more than that which the prophet adds: 'For I will make the heavens to tremble,

[1] *Tractatus Theologico-Politicus* (1670), Chap. VII. The quoted sentences are translated by J. Ratner in his *The Philosophy of Spinoza.*

and remove the earth out of her place.' " "Many occurrences in the Bible are to be regarded as Jewish expressions." "The Scripture narrates in order and style which has most power to move men and especially uneducated men . . . and therefore it speaks inaccurately of God and of events."

Asserting a subjective apperception on the part of the witnesses, a deliberate intention to impress the reader or listener with exciting descriptions, a peculiarity in the mode of expression of Hebrew penmen, Spinoza nevertheless arrives at a *non sequitur:* "Now all these texts teach most distinctly that Nature preserves a fixed and unchangeable order. . . . Nowhere does Scripture assert that anything happens which contradicts, or cannot follow from the laws of Nature," and he supports his view with a theological argument: in the Book of Ecclesiastes it is written: "I know what God does, it shall be for ever."

The events were called miracles and were explained as subjective apperceptions or as symbolic descriptions because they could not be otherwise accounted for. But apart from the events themselves, which this study endeavours to establish as historical, the words of Isaiah and of other seers and penmen of the Old Testament do not leave any room for doubt that by "stones falling from the sky" were meant meteorites; by brimstone and pitch were meant brimstone and pitch; by scorching blast of fire was meant scorching blast of fire; by storm and tempest, storm and tempest; by a darkened sun, by the earth removed from its place, by change of time and seasons, were meant just these changes in the regular processes of nature. Where is the basis for the "sure knowledge" that the earth must move without perturbation at a time when every body in the solar system more or less perturbs every other one? Until the fall of meteorites in 1803, science was sure that stones falling from the sky occurred only in legends.

The "no one imagines" of Spinoza is no longer true. The author of this book does so imagine.

CHAPTER II

# THE YEAR -687

IN ABOUT -722, after three years of siege, Samaria, the capital of the Ten Tribes, was captured by Sargon II, and the population of the Northern Kingdom, or Israel, was removed into captivity from which it never returned.

In about -701, Sennacherib, son of Sargon, undertook the third campaign of his reign; he directed it to the south, into Palestine. The record of this and other campaigns of his is written and preserved in cuneiform signs worked on the sides of prisms of baked clay. The so-called "Taylor prism" contains the narrative of eight campaigns of Sennacherib. He wrote about his road to victory: "The wheels of my war chariot were bespattered with filth and blood."

The record of the third campaign on the prism corresponds to the record preserved in II Kings 18 : 13-16. According to both sources, Sennacherib took many cities; "the proud Hezekiah, the Judean," was "closed like a bird in a cage" in his capital, Jerusalem, but Sennacherib did not capture Jerusalem; he satisfied himself with a tribute of gold and silver[1] sent to him at Lachish in southern Palestine. After that he departed with his booty.

Hezekiah had no choice but to submit; the defences of the land were inadequate. Now he used the time, which he recognized as only a respite, to build walled strongholds and to garrison them, and to prepare the brooks and the wells of the land to be stopped and destroyed at the first signal. This is described in II Chronicles (32 : 1-6).

Sennacherib, alarmed by the revolt of Hezekiah, who aligned himself with the king of Ethiopia and Egypt, Tirhakah, came again with his army and once more set up his headquarters near Lachish. One of Sennacherib's generals, Rab-sha-keh, came to Jerusalem and spoke with the emissaries of Hezekiah, loudly and in Hebrew, so that the warriors on the

[1] Thirty talents of gold in both sources; 300 talents of silver according to the Book of Kings; 800 talents of silver according to the prism.

wall could hear him, too (Isaiah 36 : 18 ff.): "Beware lest Hezekiah persuade you, saying, The Lord will deliver us. Hath any of the gods of the nations delivered his land out of the hand of the king of Assyria?" He also told them to consider the fate of Samaria, whose gods did not save it when it was stormed by the Assyrians. He informed them that Sennacherib required pledges of submission and promised that they would be exiled to a land as good as their own. Hezekiah's emissaries were ordered not to enter into any dispute. Receiving no reply, Rab-sha-keh departed for Libna where King Sennacherib had gone from Lachish. The Ethiopian king Tirhakah came against Sennacherib out of the borders of Egypt and prepared to meet him in battle. Rab-sha-keh sent again a demand to Hezekiah to submit: "Let not thy God deceive thee, saying, Jerusalem shall not be given into the hand of the king of Assyria."

It was the prophecy of Isaiah that Jerusalem would not fall into the hands of the king of Assyria and that the king who blasphemed the Lord would be destroyed by "a blast" sent by the Lord.

The story is described in detail three times in the Scriptures—in II Kings 18-20, II Chronicles 32, and Isaiah, Chapters 36-38. The first version alone contains the first part of the story about Sennacherib, who conquered all the fenced cities of Judah, and Hezekiah, king of Judah, who submitted to the Assyrian king and paid tribute to him. All three scriptural sources tell about Hezekiah's rebelling against Sennacherib and refusing to submit or to pay tribute. It is obvious that, despite the repeated mention of Lachish, there must have been two different campaigns: in the first, Hezekiah submitted and agreed to pay tribute; the second campaign was a number of years later. In the meantime, Hezekiah had built up "all the wall that was broken, and raised it up to the towers, and another wall without, and repaired Milo in the city of David, and made darts and shields in abundance. And he set captains of war over the people. And when Sennacherib came and entered Judah, Hezekiah ordered to stop all the fountains without Jerusalem, and spoke to the people in the city to be strong and courageous." And then came the miraculous destruction of the Assyrian host.

The annals of Sennacherib tell only the first part of the story: the capture of the cities of the land, the submission of Hezekiah, and the tribute he paid. The siege of Lachish is not mentioned on the prism, but an Assyrian relief of this siege is preserved. Nothing is told in the Assyrian sources about defeat in Judea, and only the epilogue, the killing of Sennacherib by his own sons, is described identically in the Scriptures and in a cuneiform inscription of Esarhaddon, son of Sennacherib.

The destruction of Sennacherib's army having taken place in a later—evidently the last—campaign of Sennacherib before his assassination, it was not inserted on the eighth-campaign prism; this must have been his ninth, or possibly his tenth, campaign. Its disastrous outcome would not have inspired the king to order a new prism which should include this campaign, too.

In the last century it was realized that the first part of the story in the Book of Kings is the counterpart of the record on the prism, and that the second part of the story in the Book of Kings, as well as the whole story in Chronicles and in the Book of Isaiah, is a separate record of a separate campaign in Palestine.[1]

The first campaign against Judah took place in -702 or -701. The date of the second campaign is established as -687, or less probably, -686.

"Of the remaining eight years of his reign [after the conclusion of the prism records] we have no information from his own annals, which now cease. Sennacherib once more arrived in the West (687 or 686?)." [2]

[1] H. Rawlinson was the first to assume two campaigns of Sennacherib against Palestine. G. Rawlinson was of the same opinion. The Taylor Cylinder covers the time down to the 20th of Adar -691. H. Winckler supported this view with the argument that Tirhakah the Ethiopian became king of Ethiopia and Egypt after -691: "It can signify only a new campaign of Sennacherib which must have taken place after the destruction of Babylon (689 B.C.) and of which we have no record by Sennacherib himself."

The reference, "in the fourteenth year of Hezekiah," in the beginning of the record in the Book of Kings, explains why the obvious fact that there were two campaigns escaped earlier commentators. Also, the mention of Lachish in both campaigns was a stumbling block. In this connection K. Fullerton remarked ("The Invasion of Sennacherib" in *Biblioteca Sacra*, 1906) that Richard Cœur de Lion also made Lachish a base of operations on two different crusades.

Modern historians support the view that Tirhakah did not become king before -689.

See also J. V. Prášek, "Sanheribs Feldzüge gegen Juda," *Mitt. d. Vorderasiat. Ges.* (1903), and R. Rogers, *Cuneiform Parallels to the Old Testament* (1926), p. 259.

[2] H. R. Hall, *Ancient History of the Near East* (1913), p. 490. "The Jewish account seems to be confused, as it stands, with that of the earlier invasion of 701 B.C. In the story of II Kings, Tirhakah is spoken of as king, which he was not till 689 B.C. at the earliest." (*Ibid.*) See also D. D. Luckenbill, *The Annals of Sennacherib* (1924), p. 12.

# IGNIS E COELO

The destruction of the army of Sennacherib is described laconically in the Book of Kings: "And it came to pass that night, that the angel of the Lord went out, and smote in the camp of the Assyrians a hundred fourscore and five thousand; and when the people arose early in the morning, behold, they were all dead corpses. So Sennacherib king of Assyria departed, and went and returned, and dwelt in Nineveh." It is similarly described in the Book of Chronicles: "And the prophet Isaiah, the son of Amoz, prayed and cried to heaven. And the Lord sent an angel which cut off all the mighty men of valour, and the leaders and captains in the camp of the king of Assyria. So he [Sennacherib] returned with shame of face to his own land."

What kind of destruction was this? *Malach*, translated as "angel," means in Hebrew "one who is sent to execute an order," supposed to be an order of the Lord. It is explained in the texts of the Books of Kings and Isaiah that it was a "blast" sent upon the army of Sennacherib.[1] "I will send a blast upon him . . . and [he] shall return to his own land," was the prophecy immediately preceding the catastrophe. The simultaneous death of tens of thousands of warriors could not be due to a plague, as it is usually supposed, because a plague does not strike so suddenly; it develops through contagion, if rapidly, in a few days, and may infect a large camp, but it does not affect great multitudes without showing a curve of cases mounting from day to day.

The Talmud and Midrash sources, which are numerous, all agree on the manner in which the Assyrian host was destroyed: a blast fell from the sky on the camp of Sennacherib. It was not a flame, but a consuming blast: "Their souls were burnt, though their garments remained intact." The phenomenon was accompanied by a terrific noise.[2]

*Arad gibil* is the Babylonian designation of *ignis e coelo* (fire from the sky).[3]

[1] II Kings 19 : 7; Isaiah 37 : 7.

[2] Tractate Shabbat 113b; Sanhedrin 94a; Jerome on Isaiah 10 : 16; Ginzberg, *Legends*, VI, 363.

[3] Cf. Winckler, *Babylonische Kultur* (1902), p. 53; *Weltmantel und Himmelszelt*, II, 451 ff.

Another version of the destruction of the army of Sennacherib is given by Herodotus. During his visit in Egypt, he heard from the Egyptian priests or guides to the antiquities that the army of Sennacherib, while threatening the borders of Egypt, was destroyed in a single night. According to this story, an image of a deity holding in his palm the figure of a mouse was erected in an Egyptian temple to commemorate the miraculous event. In explanation of the symbolic figure, Herodotus was told that myriads of mice descended upon the Assyrian camp and gnawed away the cords of their bows and other weapons; deprived of their arms, the troops fled in panic.

Josephus Flavius repeated the version of Herodotus, and added that there is another version by the Chaldeo-Hellenistic historian Berosus. Josephus wrote introductory words to a quotation of Berosus, but the quotation itself is missing in the present text of the *Jewish Antiquities*. Obviously, it was an explanation different from that of Herodotus. Josephus' own account somewhat rationalistic as usual, says a (bubonic) plague was the cause of the sudden death of one hundred and eighty-five thousand warriors in the camp of the Assyrians before the walls of Jerusalem on the very first night of the siege.

Herodotus recounts that he saw the statue of the god with a mouse in the palm of his hand, which was erected in memory of the event. Two cities in Egypt claimed the same sacred animal, the shrewmouse: Panopolis (Akhmim) in the south and Letopolis in the north. Herodotus did not travel to the south of Egypt; thus, he must have seen the statue in Letopolis. Even today many bronze mice, sometimes inscribed with the prayers of pilgrims, are found in the ground of Letopolis.

Both cities with the cult of the sacred mouse were "sacred cities of thunderbolt and meteorites." [1] The Egyptian name for Letopolis is indicated by the same hieroglyphic as "thunderbolt."

In a text dating from the New Kingdom and originating in Letopolis, it is said that a festival was established in this city in memory of "the night of fire for the adversaries." This fire was like "the flame before the wind to the end of heaven and the end of earth." [2] "I come forth and go in the devouring fire on the day of the repelling of the adversaries," says the text in

[1] G. A. Wainwright, "Letopolis," *Journal of Egyptian Archaeology*, XVIII (1932).
[2] "The devouring fire of Letopolis is reminiscent of 'the flame before the wind to the end of heaven and the end of earth' which is connected with ⟷ the primitive form of the thunderbolt sign such as that of Letopolis." *Ibid.*

the name of the god. Thus the god with the sacred mouse was a god of devouring fire.

However, interpreting the mouse as a symbol of bubonic plague,[1] the commentators agreed with Josephus that Sennacherib's army must have been destroyed by a plague.

It is peculiar that the numerous commentators of Herodotus and the no less numerous commentators of the Bible did not draw attention to a certain coincidence in these descriptions of the calamity. Hezekiah became gravely ill of some bubonic affection and was near death. Isaiah was called. He told the king that he would die, but soon he returned and offered a remedy—a lump of figs for the boil—and told the king that the Lord would deliver him from immediate death and would also deliver "this city out of the hand of the king of Assyria."

"And this shall be a sign unto thee from the Lord . . . Behold, I will bring again the shadow of the degrees, which is gone down in the sun dial of Ahaz, ten degrees backward. So the sun returned ten degrees, by which degrees it was gone down." [2]

An optical illusion is the common explanation of the meaning of this passage.[3] The sundial mentioned together with the name of Ahaz is supposed to have been a dial built by Ahaz, father of Hezekiah. But the Talmudic tradition explains that the day was shortened by ten degrees on the day when Ahaz was buried, and the day was prolonged by ten degrees when Hezekiah was ill and recovered, and this is the meaning of the "shadow of the degrees which is gone down in the sun dial of Ahaz."[4]

The rabbinical sources state in a definite manner that the disturbance in the movement of the sun happened on the evening of the destruction of Sennacherib's army by a devouring blast.[5]

[1] Cf. I Samuel 6 : 4.

[2] Isaiah 38 : 6-8; similarly in II Kings 20 : 9 ff.

[3] Schiaparelli in *Astronomy in the Old Testament*, p. 99, points to a whole literature of "curious and eccentric ideas" written on the subject of the "steps of Ahaz" and refers to Winer's *Bibl. Realwörterbuch*, I, 498-499, where "most remarkable gnomics are reviewed." "None of the explanations can be regarded as well-founded," wrote Winer, "and it will never be possible to establish the factual element that is the basis of this narrative."

[4] See the Babylonian Talmud, Sanhedrin 96a; Pirkei Rabbi Elieser 52. Other sources are mentioned by Ginzberg, *Legends*, VI, 367. M. Gaster, *The Exempla of the Rabbis* (1924), in the Chapter, "Merodach and the Sun," lists Talmudic references to the described phenomenon.

[5] Seder Olam 23. Cf. Eusebius and Jerome on Isaiah 34 : 1. See Ginzberg, *Legends*, VI, 366.

Returning to Herodotus, we shall give our attention to the following important fact neglected by the commentators. The famous paragraph of Herodotus which records, in the name of the Egyptian priests, that since Egypt became a kingdom, the sun had repeatedly changed its direction, is inserted in no other place of Herodotus' history, but directly following the story of the destruction of Sennacherib's army.

The destruction of Sennacherib's army and the disturbance in the movement of the sun are also described in two subsequent passages of the Scriptures. Now the two records seem to be in better accord.

## MARCH 23rd

It was apparently some cosmic cause that was responsible for the sudden destruction of the army of Sennacherib and brought about the perturbation in the rotating movement of the earth. Gaseous masses reaching the atmosphere could asphyxiate all breath in certain areas.

This explanation requires supporting statements from other sources; disturbances in the movement of the sun could not be confined to the sun over Palestine and Egypt. Also, other circumstances of this catastrophe, like the gaseous masses covering the sky, should have been noticed in other regions of the earth, too.

First, a more exact date for the night of the annihilation of Sennacherib's army should be established. From modern research we know that it was in the year −687 (less probably in the year −686). The Talmud and Midrash give another valuable clue: the destruction occurred during the first night of Passover. The giant host was destroyed when the people began to sing the Hallel prayer of the Passover service.[1] Passover was observed about the time of the vernal equinox.[2]

In the book of Edouard Biot, *Catalogue général des étoiles filantes et des autres météores observés en Chine après le VII^e^ siècle avant J.C.*,[3] the register begins with this statement:

"The year 687 B.C., in the summer, in the fourth moon, in

[1] The Jerusalem Talmud, Tractate Pesahim; Seder Olam 23; Tosefta Targum II Kings 19 : 35-37; Midrash Rabba, III, 221 (English ed. by H. Freedman and M. Simon.)

[2] In the last two thousand years or so, the Feast of Passover, bound to the lunar calendar, has been observed between the middle of March and the latter part of April.

[3] Paris, 1846.

the day *sin mao* (23rd of March) during the night, the fixed stars did not appear, though the night was clear [cloudless]. In the middle of the night stars fell like a rain."

The date, 23rd of March, is Biot's calculation. The statement is based on old Chinese sources ascribed to Confucius. In another translation of the text, by Rémusat,[1] the last part of the passage is rendered as follows: "Though the night was clear, a star fell in the form of rain" (*il tomba une étoile en forme de pluie*).

The annals of the *Bamboo Books* obviously refer to the same event when they inform us that in the tenth year of the Emperor Kwei (the seventeenth emperor of the Dynasty Yu, or the eighteenth monarch since Yahou) "the five planets went out of their courses. In the night, stars fell like rain. The earth shook."[2]

The words in the annals, "in the night, stars fell like rain," are the same as in the record of Confucius dealing with the cosmic event on the 23rd of March -687. The annals supply the information that the cause of this phenomena was a disturbance among the planets. The record of Confucius is a precious entry, because the time of the phenomenon—the day, the month, and the year—is given.

The sky was cloudless, so that the stars should have been visible—but were not, and this reminds us of the words of the prophets.[3]

The Biot *Catalogue*, which begins with this description of the year -687, subsequently notes only solitary meteors falling from the sky during all the following centuries up to the beginning of this era; the prodigy of the year -687 was not a pageant such as we may find again in the Chinese annals of later centuries.

The rare phenomenon occurred in that year and in that part of the year—23rd of March, -687—when, as explained above, according to modern calculations and the Talmudic data, the destruction of Sennacherib's army took place. In the Chinese record we have a short but precise account of the night, which we have recognized as the night of annihilation.

We also expect to find in Chinese sources a record of the

[1] Abel Rémusat, *Catalogue des bolides et des aérolithes observés à la Chine, et dans les pays voisins* (1819): "On a beaucoup discuté sur ce texte de Confucius" (p. 7).
[2] *The Chinese Classics* (transl. and annot. by. J. Legge, Hong Kong ed.), III, Pt. 1, 125.
[3] Joel 2 : 10; 3 : 15.

disturbance in the movement of the sun. China is forty-five to ninety degrees longitude east of Palestine, the difference in time being three to six hours.

Huai-nan-tse,[1] who lived in the second century before the present era, tells us that "when the Duke of Lu-yang was at war against Han, during the battle the sun went down. The Duke, swinging his spear, beckoned to the sun, whereupon the sun, for his sake, came back and passed through three solar mansions."

The subjective-mythological part reminds us of the primitive-subjective approach of the author of the Book of Joshua, and probably also of the contemporaries of Joshua; it is the primitive way of interpreting natural phenomena. However, it differs from what is described in the Book of Joshua in that it was not a phenomenon of a long pause by the sun, but of a short retrograde motion; in this the Chinese description corresponds with the twentieth chapter of II Kings.

The exact date of the reign of Han is not known; it is sometimes supposed, on the basis of astronomical computation, to have been in the fifth century before this era, or even later.[2] If this is true, then the event described refers to a period before the dynasty of Han became dominant in China.

The land of China is large; it was divided into many princedoms. Probably the story of Prince Tau of Yin is another description of the same event in a different part of China. Lu-Heng [3] records that Prince Tau of Yin was an involuntary guest of the king of China when the sun returned to the meridian; it was interpreted as a sign to allow the prince to return home.

The story of the Argive tyrants tells of the sun going speedily to its setting and the evening coming before its proper time; and we recognized in this the phenomenon described in the rabbinical sources as having occurred on the day of the burial of Ahaz, father of Hezekiah. The prodigy of the day of Hezekiah or of the Duke of Lu-yang and Prince Tau of Yin took place at the time of the same tyrants, or was so ascribed. "Atreus,"

[1] Huai-nan-tse VI. iv. See Forke, *The World Conception of the Chinese*, p. 86.

[2] Moyriac de Mailla (1679-1748), *Histoire général de la Chine: Tong-Kien-Kang-Mou* (1877), Vol. I, has the Han Dynasty coming to power in the last quarter of the fifth century; Forke, *The World Conception of the Chinese*, thinks that the war of the Duke of Lu-yang against Han took place in the fifth century. But these calculations are based upon an astronomical computation which may be erroneous.

[3] Lu-Heng II, 176. See Forke, *The World Conception of the Chinese*, p. 87.

says Apollodorus,[1] stipulated with Thyestes that Atreus should be king if the sun should go backward; and when Thyestes agreed, the sun set in the east."

Ovid describes this phenomenon of the days of the Argive tyrants: Phoebus broke off "in mid-career, and wresting his car about turned round his steeds to face the dawn." [2] Also in *Tristia* Ovid refers to this literary tradition [3] about "the horses of the sun turning aside." [4]

A Mayan inscription says that a planet brushed close to the earth.[5]

Three solar mansions of the Chinese must have been equal to ten degrees on the dial at the palace in Jerusalem.

According to Talmudic sources,[6] an equal perturbation, but in the opposite direction, occurred on the day Ahaz was carried to his grave: at that time the day was quickened. A case of two consecutive perturbations of a celestial body, where the second perturbation corrected the effect of the first, is recorded in the annals of modern observations. In 1875 Wolf's comet passed near the large planet Jupiter and was disturbed on its way. In 1922, when it again passed near Jupiter, it was once more disturbed, but with an effect which corrected that of the first disturbance. No perturbation was noticed in the revolution of Jupiter; its rotation probably proceeded normally, too—there was a great difference in the masses of these two bodies.

## THE WORSHIP OF MARS

The body which periodically—once in fourteen to sixteen years—approached the orbit of the earth must have been of considerable mass, for it was able to influence the rotation of

[1] Apollodorus, *The Library*, Epitome II.
[2] Ovid, *The Art of Love* (transl. J. H. Mosley, 1929), i. 328 ff.
[3] Ovid, *Tristia* (transl. A. L. Wheeler, 1924), ii. 391 ff.
[4] More about the movement of the sun toward the east instead of the west in the time of the Argive tyrants was said in the Section "East and West," and several Greek authors were quoted. More will be said when we examine oral traditions of primitive peoples in a later section on folklore.
[5] Published by Ronald Strath. I could not locate the publication. It is referred to in Bellamy's *Moons, Myths and Man* (1938), p. 258. The only other reference to the work by Strath I found in Jean Gattefossé and Claudius Roux, *Bibliographie de l'Atlantide et des questions connexes* (Lyon, 1926), under No. 1184, but these authors also were unable to trace the publication. Cf. P. Jensen, *Kosmologie*, III, R561, 5a: "A great star fell." Jupiter was known to the Babylonians as the "great star." How large was the star? Jensen asked.
[6] Tractate Sanhedrin 96a.

the earth. Apparently, however, it was much smaller than Venus, or it did not approach so closely, because the catastrophes of the days of the Exodus and the Conquest were greater than those of the time of Uzziah, Ahaz, and Hezekiah. Nevertheless, for the peoples who lived at that time, they must have been impressive experiences and must have been incorporated in their cosmogonic mythologies.

Shall we be able, when inquiring into this matter, to find guiding hints to help us obtain some data about the body which periodically approached the earth?

It would probably be the Latin people, at that time very young, just appearing on the historical scene and not loaded down with science, who would give the prodigy a prominent place in their mythology. Roman mythology was appropriated from the Greeks. Only one god of Roman mythology plays a role not comparable to that attributed to him on the Greek Olympus. It is the god Mars, whose counterpart is Ares of the Greeks.[1] Mars, the lord of war, was second to Jupiter-Zeus. He personified the planet Mars, to him was dedicated the month of March (Mars), and as a god he was supposed to be the father of Romulus, the founder of Rome. He was the national god of the Romans. Livy wrote in the preface to his history of Rome, "the mightiest of empires, next after that of Heaven": "The Roman people . . . profess that their Father and the Father of their Empire was none other than Mars."

Placing the time of Mars' activity as late as the foundation of Rome indicates that the Romans had a tradition that the city on the Tiber came into existence during a generation which witnessed some great exploit of their god-planet.

The founding of Rome took place close in time to the great perturbations of nature in the days of Amos and Isaiah. According to the calculations of Fabius Pictor, Rome was founded in the latter half of the first year of the eighth Olympiad, or the year –747; other Roman authorities differ by a few years only.[2] The year –747 is the beginning of an astronomical

[1] Besides Ares, Hercules also represents the planet Mars. Eratosthenes (*Eratosthenis catasterismorum reliquiae*, ed. C. Robert, 1878): "Tertia est stella Martis quam alii Herculis dixerunt" (Mars is the third star, which others say is Hercules). Similarly, Macrobius (*Saturnalia* iii. 12. 5-6), whose authority is Varro.

[2] Polybius dated the foundation of Rome in the second year of the seventh Olympiad (–750); Porcius Cato, in the first year of the seventh Olympiad (–751); Verrius Flaccus, in the fourth year of the sixth Olympiad (–752); Terentius Varro, in the third year of the sixth Olympiad (–753); Censorinus followed Varro.

era in the Middle East; and the "commotion of Uzziah" took place, apparently, in the same year.

According to a persistent Roman tradition, the conception of Romulus by his mother, the foundation of Rome, and the death of Romulus occurred in years of great commotions accompanied by celestial phenomena and disturbances in solar movement. These changes were connected in some way with the planet Mars. Plutarch wrote: "To the surname of Quirinus bestowed on Romulus some give the meaning of Mars." [1] The legend says that Romulus was conceived in the first year of the second Olympiad (−772) when the sun was totally eclipsed. According to Latin historians, on the very day of Rome's foundation, the sun was disrupted in its movement and the world was darkened.[2] In Romulus' time "a plague fell upon the land, bringing sudden death without previous sickness," and "a rain of blood" and other calamities. Earthquakes convulsed the earth for a long period. Jewish tradition knows that "the first settlers of Rome found that the huts collapsed as soon as built." [3]

The death of Romulus occurred when, according to Plutarch, "suddenly strange and unaccountable disorders with incredible changes filled the air; the light of the sun failed, and night came down upon them, not with peace and quiet, but with awful peals of thunder and furious blasts," and amidst this storm Romulus disappeared.[4]

Ovid's description of the phenomena on the day of Romulus' death is this: "Both the poles shook, and Atlas shifted the burden of the sky. . . . The sun vanished and rising clouds obscured the heaven . . . the sky was riven by shooting flames. The people fled and the king [Romulus] upon his father's [Mars'] steeds soared to the stars."[5]

Hezekiah was a contemporary of Romulus and Numa; this was known to Augustine: "Now these days extend . . . down to Romulus king of Romans, or even to the beginning of the reign of his successor Numa Pompilius. Hezekiah king of Judah certainly reigned till then." [6]

[1] Plutarch, *Lives*, "The Life of Romulus" (transl. B. Perrin, 1914).
[2] Cf. F. K. Ginzel, *Spezieller Kanon der Sonnen- und Mondfinsternisse* (1899), and T. von Oppolzer, *Kanon der Finsternisse* (1887).
[3] Literature in Ginzberg, *Legends*, VI, 280.
[4] Plutarch, *Lives*, "The Life of Romulus."
[5] Ovid *Fasti* (transl. Frazer, 1931), II ll. 489 ff.
[6] Augustine, *The City of God*, Bk. XVIII, Chap. 27.

If Mars really was the deified cosmic visitor of the days of Hezekiah and Sennacherib, then one might expect not only that the activities of Mars would have been ascribed to the generation of Romulus and the foundation of Rome, but that the very date of the perturbation would have been a celebrated date in the cult of Mars.

The year of the second campaign of Sennacherib against Palestine is established by modern research as –687. The Talmud helps to set the time of the year: it was the night of the feast of spring, Passover. Chinese sources give the exact date, midnight of the 23rd of March, –687, as the date of a great cosmic activity.

The main festival in the cult of Mars took place in the month dedicated to this god-planet. "The *ancilia*, or sacred shields . . . were carried in procession by the Salii, or dancing warrior-priests of Mars on several occasions during the month of March up to the 23rd (*tubilustrium*), when the military trumpets (*tubae*) were lustrated; and again in October to the 19th (*armilustrium*), when both the *ancilia* and the arms of the exercitus were purified and put away for the winter. . . . It is only at the end of February that we find indications of the coming Mars-cult." [1] "The most important role in the cult of Mars appears to be played by the festival of *tubilustrium* on the twenty-third day of March." [2]

The date, the 23rd of March, taken with all the other circumstances mentioned above, must impress us. The fact that Mars had festivals on two dates (the other date, the 19th of October, is almost a month after the autumnal equinox) is easily understandable if one remembers that there was more than one perturbation connected with the same cosmic cause.

The disturbance in the movement of the sun a few hours before the Assyrian host perished occurred on the first day of Passover. The cataclysm of the days of the Exodus was caused by the planet Venus. Therefore, about the time of the vernal equinox there were two festivals, one for the planet Mars, the other for the planet Venus, which coincided in time. The festival of Minerva lasted from the nineteenth to the twenty-third of March, and on March 23rd, Mars, and also Minerva-Athene, were the honoured deities.[3]

[1] Quoted from W. W. Fowler, "Mars," *Encyclopaedia Britannica*, 14th ed.
[2] Roscher, "Mars," in Roscher's *Lexikon der griech. und röm. Mythologie.*
[3] *Ibid.*, Col. 2402.

## MARS MOVES THE EARTH FROM ITS PIVOT

Venus was a comet, and in historical times it became a planet. Was Mars a comet in the eighth century before this era? There is evidence that long before the eighth century Mars was a planet in the solar system. A four-planet system was known to Chaldean astronomy, in which Venus was absent but Mars was present.

There does not exist, at least in the extant material, any mention of the first appearance of Mars, whereas expressions referring to the birth of the planet Venus have been found in literary sources of the peoples of both hemispheres.

The Babylonian name of the planet Mars is Nergal.[1] This name is referred to in early times, many centuries prior to the eighth century. But it was in that latter century that this planet became a most important deity. Many prayers to it were composed. "Radiant abode, that beams over the land . . . who is thy equal?" Temples were built to this planet and statues erected. When Samaria was conquered by Sargon, father of Sennacherib, and new settlers were brought to live there, they erected in Samaria a shrine to the planet Mars.[2]

The planet Mars was feared for its violence. "Nergal, the almighty among the gods, fear, terror, awe-inspiring splendour,"[3] wrote Esarhaddon, son of Sennacherib. Shamash-shum-ukin, king of Babylonia and grandson of Sennacherib, wrote: "Nergal, the most violent among the gods."

It is characteristic that Nergal was regarded by the people of Assyria, as a god who brought defeat. Another grandson of Sennacherib, Assurbanipal, king of Assyria, wrote: "Nergal, the perfect warrior, the most powerful one among the gods, the pre-eminent hero, the mighty lord, king of battle, lord of power and might, lord of the storm, who brings defeat." [4]

It is also a conspicuous fact that the name of Nergal became very common as a component of personal names in the seventh and sixth centuries. Two generals, both by the name of

[1] Böllenrücher, *Gebete und Hymnen an Nergal* (1904), p. 3.
[2] II Kings 17 : 30.
[3] Luckenbill, *Records of Assyria*, II, Sec. 508.
[4] *Ibid.*, Sec. 922.

Nergalsharezer, were among Nebuchadnezzar's marshals;[1] a king by the name of Nergilissar ruled in Babylon.[2] Priests, warriors, traders in cattle, criminals bearing the name of Nergalsharezer, are familiar figures in the documents of the seventh century.

In the eighth century in Babylonia, the planet Mars was called "the unpredictable planet."[3]

Historical inscriptions of the eighth century speak of the oppositions of the star Mars (Nergal). These together with conjunctions were carefully watched. "The movements of Mars were extremely important in Babylonian astrology—its rise and setting, its disappearance and return . . . its position in relation to the equator, the change in its illuminating power, its relation to Venus, Jupiter and Mercury."[4] In India, also, "the various phases of the retrograde motion of the planets and especially of Mars seem to have been objects of great attention."[5]

Prayers were addressed to Nergal with the lifting of hands toward the star Mars.[6] "Thou who walkest in the sky . . . with splendour and terror . . . king of battle, the raging fire-god, god Nergal." Nergal-Mars was called by the Babylonians the "fire-star."[7] Nergal, the fire-star, comes like a raging storm. He is also called Sharappu, "the burner," and "light that flames from heaven," and "lord of destruction."[8] Mars was generally regarded by other peoples, too, as a "fire-star."[9] Ying-Huo, or the fire planet, is the name of Mars in Chinese astronomical charts.[10] Sargon (−724 to −705), father of Sennacherib, wrote on one occasion: "In the month of Abu, the month of descent of the fire-god."[11]

But we ask for a direct statement that the planet Mars-Nergal was the immediate cause of the cataclysms in the eighth and seventh centuries, when the world, in the language

[1] Jeremiah 39 : 3.
[2] The order of succession of the kings of the Neo-Babylonian Empire will be discussed in *Ages in Chaos.*
[3] Schaumberger, in Kugler, *Sternkunde und Sterndienst in Babel*, 3rd supp., p. 307.
[4] Bezold, in Boll's *Sternglaube und Sterndeutung*, p. 6.
[5] Thibaut, "Astronomie, Astrologie und Mathematik," *Grundriss der indo-arischen Philologie und Alterthumskunde, III* (1899).
[6] Böllenrücher, *Gebete und Hymnen an Nergal*, pp. 9, 19 ("Zauberspruch mit Handerhebung an den Mars-Stern").
[7] Schaumberger in Kugler's *Sternkunde*, p. 304; Böllenrücher, *Gebete und Hymnen an Nergal*, pp. 21 ff.
[8] Langdon, *Sumerian and Babylonian Psalms* (1909), p. 85.
[9] Apuleius, *Tractate of the World;* literature in Chwolson, *Die Ssabier und Ssabismus*, II, 188.
[10] Rufus and Hsing-chih-tien, *The Soochow Astronomical Chart.*
[11] Luckenbill, *Records of Assyria*, II, Sec. 121.

of Isaiah, was "moved exceedingly" and "became removed from its place." This very action is ascribed to the planet Mars-Nergal: "The heaven he makes dark, he moves the Earth off its hinges."[1] And again: "Nergal . . . on high stills the heavens . . . causes the earth to shudder."[2]

[1] Böllenrücher, *Gebete und Hymnen an Nergal*, p. 9.
[2] Langdon, *Sumerian and Babylonian Psalms*, p. 79.

## CHAPTER III

# WHAT CAUSED VENUS AND MARS TO SHIFT THEIR ORBITS?

When Venus became a new member of the solar system, it moved on a stretched ellipse, and for centuries imperilled the other planets. Because of its dangerous circling, Venus was diligently observed in both hemispheres, and records were kept of its movement.

In the last centuries before this era, the 225-day year of Venus, and apparently also its orbit, were practically the same as in modern times. As early as the second half of the seventh century before this era, Venus, watched until then with anxiety, had already ceased to be a cause of dreadful expectation; it probably reached then the orbital stage in which it was found in the last centuries before this era, and where we still find it today. What caused the change in the orbit of Venus?

I shall pose another problem besides the first. Mars did not arouse any fears in the hearts of the ancient astrologers, and its name was seldom mentioned in the second millennium. In Assyria-Babylonia, in inscriptions made before the ninth century, the name of Nergal is found only on rare occasions. On the astronomical ceiling of Senmut Mars does not appear among the planets. It did not play any conspicuous part in the early mythology of the celestial gods.

But in the ninth or eighth century before this era, the situation changed radically. Mars became the dreaded planet. Accordingly, Mars-Nergal rose to the position of the frightful storm and war god. The question must then present itself: Why, previous to that time, did Mars signify no danger to the earth, and what caused Mars to shift its orbit nearer to the earth?

The planets of the solar system move in nearly the same plane, and if one planet were to revolve along a stretched ellipse, it would endanger the other planets. The two problems—what caused Venus to change its orbit, and what caused Mars to change its orbit—may have a common explanation. The common cause may have been some comet which changed the orbits of Venus and Mars; but it is simpler to suppose that two planets, one of which had a greatly elongated orbit, collided, and that no third agent was necessary to bring about that result.

A conflict between Venus and Mars, if it occurred, might well have been a spectacle observable from the earth. It is not impossible that the two planets came repeatedly into contact, each time with different results.

If a contact between Venus and Mars really occurred and was observed from the earth, it must have been commemorated in traditions or literary documents.

## WHEN WAS THE *ILIAD* CREATED?

> A mighty strife had waxen great
> Within the members of the sphere.
>
> —EMPEDOCLES[1]

To this day it has not been established at what date the *Iliad* and *Odyssey* were composed. Even ancient authors differed greatly in reckoning the time when Homer lived. It was estimated to be as late as –685 (the historian Theopompus) and as early as –1159 (certain authorities quoted by Philostratus). Herodotus wrote that "Homer and Hesiod" created the Greek pantheon "not more than 400 years before me," which would mean not prior to –884, –484 being regarded as the year of Herodotus' birth. The question is still debated. Some authors argue that there was a long interval between the time when the epic works of Homer were composed and the time when they were put into writing; others think that these works must have been created not long before the Greeks acquired the art of writing, about –700.[2] It is also argued that the Greeks must have known this art long before –700 on the assumption that the Homeric works were created much before that date. It is generally assumed that the fall of Troy antedated Homer by several generations, and also that the great epic works were the creation of generations. The fall of Troy is

[1] *The Fragments of Empedocles* (transl. W. E. Leonard, 1908), p. 30.

[2] See R. Carpenter, "The Antiquity of the Greek Alphabet" and B. Ullman, "How Old Is the Greek Alphabet?" in *American Journal of Archaeology*, XXXVII (1933) and XXXVIII (1934), respectively.

sometimes thought to have taken place in the twelfth century.[1]

On the other hand, it has been shown that the cultural background of the Homeric epos is that of the eighth or even the seventh century; the age of iron was well under way, and many other details would preclude an earlier scene.[2] It is highly probable that the Homeric works were created at that time or shortly thereafter. Whether these poems were first sung by a bard who lived centuries after the destruction of Troy depends on the time when Troy was destroyed. The tradition about Aeneas who, saved when Troy was captured, went to Carthage (a city built in the ninth century) and from there to Italy, where he founded Rome (a city first built in the middle of the eighth century), implies that Troy was destroyed in the eighth or late in the ninth century.

But for what purpose do I burden my present work with this question? It may seem that the two problems—how Venus changed its orbit to a circle, and how Mars changed its orbit so as to come in contact with the earth—are weighted with a third problem from a far-removed field and in itself complicated. And even if these matters have something in common, how can a problem with three unknowns be solved?

We shall come closer to a solution of the astronomical problem with which we are concerned and the problem of the epics of Troy if we recognize the cosmic scene of these epics.

A simple test can be made. If Ares, the Mars of the Greeks, is not mentioned in the creations of Homer, this would support the view that the *Iliad* and *Odyssey* were created in the tenth century or earlier, or at least that the drama they describe had taken place not later than this time. But if Ares is presented as a war god in these epics, it would indicate that they were composed in the eighth century or thereafter. It was in the eighth century that Mars-Nergal, an obscure deity, became a prominent god. Epic poems, rich in mythology, that originated in the eighth or seventh century, would not be silent about Mars-Ares, who became "outrageous" at that time.

With this yardstick at hand, the epic poems of Homer must

[1] When the ancient site was discovered, Schliemann identified the ruins of the second city (from the bottom) as those of the Ilium of the *Iliad;* but later explorers disagreed and pronounced the ruins of the sixth city as those of Homeric Troy.

[2] G. Karo, "Homer" in Ebert's *Reallexikon der Vorgeschichte*, Vol. V.

be re-examined. The task will not be difficult; the *Iliad* is full of descriptions of the violent deeds of Ares.

In this epic the story is told of the battles which the Greeks, besieging Troy, waged against the people of Priam, king of Troy. Deities took a prominent part in these battles and skirmishes. Two of them—Athene and Ares—were by far the most active. Athene was the protectress of the Greeks; Ares was on the side of the Trojans. They were the chief antagonists throughout the epopee.

At first Athene removed Ares from the battlefield:

> And flashing-eyed Athene took furious Ares by the hand and spake to him, saying: "Ares, Ares, thou bane of mortals, thou blood-stained stormer of walls, shall we not now leave the Trojans and Achaeans to fight?" . . . [She] led furious Ares forth from the battle.[1]

But they met together again in the field; "furious Ares" was "abiding on the left of the battle."

Aphrodite, the goddess of the moon, wished to participate in the war also, but Zeus, presiding in heavenly Olympus, told her:

> "Not unto thee, my child, are given works of war; nay, follow thou after the lovely works of marriage, and all these things shall be the business of swift Ares and Athene."

Thus the god of the planet Jupiter admonished the goddess of the moon to leave the combat that it might be fought out by the god of the planet Mars and the goddess of the planet Venus. Phoebus Apollo, the god of the sun, spoke to the god of the planet Mars:

> Then unto furious Ares spake Phoebus Apollo: "Ares, Ares, thou bane of mortals, thou blood-stained stormer of walls, wilt thou not now enter into the battle?" . . .
>
> And baneful Ares entered amid the Trojan's ranks. . . . He called: . . . "How long will ye still suffer your host to be slain by the Achaeans?"

The battlefield was darkened by Ares:

> And about the battle furious Ares drew a veil of night to aid the Trojans . . . he saw that Pallas Athene was departed, for she it was that bare aid to the Danaans.

[1] *The Iliad*, Bk. V (transl. A. T. Murray; Loeb Classical Library, 1924-1925).

Hera, the goddess of the earth, "stepped upon the flaming car" and "self-bidden groaned upon their hinges the gates of heaven which the Hours had in their keeping, to whom are entrusted great heaven and Olympus." She spoke to Zeus:

"Zeus, hast thou no indignation with Ares for these violent deeds, that he hath destroyed so great and so goodly a host of the Achaeans recklessly? . . . Wilt thou in any wise be wroth with me if I smite Ares?"

And Zeus replied:

"Nay, come now, rouse against him Athene . . . who has ever been wont above others to bring sore pain upon him."

So came the hour of battle.

Then Pallas Athene grasped the lash and the reins, and against Ares first she speedily drave. . . . Athene put on the cap of Hades, to the end that mighty Ares should not see her.

Ares, "the bane of mortals," was attacked by Pallas Athene, who sped the spear "mightily against his nethermost belly."

"Then brazen Ares bellowed loud as nine thousand warriors or ten thousand cry in battle, when they join in the strife of the War-god."

Even as a black darkness appeareth from the clouds when after heat a blustering wind ariseth, even in such wise . . . did brazen Ares appear, as he fared amid the clouds unto broad heaven.

In heaven he appealed to Zeus with bitter words of complaint against Athene:

"With thee are we all at strife, for thou art father to that mad and baneful maid, whose mind is ever set on deeds of lawlessness. For all the other gods that are in Olympus are obedient unto thee . . . but to her thou payest no heed . . . for that this pestilent maiden is thine own child."

And Zeus replied:

"Most hateful to me art thou of all gods that hold Olympus, for ever is strife dear to thee and wars and fightings."

The first round was lost by Ares. "Hera and Athene . . . made Ares, the bane of mortals, to cease from his manslaying."

In this vein the poem proceeds, its allegorical features being only too readily overlooked. In the fifth book of the *Iliad* Ares is called by name more than thirty times, and throughout the poem he never disappears from the scene, whether in the sky or on the battleground. The twentieth and twenty-first books describe the climax of the battle of the gods at the walls of Troy.

[Athene] would utter her loud cry. And over against her spouted Ares, dread as a dark whirlwind, calling with shrill tones to the Trojans.

Thus did the blessed gods urge on the two hosts to clash in battle, and amid them made grievous strife to burst forth. Then terribly thundered the father of gods and men from on high; and from beneath did Poseidon cause the vast earth to quake, and the steep crests of the mountains. All the roots of many-fountained Ida were shaken, and all her peaks, and the city of the Trojans, and the ships of the Achaeans. And seized with fear in the world below was Aidoneus, lord of the shades . . . lest above him the earth be cloven by Poseidon, the Shaker of Earth, and his abode be made plain to view for mortals and immortals . . . so great was the din that arose when the gods clashed in strife.

In this battle of gods above and beneath, Trojans and Achaeans clashed together and the whole universe roared and shivered. The battle was fought in gloom; Hera spread a thick mist. The river "rushed with surging flood, and roused all his streams tumultuously." Even the ocean was inspired with "fear of the lightning of great Zeus and his dread thunder, whenso it crasheth from heaven." Then rushed into the battle a "wondrous blazing fire. First on the plain was the fire kindled, and burned the dead . . . and all the plain was parched." Then to the river turned the gleaming flame. "Tormented were the eels and the fish in the eddies, and in the fair streams they plunged this way and that. . . . The fair streams seethed and boiled." Nor had the river "any mind to flow onward, but was stayed," unable to protect Troy.

Upon the gods "fell strife heavy and grievous." "Together then they clashed with a mighty din, and the wide earth rang, and round about great heaven pealed as with a trumpet. . . . Zeus—the heart within him laughed aloud in joy, as he beheld the gods joining in strife."

Ares . . . began the fray, and first leaping upon Athene, brazen spear in hand, and spake a word of reviling: "Wherefore now again, thou dog-fly, art thou making gods to clash with gods in strife . . . ? Rememberest thou not what time . . . thyself in sight of all didst grasp the spear and let drive straight at me, and didst rend my fair flesh?"

This second encounter between Ares and Athene was also lost by Ares.

He [Ares] smote upon her tasselled aegis. . . . Thereon blood-stained Ares smote with his long spear. But she gave ground, and seized with her stout hand a stone that lay upon the plain, black and jagged and great. . . . Therewith she smote furious Ares on the neck, and loosed his limbs. . . .

Pallas Athene broke into a laugh. . . . "Fool, not even yet hast thou learned how much mightier than thou I avow me to be, that thou matchest thy strength with mine."

Aphrodite came to wounded Ares, "took [him] by the hand, and sought to lead [him] away." But "Athene sped in pursuit. . . . She smote Aphrodite on the breast with her stout hand . . . and her heart melted."

These excerpts from the *Iliad* show that some cosmic drama was projected upon the fields of Troy. The commentators were aware that originally Ares was not merely the god of war, and that this quality is a deduced and secondary one. The Greek Ares is the Latin planet Mars; it is so stated in classic literature a multitude of times. In the so-called Homeric poems, too, it is said that Ares is a planet. The Homeric hymn to Ares reads:

Most mighty Ares . . . chieftain of valour, revolving thy fiery circle in ether among the seven wandering stars [planets,] where thy flaming steeds ever uplift thee above the third chariot.[1]

But what might it mean, that the planet Mars destroys cities, or that the planet Mars is ascending the sky in a darkened cloud, or that it engages Athene (the planet Venus) in battle? Ares must have represented some element in nature, guessed the commentators. Ares must have been the personification of the raging storm, or the god of the sky, or the god of light, or a sun-god, and so on.[2] These explanations are futile. Ares-Mars is what his name says—the planet Mars.

[1] *The Odyssey of Homer with the Hymns* (transl. Buckley), p. 399. The translation by H. Evelyn-White (Hesiod volume in the Loeb Classical Library) is: "Who whirl your fiery sphere among the planets in their sevenfold courses through the ether wherein your blazing steeds ever bear you above the third firmament of heaven." Allen, Holliday, and Sikes, *The Homeric Hymns* (1936), p. 385, regard the hymn to Ares as post-Homeric.

[2] These divergent views are offered by L. Preller (*Griechische Mythologie* [1894] ), G. F. Lauer (*System der griechischen Mythologie* [1853], p. 224), F. G. Welcker (*Griechische Götterlehre*, I [1857], 415), and H. W. Stoll (*Die ursprüngliche Bedeutung des Ares* [1855] ).

I find in Lucian a statement which corroborates my interpretation of the cosmic drama in the *Iliad.* This author of the second century of the present era writes in his work *On Astrology* this most significant and most neglected commentary on the Homeric epics:

"All that he [Homer] hath said of Venus and of Mars his passion, is also manifestly composed from no other source than this science [astrology]. Indeed, it is the conjuncture of Venus and Mars that creates the poetry of Homer."[1]

Lucian is unaware that Athene is the goddess of the planet Venus,[2] and yet he knows the real meaning of the cosmic plot of the Homeric epic, which shows that the sources of his instruction in astrology were cognizant of the facts of the celestial drama.

My interpretation of the Homeric poem, I find, has been anticipated by still others. Who they were, it is impossible to say. However, Heraclitus, a little known author of the first century, who should not be confused with the philosopher, Heraclitus of Ephesus, wrote a work on Homeric allegories.[3] In his opinion Homer and Plato were the two greatest spirits of Greece, and he tried to reconcile the anthropomorphic and satiric description of gods by Homer with the idealistic and metaphysical approach of Plato. In Paragraph 53 of his *Allegories,* Heraclitus confutes those who think that the battles of the gods in the *Iliad* signify collisions of the planets. Thus I find that some of the ancient philosophers must have held the same opinion at which I arrived independently after a series of deductions.

The problem of the date when the Homeric epics originated was raised here, to be solved with the help of this criterion: If the cosmic battle between the planets Venus and Mars is mentioned there, then the epics could not have originated much before the year –800. If the earth and the moon are involved in this struggle, the time of the birth of the *Iliad* must be lowered to –747 at least and probably to an even later date. The first earthshaking contact with our planet had already taken place, and for this reason Ares is repeatedly called "bane of mortals, blood-stained stormer of walls."

[1] Lucian, *Astrology* (transl. A. M. Harmon, 1936), Sec. 22.
[2] In the same sentence Lucian identifies Venus with Aphrodite of the *Iliad.*
[3] *Heracliti questiones Homericae* (Teubner's ed. 1910). Cf. F. Boll, *Sternglaube und Sterndienst* (ed. W. Gundel, 1926), p. 201.

Homer was thus, at the earliest, a contemporary of the prophets Amos and Isaiah, or more likely he lived shortly after them. The Trojan War and the cosmic conflict were synchronous; the time of Homer was not separated from the time of the Trojan war by several centuries, possibly not even by a single one.

The statement by Lucian regarding the inspiring drama of the Homeric epics—the conjunction of the planets Venus and Mars—can be refined. There was more than one fateful conjunction between Venus and Mars—at least two are described in the *Iliad*, in the fifth and the twenty-first books. The conjunctions were near contacts; the mere passage of one planet in front of another could not have provided material for a cosmic drama.

## HUITZILOPOCHTLI

The Greeks chose Athene, the goddess of the planet Venus, as their patron, but the people of Troy looked to Ares-Mars as their protector. A similar situation existed in ancient Mexico. Quetzal-cohuatl, known as the planet Venus, was the patron of the Toltecs. But the Aztecs who later came to Mexico and supplanted the Toltecs, revered Huitzilopochtli (Vitchilupuchtli) as their protector-god.[1]

Sahagun says that Huitzilopochtli was "a great destroyer of towns and killer of people." The epithet "blood-stained stormer of walls" is familiar to us from the *Iliad*, where it is regularly applied to Mars. "In warfare he [Huitzilopochtli] was like live-fire, greatly feared by his enemies," writes Sahagun.[2]

In his work on the Indians of America, H. H. Bancroft writes:

"Huitzilopochtli had, like Mars and Odin, the spear or a bow in his right hand, and in the left, sometimes a bundle of arrows, sometimes a round white shield. . . . On these weapons depended the welfare of the state, just as on the ancile of the Roman Mars, which had fallen from the sky, or on the palladium of the warlike Pallas Athene. Bynames also point out Huitzilopochtli as war god; so he is called the terrible god Tetzateotl, or the raging Tetzahuitl."[3] Bancroft proceeds:

[1] J. G. Müller, *Der mexikanische Nationalgott Huitzilopochtli* (1847).
[2] Sahagun, *A History of Ancient Mexico* (transl. F. R. Bandelier, 1932), p. 25.
[3] H. H. Bancroft, *The Native Races of the Pacific States* (1874-1876), III, 302.

"One might be led to compare the capital of the Aztecs with ancient Rome, on account of its warlike spirit, and therefore it was right to make the national god of Aztecs a war god like the Roman Mars."[1]

But Huitzilopochtli was not like Mars, he was Mars. The identity of their appearance, character, and action is dictated by the fact that Mars and Huitzilopochtli were one and the same planet-god.

The conflict between Venus and Mars was also symbolized in religious ceremonies of the ancient Mexicans. In one of these ceremonies the priest of Quetzal-cohuatl shot an arrow into an effigy of Huitzilopochtli, which penetrated the god, who was then considered dead.[2] This appears to have been a symbolic repetition of the electrical discharge that Venus ejected toward Mars.

But the Aztecs would not concede the death of Mars, the bellicose destroyer of towns, the god of sword and pestilence, and carried on their wars against the Toltecs, the people who looked to the planet Venus. These wars between the Toltecs and the Aztecs must have taken place earlier than is generally supposed; they might have occurred before the present era, when there was rivalry between the peoples devoted to Venus and those devoted to Mars, and when the memory of the cosmic conflict was still vivid.

## TAO

> What is it that we call the Tao?
> There is the Tao, or Way of Heaven;
> and there is the Tao, or Way of Man.
> —KWANG-TZE

Planets of the solar system were disturbed by the contacts of Venus, Mars, and the earth. We have already referred to the annals of the *Bamboo Books*, where it is written that in the tenth year of the Emperor Kwei, the eighteenth monarch since Yahou, "the five planets went out of their courses. In the night, stars fell like rain. The earth shook." [3] The disturbances in the family of planets were caused by collisions between

[1] H. H. Bancroft, *The Native Races of the Pacific States* (1874–1876), III, p. 301.
[2] Sahagun, *Historia general de las cosas de la Nueva España*, III, Chap. I, Sec. 2.
[3] James Legge (ed.), *The Chinese Classics*, III, Pt. 1, 125.

Venus and Mars. The battles of two stars appearing as bright as suns are mentioned in another Chinese chronicle as having occurred in the days of the same Emperor Kwei (Koei-Kie):

"At this time the two suns were seen to battle in the sky. The five planets were agitated by unusual movements. A part of Mount T'aichan fell down."[1]

The two battling stars are recognized by us as Venus and Mars. In the language of Eratosthenes, the Alexandrian librarian of the third century before this era: "In the third place is the star [stella] of Mars. . . . It was pursued by the star [sidus] Venus; then Venus took hold of him and inflamed him with an ardent passion."[2]

In an astronomical chart dating from the Middle Ages (1193), used in the education of emperors and known as the Soochow Astronomical Chart,[3] it is asserted on the authority of the ancients that it happened that planets went off their courses. It is said that once Venus ran far off the zodiac and attacked the "Wolf-Star." A change in the course of the planets was regarded as a sign of heavenly wrath, since it occurred when the emperor or his ministers sinned.

In the old Chinese cosmology "Earth is represented as a body suspended in air, moving eastward,"[4] and thus was understood as one of the planets.

The following passage from the Taoist text of Wen-Tze[5] contains a description of calamities which, as we have found, belong together:

"When the sky, hostile to living beings, wishes to destroy them, it burns them; the sun and the moon lose their form and are eclipsed; the five planets leave their paths; the four seasons encroach one upon another; daylight is obscured; glowing mountains collapse; rivers are dried up; it thunders then in winter, hoarfrost falls in summer; the atmosphere is thick and human beings are choked; the state perishes; the aspect and the order of the sky are altered; the customs of the age are disturbed [thrown into disorder] . . . all living beings harass one another."

Hoei-nan-tze, a Taoist author of the third century of this

[1] L. Wieger, *Textes historiques* (2nd Ed., 1922-1923), I, 50.
[2] Eratosthenes, ed. Robert, p. 195.
[3] *The Soochow Astronomical Chart* (transl. and ed. by Rufus and Hsing-chih tien).
[4] J. C. Ferguson, *Chinese Mythology* (1928), p. 29.
[5] Wen-Tze in *Textes Taoistes*, transl. C. de Harlez (1891).

era, speaks of the sun and the earth leaving their paths; he transmits the tradition that "if the five planets err on their routes," the State and the provinces are overcome by a flood.[1]

Taoism is the dominant religion of China. "The term Tao originally meant the revolution of the way of the heavens about the earth. This movement of the heavens was regarded as the cause of the phenomena on earth. The Tao was located about the celestial pole which was considered to be the seat of power because all revolves about it. In the course of time Tao was viewed as the universal cosmic energy behind the visible order of nature."[2]

## YUDDHA

In an old textbook on Hindu astronomy, the *Surya-Siddhanta*, there is a chapter, "Of planetary conjunctions." Modern astronomy knows only one kind of conjunction between planets, when one planet (or sun) stands between the earth and another planet (differentiated only as superior and inferior conjunction and opposition). But ancient Hindu astronomy distinguished between many different conjunctions, translated as follows: *samyoga* (conjunction), *samagama* (coming together), *yoga* (junction), *melaka* (uniting), *yuti* (union), *yuddha* (encounter, in the meaning of conflict, fight).[3]

The first paragraph of this chapter, "Of planetary conjunctions," of the *Surya-Siddhanta* tells us that between planets there occur encounters in battle (*yuddha*) and simple conjunction (*samyoga samagama*). The force of the planets, which manifests itself in conjunctions, is called *bala*. A planet can be vanquished (*jita*) in an "*apasvya* encounter," struck down (*vidhvasta*), utterly vanquished (*vijita*). A powerful planet is called *balin*, and the victor-planet in an encounter, *jayin*. "Venus is generally victor."

To the last sentence the translator of *Surya-Siddhanta* wrote: "In this passage we quit the proper domain of astronomy, and trench upon that of astrology." Aside from the introductory lines in which the work is presented as a revelation of the sun (a common introduction in many astronomical works of the Hindus), it is written in very sober terms. It makes use of square roots and geometrical figures, and speaks in algebraic

[1] Hoei-nan-tze in *Textes Taoïstes*.

[2] L. Hodous, "Taoism," *Encyclopaedia Britannica*, 14th ed.

[3] *Surya-Siddhanta*, Chap. VII (transl. Burgess).

terms; every sentence of the work is in scientific language, very precious, indeed.[1]

This manual of the *Surya* contains also the correct notion of the earth as a "sphere" or "globe in the ether," showing that the Hindus of early times knew that the earth is one of the planets, though they thought it to be situated in the centre of the universe.[2] Aryabhatta held the opinion that the earth revolves on its axis.[3] Like the author of the Book of Job, who wrote that the earth hangs "upon nothing" (26 : 7), the *Surya* knew that "above" and "beneath" are only relative: "And everywhere upon the globe of the earth, men think their own place to be uppermost—but since it is a globe in the ether, where should there be an upper, or where an under side of it?"[4]

The strange chapter of *Surya-Siddhanta* dealing with the conjunctions of planets and with their conflicts when in close proximity made modern scholars think that this portion did not have the scientific value of the rest of the work, and was a product of astrological invention, or even an interpolation. We know now that this chapter has equal scientific value with other chapters of the work and that encounters between planets actually took place a number of times in the solar system.

In Hindu astronomy a junction of the planets is called *yoga* [yuga]. Very revealing is the fact that the world ages are also called *yogas*, planetary conjunctions[5] (or more precisely, junctions).

## THE BUNDAHIS

Theomachy, the battle of the gods, described in the Homeric epics, in the *Edda*, and in the Huitzilopochtli epos, is related also in the Indo-Iranian text of the *Bundahis*.[6] "The planets ran against the sky and created confusion" in the entire cosmos.[7]

[1] The following formula may serve as an example of the *Surya* method: "Multiply the earth's circumference by the sun's declination in degrees, and divide by the number of degrees in a circle; the result, in *yojanas*, is the distance from the place of no latitude where the sun is passing overhead." (Chap. xii).
[2] Tycho Brahe, in post-Copernican times, still adhered to this view.
[3] *Surya-Siddhanta*, note to p. 13.
[4] *Ibid.*, p. 248.
[5] Bentley, *A Historical View of the Hindu Astronomy* (1825), p. 75: "The periods themselves were named Yugas, or conjunctions."
[6] *The Bundahis, Pahlavi Texts* (transl. West).
[7] "Die Planeten rannten, Verwirrung stiftend, gegen den Himmel an." J. Hertel, "Der Planet Venus in Avesta," *Berichte der Sächsischen Akademie der Wissenschaften, Phil. hist. Klasse*, LXXXVII (1935).

In the long battle of the celestial bodies, one of them made the world entirely dark, disfigured creation, and filled it with vermin. This act of the cosmic drama was recognised by us as the first contact of the earth with the comet Typhon, the same as Pallas Athene. Other acts of the drama followed. The planetary disturbances lasted for a long time. "The celestial sphere was in revolution. . . . The planets, with many demons, dashed against the celestial sphere, and mixed the constellations; and the whole creation was as disfigured as though fire disfigured every place and smoke arose over it."[1]

The planet named Gokihar or "Wolf-progeny" and "special disturber of the moon,"[2] and a celestial body called Mievish-Muspar, "provided with tails," or a comet,[3] brought confusion to the sun, moon, and stars. But in the end "the sun has attached Muspar to its own radiance by mutual agreement, so that he may be less able to do harm."[4]

In this description of "the battle of the planets," we recognize the wolf-progeny and disturber of the moon, the planet Gokihar, as Mars; Muspar with tails apparently is Venus, called also Tistrya, or "the leader of the stars against the planets." As the final result of these battles, the sun made Venus into an evening-morning star or put Lucifer lower down so that it could do no harm. In the *Bundahis* the conflicting forces are called, not "gods," but merely "planets."

## LUCIFER CUT DOWN

It can be said that the planet Mars saved the terrestrial globe from a major catastrophe by colliding with Venus. Since the days of Exodus and Joshua, Venus was dreaded by the peoples of the earth. For about seven hundred years this terror hung over mankind like the sword of Damocles. Human sacrifices were made to Venus in both hemispheres in order to propitiate her.

After centuries of terror, the sword of Damocles was removed from above the heads of mankind, only to be replaced by another. Mars became the dread of the peoples and its return was feared every fifteen years. Before this, Mars had absorbed

[1] *Bundahis*, Chap. 3, Secs. 19-25.
[2] See infra the Section "Fenris-Wolf," note 5.
[3] Olrik, *Ragnarok*, p. 339.
[4] *Bundahis*, Chap. V, Sec. 1.

the blow, even the repeated blows of Venus, and had saved the earth.

Venus, which collided with the earth in the fifteenth century before the present era, collided with Mars in the eighth century. At that time Venus was moving at a lower elliptical velocity than when it first encountered the earth; but Mars, being only about one-eighth the mass of Venus, was no match for her. It was therefore a notable achievement that Mars, though thrown out of the ring, nevertheless was instrumental in bringing Venus from an elliptical to a nearly circular[1] orbit. Looked at from the earth, Venus was removed from a path that ran high to the zenith and over the zenith to its present path[2] in which it never retreats from the sun more than 48 degrees, thus becoming a morning or an evening star that precedes the rising sun or follows the setting sun. The awe of the world for many centuries, Venus became a tame planet.

Isaiah, referring figuratively to the king of Babylon who destroyed cities and made the land into a wilderness, uttered his remarkable words about Lucifer that fell from heaven and was cut down to the ground. The commentators recognized that behind these words applied to the king of Babylon must have been some legend about the Morning Star. The metaphor regarding the king of Babylon implied that his fate and the fate of the Morning Star were not dissimilar; both of them fell from on high. But what could it mean that the Morning Star fell from the heights? asked the commentators.

Significant are the words of Isaiah about the Morning Star, that it "weakened the nations" before it was cut down to the ground. It weakened the nations in two collisions with the earth, and it weakened the nations by keeping them in constant fear for centuries.

The Book of Isaiah, in every chapter, provides abundant evidence that with the removal of Venus, so that it no longer crossed the orbit of the earth, danger was not eliminated, but became even more threatening.

[1] Eccentricity of Venus' orbit is .007.

[2] Inclined 3° 4′ to the plane of the ecliptic (Duncan, 1945).

CHAPTER IV

# SWORD-GOD

IN BABYLON OF the eighth century the planet Mars became a great and feared god, to whom many prayers were composed and hymns and invocations were sung and magic formulas were whispered. Such formulas are referred to as "magic words with raising the hand to the planet Nergal [Mars]." These prayers were addressed directly to the planet Mars.[1] Like the Greek Ares, Nergal is called "king of battle, who brings the defeat, who brings the victory." Nergal could not be regarded as favouring the people of the Double Streams; on a most fateful night he inflicted a defeat on Sennacherib.

> Shine of horror, god Nergal, prince of battle,
> Thy face is glare, thy mouth is fire,
> Raging Flame-god, god Nergal.
>
> Thou art Anguish and Terror,
> Great Sword-god,
> Lord who wanderest in the night,
> Horrible, raging Flame-god . . .
> Whose storming is a storm flood.

In one of its great conjunctions, Mars' atmosphere was stretched so that it appeared like a sword. Often before and later, too, celestial prodigies assumed the shape of swords. Thus, in the days of David a comet appeared in the form of a human being "between the earth and the heaven, having a drawn sword in his hand stretched out over Jerusalem." [2]

The Roman god Mars was pictured with a sword; he became the god of war. The Chaldean Nergal is called "Sword-god." Of this sword Isaiah spoke when he predicted the repetition of the catastrophe, a stream of brimstone, flame, storm, and

[1] Böllenrücher, *Gebete und Hymnen an Nergal*, p. 19. Bezold in Boll, *Sternglaube und Sterndeutung*, p. 13: "Gebete der Handerhebung: von denen eine Anzahl an Planetengötter andere dagegen ausdrücklich an die Gestirne selbst (Mars) gerichtet sind" (prayers with the lifting of the hand: some of them are directed to the planetary gods and others expressly to the planets themselves).
[2] I Chronicles 21 : 16.

reeling of the sky. "Then shall the Assyrian fall with the sword, not of a mighty man; and the sword, not of a mean man, shall devour him . . . and his princes shall be afraid of the ensign."[1] "And all the host of heaven shall be dissolved . . . for my sword shall be bathed in heaven." [2]

The ancients classified the comets according to their appearance. In the old astrological texts, as in the book of *Prophecies of Daniel*, comets that took the form of a sword were originally related to the planet Mars.[3]

Besides the swordlike appearance of the atmosphere of Mars, elongated on its approach to the earth, there was also another reason to make the planet Mars the god of war. A bellicose or martial character was ascribed to the planet because of the great excitement it caused, excitement that brought anxiety to many peoples, that led to migrations and to wars. Since early times celestial prodigies have been regarded as portents that forecast great commotions and great wars.

A planet that collided with other planets in the sky and rushed against the earth as if with a fire-sword became the god of battle, wresting this title from the hands of Athene-Ishtar.

"The gods of heaven put themselves in war against thee," the hymns of the planet Nergal say, and this is the war that was recounted in the *Iliad*.

Nergal was named *quarradu rabu*, "the great warrior"; he waged war against gods and the earth. The most frequent ideogram for Nergal in Semitic cuneiform is read *namsaru*, which means "sword"; [4] the planet Mars, in the Babylonian inscriptions of the seventh century, was called "the most violent among the gods."

Herodotus said that the Scythians worshipped Ares (Mars), and that a scimitar of iron was their image of him; to him they made human sacrifices and poured the blood on the scimitar.[5] Solinus wrote of the people of Scythia: "The god of this people is Mars; instead of images they worship swords." [6]

War in heaven among the colliding planets, war on earth among the nations wandering in unrest, a planet running toward the earth with an outstretched flaming sword, attacking

[1] Isaiah 31 : 8-9. [2] Isaiah 34 : 4-5.
[3] Gundel, "Kometen," in Pauly-Wissowa, *Real-Encyclopädie*, XI, Col. 1177, with reference to *Cat. cod. astr.*, VIII, 3, p. 175.
[4] Böllenrücher, *Gebete und Hymnen an Nergal*, p. 8.
[5] Herodotus iv. 62.
[6] Solinus *Polyhistor* (transl. A. Golding, 1587), Chap. xxiii.

land and sea, participating in the wars among the nations—all these made Mars the god of war.

The sword of the god of battle was not like the sword "of a mighty man"; it was not thrust into the belly, and yet it caused sickness and death. The god of war scattered pestilence. In a prayer to the planet Mars (Nergal) it is said: [1]

> Radiant abode, that beams over the land . . .
> Who is thy equal?
> When thou ridest in the battle,
> When thou throwest down,
> Who can escape thy look?
> Who can run away from thy storming?
> Thy word is a mighty catch net,
> Stretched over Sky and Earth. . . .
>
> His word makes human beings sick,
> It enfeebles them.
> His word—when he makes his way above—
> Makes the country sick.

The outbreak of pestilence that appears to have accompanied the first contact with the planet Mars was repeated on each subsequent contact. Amos uttered these words: "I have smitten you with blasting and mildew. . . . I have sent among you the pestilence after the manner of Egypt."

The planet Nergal was regarded by the Babylonians as the god of war and pestilence; thus, too, did the Greeks regard the planet Ares and the Romans the planet Mars.

## FENRIS-WOLF

In the Babylonian astrological texts it is said that "a star takes the shape of divers animals: lion, jackal, dog, pig, fish." [2] This, in our opinion, explains the worship of animals by ancient peoples, notably the Egyptians.

The planet Mars, its atmosphere distorted by its approaches to other celestial bodies—Venus, earth, moon—took on different shapes. The Mexicans narrated that Huitzilopochtli, the bellicose destroyer of cities, took the form of various birds and beasts.[3] On one occasion Mars very characteristically resembled

[1] Böllenrücher, *Gebete und Hymnen an Nergal*, p. 36.
[2] Kugler, *Babylonische Zeitordnung*, Vol. II of *Sternkunde und Sterndienst in Babel*, 91.
[3] Sahagun, *Historia general de las cosas de Nueva España*, Vol. I.

a wolf or a jackal. In Babylonia Mars had seven names—Jackal was one of them.[1] Also, the god with the head of a jackal or wolf in the Egyptian pantheon was apparently Mars. Of him it is said that he is a "prowling wolf circling this land." [2]

In the Chinese Chart of Soochow, in which it is related on the authority of more ancient sources that "Once Venus suddenly ran into the Wolf-Star," Wolf-Star apparently means Mars.[3]

Wolf or Lupus Martius was the animal symbol for Mars of the Roman religion.[4] It gave rise to the legend about Romulus, son of Mars, who was fed by a she-wolf. According to the tradition, the conception of Romulus took place during a prolonged eclipse.

The Slavic Vukadlak, who followed the clouds and devoured the sun or the moon, had the shape of a wolf.[5] The North-Germanic tribes, too, spoke of the wolf Sköll that pursued the sun.[6] In the *Edda*, the planetary god that darkened the sun is called Fenris-Wolf. "Whence comes the sun to the smooth sky back, when Fenris has swallowed it forth?" The battle of Mars and Venus is presented, in the Icelandic epos, as a fight between the wolf Fenris and the serpent Midgard.

"The bright snake gaping in the heaven above" and "the foaming wolf" battle in the sky. Storms come in summer. Then comes the day, and "dark grows the sun"; in a great upheaval "the heaven is cloven." "In anger smites the warden of earth, forth from their homes must all men flee. . . . The sun turns black, earth sinks in the sea, the hot stars down from the heaven are whirled, fierce grows the stream . . . till fire leaps high above heaven itself." [7]

[1] Bezold, in Boll's *Sternglaube und Sterndeutung*, p. 9.
[2] Breasted, *Records of Egypt*, III, Sec. 144.
[3] The translators of the chart surmised that by Wolf-Star Sirius is meant.
[4] Cf. Virgil *Aeneid* iv. 566; Livy, *History of Rome*, Bk. XXII, i. 12. A statue of Mars on the Appian Way stood between figures of wolves. "Among the animal symbols of Mars, the wolf holds first place. . . . The wolf belonged so definitely to Mars that Lupus Martius or Martialis became its usual name. As to the meaning of this symbol, it is difficult to understand it." Roscher in Roscher's *Lexikon d. griech. und röm. Myth., s. v.* "Mars," Col. 2430.
[5] J. Machal, *Slavic Mythology* (1918), p. 229.
[6] L. Frobenius, *Das Zeitalter des Sonnengottes* (1904), I, 198.
[7] *The Poetic Edda: Völuspa* (transl. Bellows, 1923).

# SWORD-TIME, WOLF-TIME

> Quaking of places,
> tumult of peoples,
> scheming of nations,
> confusion of leaders.
>
> —IV Ezra 9

The fear of the Judgment Day not only did not pacify the nations, but on the contrary, uprooted them, impelling them to migration and war.

The Scythians came down from the plains of the Dnieper and Volga and moved southward. The Greeks left their home in Mycenae and on the islands of the Aegean and carried on the siege of Troy through years of cosmic disturbances. Assyrian kings waged war in Elam, Palestine, Egypt, and beyond the Caucasus.

Civil war in the nations, tribal strife, and strife between members of households became so widespread that the same complaint was heard in many parts of the world. As I have already said, Mars was named the war god not only because of his swordlike appearance, but also because of these conflicts.

". . . The land [is] darkened, and the people shall be as the fuel of the fire; no man shall spare his brother," said Isaiah (9 : 19). In Egypt an inscription of the eighth century that refers to the moon disturbed in its movement, mentions incessant fighting in the land: "While years passed in hostility, each one seizing upon his neighbour, not remembering his son to protect."[1] Isaiah, speaking of the Day of Wrath, says: "And I will set the Egyptians against the Egyptians: and they shall fight every one against his brother, and every one against his neighbour; city against city, and kingdom against kingdom."[2] It was no different seven hundred years earlier, in the days of the catastrophes caused by Venus. At that time an Egyptian sage complained: "I show thee the land upside down; the sun is veiled and shines not in the sight of men. I show thee the son as enemy, the brother as foe, a man slaying his father."[3]

The Icelandic *Völuspa* says: "Dark grows the sun. . . .

[1] Breasted, *Records of Egypt*, IV. Sec. 764. [2] Isaiah 19 : 2.
[3] Gardiner, "New Literary Works from Ancient Egypt," *Journal of Egyptian Archaeology*, I (1914).

Brothers shall fight and fell each other. . . . Axe-time, sword-time, shields are sundered, wind-time, wolf-time, ere the world falls; nor ever shall men each other spare." [1]

The wars of Shalmaneser IV, Sargon II, and Sennacherib were carried on in the intervals between the catastrophes and at the very time of their occurrence. The campaigns were repeatedly interrupted by the forces of nature. Of his second campaign Sennacherib wrote: "The month of rain set in with extreme cold and the heavy storms sent down rain upon rain and snow. I was afraid of the swollen mountain streams; the front of my yoke I turned and took the road to Nineveh." [2] Before Sennacherib set out on his last campaign to Palestine, his astrologers told him that he had to hurry if he would escape calamity; [3] as we know, he did not escape it. At the same time Isaiah, who encouraged Hezekiah to resist Sennacherib, reckoned with the possibility of a disaster in the year of the opposition of Mars, and thus built his hope on the intervention of the forces of nature.

The Babylonians called the year of the close opposition of Mars "the year of the fire-god," and the month "the month of descent of the fire-god," as, for instance, in an inscription of Sargon.[4]

In *The Birth of the War-God*, the Hindu poet Kalidasa gives a vivid picture of the wars above and on the earth, weaving them into one great battle.

"Foul birds came, a horrid flock to see . . . and dimmed the sun. . . . And monstrous snakes, as black as powdered soot, spitting hot poison high into the air, brought terror to the army underfoot. . . . The sun a sickly halo round him had; coiling within it frightened eyes could see great, writhing serpents . . . and in the very circle of the sun were phantom jackals."

There fell, with darting flame and blinding flash
Lighting the farthest heavens, from on high
A thunderbolt whose agonising crash
Brought fear and shuddering from a cloudless sky.

There came a pelting rain of blazing coals
With blood and bones of dead men mingled in;
Smoke and weird flashes horrified their souls;
The sky was dusty grey like asses' skin.

[1] *The Poetic Edda: Völuspa* (transl. Bellows).
[2] Luckenbill, *Records of Assyria*, II, Sec. 250.
[3] Ginzberg, *Legends*, IV, 267, n. 53.
[4] Luckenbill, *Records of Assyria*, II, Sec. 121.

The elephants stumbled and the horses fell,
The footmen jostled, leaving each his post,
The ground beneath them trembled at the swell
Of ocean, when an earthquake shook the host.[1]

Lightning is usually discharged between two clouds or a cloud and the ground. But if for some reason the charge of the ionosphere, the electrified layer of the upper atmosphere, should be sufficiently increased, a discharge would occur between the upper atmosphere and the ground, and a thunderbolt would crash from a cloudless sky.

The planet-god Shiva, Kalidasa says, "deposited his seed in fire" and gave birth to Kumara who battled the great demon named Taraka that "troubled the world."

The Babylonian astrologers ascribed to their planet-gods the ability to emit the sounds of different animals—lion, pig, jackal, horse, ass—and of two species of birds.[2] The ancient Chinese likewise asserted that planets emit animal sounds when they approach the earth with a rain of stones.[3] It is fairly probable that on some occasion the crash of the discharge "from the cloudless sky" sounded like Ta-ra-ka, the name of the demon who battled the planets.

The Ethiopian king who went up against Sennacherib called himself Taharka or Tirhakah.[4] In many places in the Near and Middle East this or similar names suddenly became very popular at the close of the eighth century before the present era; before that time it was unknown.

Taraka troubled the world so that

The seasons have forgotten how
to follow one another now;
they simultaneously bring
flowers of autumn, summer, spring.

The night when Sennacherib's army was destroyed, he survived, but according to rabbinical sources, was badly burned. Some time after his inglorious return from Palestine without his army, he was killed by two of his sons as he knelt in a temple; Esarhaddon pursued his brother-patricides, killed them, and became king. On one of his campaigns against Egypt, his armies became so panicky at some natural phenomenon that they scattered and fled from Palestine where

[1] Translated by A. W. Ryder (1912). [2] Kugler, *Babylonische Zeitordnung*, p. 91.
[3] F. Arago, *Astronomie populaire*, IV, 204. [4] Isaiah 37 : 9.

Sennacherib had lost his army to the storm-god Nergal. The laconic cuneiform chronicles, composed in the days of Nabonidus, the last Babylonian king, who lived in the sixth century, record the main events of Esarhaddon's war: "In the sixth year the troops of Assyria went to Egypt. They fled before a great storm." [1] An army as disciplined as the Assyrian army under one of its famous kings would not have run away from a cloudburst. The event mentioned in this inscription suggested to its modern publisher that the scriptural story of a blast that destroyed the Assyrian host refers, not to Sennacherib's army, but to that of his successor-son; otherwise one must think that on two similar occasions a natural cause subdued the Assyrian army. However, it is probable that after the army of Sennacherib was annihilated, violent atmospheric discharges and some portents in the sky, so numerous in those years, threw the Assyrian troops into a panic so that they fled.

The trembling earth, the displacement of the poles, the change in the climate, the frightening prodigies in the sky, caused great movements of peoples. The Aztecs changed their homeland. "These Mexicans carried with them an idol which they called Huitzilopochtli. . . . They asserted that this idol commanded them to leave their country, promising to make them lords and masters of all the lands . . . which abounded with gold, silver, feathers . . . and all the things necessary for life. The Mexicans departed like the children of Israel in their search of a promised land." [2] In India the patron of the invading Aryan race was Indra, the god of war, the Hindu Mars.

The Ionians and Dorians spread to the islands, the Latins were pressed by newcomers to the Apennine Peninsula, the Cimmerians wandered from Europe across the Bosphorus into Asia Minor, the Scythians crossed the Caucasus into Asia.

## SYNODOS

We remember that Josephus Flavius, after giving Herodotus' account of the destruction of Sennacherib's army, intended to quote a divergent account of Berosus, and introduced it with the words, "Here is what wrote Berosus," but the account is not preserved. Now, if we know what happened on the night

[1] Sidney Smith, *Babylonian Historical Texts* (1924), p. 5.
[2] *Manuscrit Ramirez* (of the 16th century) translated by D. Charnay, *Histoire de l'origine des Indiens qui habitent la Nouvelle Espagne selon leurs traditions* (1903), p. 9.

of March 23, -687, are we not able to find out what the missing account of Berosus was?

We can assume that Berosus knew that the catastrophe was caused by a planet in contact with the earth. Seneca, in his work, *Naturales quaestiones*, described the cataclysms of water and fire that visited this world and brought it to the brink of destruction. He also presented the opinion of Berosus, which is remarkable in that it reflects ancient knowledge similar to that at which we arrived after a long series of deductions and conclusions. Seneca wrote: "Berosus, the translator of Bel, attributed to the planets the cause of these perturbations." And he added: "His certainty in this matter was so great as to fix the dates of the universal conflagration and deluge. Everything terrestrial, he says, will be burned, when the stars which now follow different orbits will reunite in the sign of Cancer, and will place themselves in one line, so that a straight line would pass through the centres of all these globes. The deluge will come when the same planets will have conjunction in Capricorn." [1]

Disregarding the specific details of this assumption, there still remains a kernel of truth. The catastrophes of flood and of conflagration were ascribed to the influence of planets, and the conjunction was called the fatal moment. Such being the opinion of Berosus on the cause of the world catastrophes, the catastrophe that befell Sennacherib was probably explained by him in the same way. We are thus able to reconstruct Berosus' record which was omitted in Josephus.

Chaldean scholars were aware that the planetary system is not rigid and that the planets undergo changes. We find in Diodorus of Sicily: "Each of the planets, according to them [the Chaldeans] has its own particular course, and its velocities and periods of time are subject to change and variation." [2] They counted the earth among the planets, for Diodorus wrote that the Chaldeans stated "that the moon's light is reflected and her eclipses are due to the shadow of the earth". [3] This

[1] The same idea, but with varying positions of the stars as the cause of the catastrophes, is found in Nigidius, quoted by Lucan, and in Olympiodor, *Commentary to Aristotle*. See Boll, *Sternglaube*, p. 201, and *idem*, *Sphaera*, p. 362; Gennadius (George Scholarius, patriarch at Constantinople), *Dialogus Christiani cum Judaeo* (1464). A French edition of the works of Gennadius was printed in 1930.

[2] Diodorus of Sicily, *The Library of History*, ii. 31 (transl. Oldfather).

[3] *Ibid.*

implies that they knew the earth is a sphere in space, a fact known also to a number of Greek philosophers.[1]

A few Greek philosophers were aware that planets, on close contact, are greatly disturbed, and that out of their agitated atmospheres comets are born. The perturbations in such contacts may be so strong that, when the earth is involved, deluge or world conflagration may take place.

Zeno, the founder of the Stoic school of thought,[2] and likewise Anaxagoras (-500 to -428) and Democritus (-460 to -370), declared that planets at conjunction may become coalescent, thus taking the form of comets. Aristotle, who misunderstood their teaching, declared: "We have ourselves observed Jupiter coinciding with one of the stars of the Twain and hiding it, and yet no comet was formed." [3]

Diogenes Laërtius recorded that Anaxagoras thought that comets are "a conjunction of planets which emit flames"; [4] and Seneca, without naming Anaxagoras and Democritus, wrote: "Here is the explanation which is given by some ancient authors. When a planet enters in conjunction with another, they confound their lights into one light, and they have the appearance of an elongated star. . . . The interval which separates them is illuminated by both of them, inflames and transforms into a trail of fire." [5] Seneca, who regarded this as an explanation of the nature of comets, questioned it, reasoning that "planets cannot remain for a long time in conjunctions, because by necessity of the law of velocity they would separate."

Plato, on the authority of the Egyptian sages, ascribed the deluge and conflagration of the world to the action of a celestial body that, changing its path, passed close by the earth, and he even pointed to the planets as the cause of periodic world catastrophes.[6] The Greek term for the collision of planets is *synodos*, which, in the words of a modern interpreter, requires a meeting in space and also a collision of planets.[7]

The Romans knew that the earth is one of the planets; Pliny,

[1] Aristarchus of Samos recognized that the earth revolves together with other planets around the sun.
[2] Seneca, *De Cometis*.
[3] Aristotle, *Meteorologica*, i. 6 (transl. E. W. Webster, 1931).
[4] Diogenes Laërtius, *Lives*, "Life of Anaxagoras."
[5] Seneca, *De cometis*.
[6] Plato, *Timaeus*, 22C, 39D.
[7] Boll, *Sternglaube*, pp. 93 and 201. The Greek term "requires a meeting in the same horizontal and vertical planes and a collision. The planets thrust one another and cause the destruction of the world" ("ein Zusammentreffen und auch ein Zusammenstossen auf derselben Ebene, also nach Breite und Höhe stossen die Planeten ineinander und lösen dadurch das Weltende aus").

for instance, wrote: "Human beings are distributed all round the earth and stand with their feet pointing toward each other. . . . Another marvel, that the earth herself hangs suspended and does not fall and carry us with it." [1]

The earth, one of the planets, had been subject to conflicts with other planets, and traces of knowledge of these occurrences may be found in the early writers. Origen writing against Celsus stated: "We do not refer either the deluge or the conflagration to the cycles and planetary periods; but the cause of them we declare to be the extensive prevalence of wickedness, and its (consequent) removal by a deluge or a conflagration." [2] Celsus and Origen were familiar with the view that the deluge and world conflagration were caused by planets, and that these world catastrophes could be calculated in advance.

Pliny wrote: "Most men are not acquainted with a truth known to the founders of the science from their arduous study of the heavens," namely, that thunderbolts "are the fires of the three upper planets." [3] He differentiated them from lightning caused by the dashing together of two clouds. Seneca, his contemporary, also distinguished lightnings that "seek houses" or "lesser bolts" and the bolts of Jupiter "by which the threefold mass of mountains fell." [4]

A vivid picture of an interplanetary discharge is given by Pliny: "Heavenly fire is spit forth by the planet as crackling charcoal flies from a burning log." [5] If such a discharge falls on the earth, "it is accompanied by a very great disturbance of the air," produced "by the birth-pangs, so to speak, of the planet in travail." [6]

Pliny says also that a bolt from Mars fell on Bolsena, "the richest town in Tuscany," and that the city was entirely burned up by this bolt. [7] He refers to Tuscan writings as the source of his information. By Tuscan writings are meant Etruscan books.

Bolsena, or the ancient Volsinium, was one of the chief cities of the Etruscans, the people whose civilization preceded that of the Latin Romans on the Apennine Peninsula. The Etruscan states occupied the area of what was later known as Tuscany, between the Tiber and the Arno.

[1] Pliny, *Natural History*, ii. 45.
[2] Origen, *Against Celsus*, Bk. iv, Chap xii, in Vol. IV of *The Ante-Nicene Fathers* (ed. A. Robert and J. Donaldson, 1890).
[3] Pliny, *Natural History*, ii. 18.
[4] Seneca, *Thyestes*.
[5] Pliny, ii. 18. [6] *Ibid.* [7] *Ibid.*, ii. 53.

Near Bolsena, or Volsinium, is a lake of the same name. This lake fills a basin nine miles long, seven miles wide, and 285 feet deep. For a long time this basin was regarded as the water-filled crater of a volcano. However, its area of 117 square kilometres exceeds by far that of the largest known craters on the earth—those in the Andes in South America and those in the Hawaiian (Sandwich) Islands in the Pacific. Hence, the idea that the lake is the crater of an extinct volcano has recently been questioned. Moreover, although the bottom of the lake is of lava, and the ground around the lake abounds with ashes and lava and columns of basalt, the talus of a volcano is lacking.

Taking what Pliny said of an interplanetary discharge together with what has actually been found at Volsinium, one may wonder whether the cinders and the lava and the columns of basalt could possibly be the remains of the contact Pliny mentions. Again, if the discharge was caused by Mars, it would probably have occurred in the eighth pre-Christian century. The catastrophes of that century brought the great Etruscan civilization into sudden decline and launched the migration of newcomers to Italy leading to the founding of Rome. The Etruscans, as cited by Censorinus and quoted in the Section on "The World Ages," thought that celestial prodigies augured the end of each age. "The Etruscans were versed in the science of the stars, and after having observed the prodigies with attention, they recorded these observations in their books."

## THE STORMER OF THE WALLS

Following the upheavals in which, in the words of the Babylonians, Mars-Negal "moved the earth off its hinges," and, in the words of Isaiah, "the earth moved exceedingly" and was "removed out of her place," mighty and repeated earthquakes devastated whole countries, destroyed cities, and shattered the walls of strongholds. "Bloodstained stormer of walls" is the ever repeated epithet of Ares in Homer. Hesiod, too, calls Ares "sacker of towns."[1] "Behold," said Amos, "the Lord commandeth, and he will smite the great house with breaches [into pieces]." Then came the "commotion" of the days of Uzziah, and of the days of Ahaz, and of the days of Hezekiah, when "the bricks are fallen down" (Isaiah 9 : 10) and only "a very

[1] Hesiod, *Theogony*, ll. 935 ff. *Purandara* or "town destroying" is the usual appellative of Indra.

small remnant" of the people remained (Isaiah 1 : 9). Those were days of "trouble, and of treading down, and of perplexity by the Lord God of Hosts" and "breaking down the walls" (Isaiah 22 : 5).

Recurrent displacement of the terrestrial globe, torsion of the lithosphere, and migration of the inner parts of the globe must have caused a succession of earthquakes over a prolonged period. But in comparison with the great catastrophes, when "heaven reeled," the local earthquakes received only slight attention.

In the reports of the astrologers of Nineveh and Babylon, earthquakes are often mentioned in just a single line, as in the following message: "Last night there was an earthquake." The frequent trembling of the earth became a source of omens for the magicians, which were reduced to formulas: "When the earth quakes in the month of Shevat," or "When the earth quakes in the month of Nisan," then one or another event will take place. As in the following sentence, the observation could be basically correct: "When the earth quakes through the whole day, there will be a destruction of the land. When it quakes continually, there will be an invasion of the enemy."[1]

Reports concerning earthquakes in Mesopotamia in the eighth and seventh centuries are very numerous, and they are dated.[2] Nothing comparable is known in modern times. In some of these reports, Nergal (Mars) is mentioned as the cause of the calamity. "The earth shook; a collapsing catastrophe was all over the country; Nergal strangles the country."[3] Temples constructed with great care, so that the foundations might absorb shocks and resist them, were often destroyed by the catastrophes, and the cause was again the planet Nergal. Thus Nergal is referred to in connection with the collapse of the temple in Nippur that was destroyed in an earthquake.[4]

The kings of Babylon, the successors to Sennacherib, record in many inscriptions the repairing of breaches in the palaces and temples of the land. Sometimes the same temples or palaces were repaired by two kings in close succession, as in the case of Nergilissar (Neriglissar) and Nebuchadnezzar.[5] In the great

[1] R. C. Thompson (ed.), *The Reports of the Magicians and Astrologers of Nineveh and Babylon in the British Museum* (1900), Vol. II, Nos. 263, 265.
[2] See Kugler, *Babylonische Zeitordnung*, p. 116. [3] *Ibid.*
[4] Langdon, *Sumerian and Babylonian Psalms*, p. 99.
[5] See p. 235, note 2.

catastrophes of the eighth to the seventh centuries, practically no structure escaped damage, and new buildings were erected so as to absorb frequent shocks. At the close of the seventh century, Nebuchadnezzar described the precautions taken in placing the foundations of the palaces "on the breast of the netherworld"; these foundations of large stones with joints fitting one into the other have been unearthed in excavations.[1] The Babylonians also found that walls of burnt bricks were of greater elasticity than walls of stones; they were built on foundations of great blocks of stone.[2]

These ever recurrent earthshocks in a country as rich in oil as Mesopotamia also caused eruptions of earth deposits: "The earth threw oil and asphalt," observed the official astrologers, as the effect of an earthquake.[3]

The Scriptures and the rabbinical sources recorded repeatedly the repairing of breaches in the House of the Lord. On the day of the "commotion" of Uzziah the temple suffered a great breach.[4] References to breaches in houses, large palaces, and small dwellings are very numerous in the prophets of the eighth century. Isaiah speaks of "breaches of the city of David that they are many."[5] Repair of breaches in the Temple was the permanent concern of the kings of Jerusalem, also "the wall that was broken" of the city's outer bulwark.[6]

Since in modern times earthquakes occur only very seldom in Palestine, the frequent reference of the prophets and psalmists to them caused perplexity: "The earthquake held a place in the religious conceptions of the Israelites quite out of proportion to its slight and relatively rare occurrence in Palestine."[7]

Troy, the scene of the Homeric epos, was destroyed by an earthquake. The famous "sixth city" at Hissarlik, recognized as the fortress of Priam, king of the Trojans, fell because of

[1] R. Koldeway, *The Excavations at Babylon* (1914); *idem*, *Das wieder entstandene Babylon* (4th ed., 1925).

[2] Koldeway, *Die Königsburgen von Babylon* (1931-1939), Vols. I and II. Cf. Pliny, ii. 84: "The solidly built portion of the city being specially liable to collapses of this nature . . . walls built of clay bricks suffer less damage from being shaken."

[3] Kugler, *Babylonische Zeitordnung*, p. 117.

[4] Josephus, *Antiquities*, IX, x. 4. See Ginzberg, *Legends*, VI, 358.

[5] Isaiah 22 : 9.

[6] II Kings 12 : 5; 22 : 5; II Chronicles 32 : 5; Amos 6: 11; 9 : 11.

[7] A. Lods, *Israel: From Its Beginnings to the Middle of the Eighth Century* (transl. S. H. Hooke, 1932), p. 31.

earthshocks, a fact established in the excavation by the archaeological expedition of the University of Cincinnati.[1]

There are a number of theories concerning the cause of the earthquakes, but none of them is generally accepted. One connects the cause of earthquakes with the process of mountain building. Mountains are supposed to have their origin in the cooling of the earth and contraction of its crust.[2] This theory is based on the assumption that originally the earth was liquid. The folding of the crust creates mountains and causes earthquakes.

Another theory sees the cause of earthquakes in the migration of land masses, even of entire continents. This theory, too, is based on the concept of a thin crust resting on a viscous substratum. Geological and faunal similarities of South America and West Africa suggested their separation in recent geological times, and their migration in opposite directions. According to this theory, thermal convection is the mechanical cause of this migration, with magma supplying the heat.

Still another theory supposes that there are great mountains and deep valleys on the inner surface of the crust, facing the magma. The sliding of huge rocks along these mountainous slopes under the pull of gravity is presumed to be the cause of earthquakes.

The mountainous western coast of North and South America, or the shore of the Cordilleras, and the eastern coast of Asia stretching into the East Indies form the area of greatest earthquake activity, with 80 per cent of the entire mechanical force released in earthquakes concentrated there. Another area stretches from the Mediterranean toward the highland of Asia.

In an attempt to find the relation of earthquakes to other natural phenomena, a statistical investigation of the earthquakes of the middle of the nineteenth century was conducted, and the results suggested that earthquakes are more numerous when the moon is new and again when it is full, or when the pull of the moon acts in the same direction as the pull of the sun or when it acts in the opposite direction. The time when the moon is in perigee, or closest to the earth, was also found to be

[1] C. W. Blegen, "Excavation at Troy," *American Journal of Archaeology*, XXXIX (1935), 17.
[2] See the discussion of the problem of mountain building in the Section "The Planet Earth."

favourable for earthquakes.[1] These observations were challenged as to their general validity.

However, mountain building is a process the causes of which have not been established; the migration of continents is but a hypothesis; and the crumbling of the earth's crust must have some additional cause besides the force of gravity, because this force was active when the crust was built and made possible the formation of the crust in its present shape. Hence, all these theories are only hypotheses about unknown causes of known phenomena.

On the basis of the material offered in the foregoing pages, the assumption is made here that earthquakes result from torsion of the crust following a change in the position of the equator and the displacement of matter inside the globe caused by the direct attraction of a cosmic body when in a close contact. Pull, torsion, and displacement were responsible for mountain building, too.

If this conception of the causes of earthquakes is correct, then there must have been fewer and fewer earthquakes during the course of time since the last cosmic catastrophe. The regions of the Apennine Peninsula, the eastern Mediterranean, and Mesopotamia, for which we have reliable records, can be compared in this respect with the same regions today.

Earthquakes in Asia Minor, Greece, and Rome are described or mentioned by many classic authors. For the purpose of comparison with the earth-tremor activity of the present day, it is enough to point to fifty-seven earthquakes reported in Rome in a single year[2] during the Punic wars (-217).

If our interpretation of the cause of earthquakes is correct, then not only must more tremors and stronger shocks have been experienced in olden times, but also their cause must have been known to the ancients.

Pliny wrote: "The theory of the Babylonians deems that even earthquakes and fissures in the ground are caused by the force of the stars that is the cause of all other phenomena, but only by that of those three stars (planets) to which they assign thunderbolts." [3]

[1] Cf. the scientific publications of A. Perrey.
[2] Pliny, ii. 86.
[3] Pliny, ii. 81.

CHAPTER V

# THE STEEDS OF MARS

THE CASE OF Abraham Rockenbach and David Herlicius, who wrote about the year 1600, and who were informed on the matter of the comets of antiquity,[1] shows that the contents of some old manuscripts were known to the scholarly world then, though not to modern scholars.

A scholar and pamphleteer, Jonathan Swift, in his *Gulliver's Travels* (1726), wrote that the planet Mars had two satellites, very small ones. "Certain astrologers . . . have likewise discovered two lesser stars, or satellites, which revolve about Mars, whereof the innermost is distant from the centre of the primary planet exactly three of its diameters, and the outermost five; the former revolves in the space of ten hours, and the latter in twenty-one and a half . . . which evidently shows them to be governed by the same law of gravitation, that influences the other heavenly bodies."[2]

Actually Mars has two satellites, mere rocks, one being as small as about ten (?) miles in diameter, the other only five (?) miles.[3] One travels around Mars in 7 hours 39 minutes, the other in 30 hours 18 minutes. Their distance from the centre of Mars is even less than Swift said it was.[4] They were discovered by Asaph Hall in 1877. With the optical instruments of the days of Swift, they could not have been seen, and neither Newton nor Halley, the contemporaries of Swift, nor William Herschel in the eighteenth or Leverrier in the nineteenth century suspected their existence.[5] It was bold of Swift to assume their very short periods of revolution (months), measured only in hours; it was a very rare coincidence, indeed, if Swift invented these satellites, guessing correctly not only their existence, but also their number (two), and especially

[1] See the Section, "The Comet of Typhon."

[2] *Travels into Several Remote Nations of the World*, by Lemuel Gulliver (London, 1726), II, 43.

[3] The diameters of these satellites are not exactly known (Russell, Dugan and Stewart. 1945).

[4] Phobos is distant from the planet's surface less than one diameter of the planet (from the planet's centre less than one and a half diameters of the planet).

[5] Leverrier died one month after Asaph Hall made his discovery.

their very short revolutions. This passage of Swift aroused the literary critics' wonder.

It is an even chance that Swift invented the two satellites of Mars and thus by a rare accident came close to the truth. But it may also have been that Swift had read about the trabants in some text not known to us or to his contemporaries. The fact is that Homer knew about the "two steeds of Mars" that drew his chariot; Virgil also wrote about them.[1]

When Mars was very close to the earth, its two trabants were visible. They rushed in front of and around Mars; in the disturbances that took place, they probably snatched some of Mars' atmosphere, dispersed as it was, and appeared with gleaming manes.[2] The steeds were yoked when Mars (Ares) prepared to descend to the earth on a punitive expedition.

When Asaph Hall discovered the satellites, he gave them the names of Phobos (Terror) and Deimos (Rout), the two steeds of Mars;[3] without fully realizing what he did, he gave the satellites the same names by which they were known to the ancients.

Whether or not Swift borrowed his knowledge of the existence of two trabants of Mars from some ancient astrological work, the ancient poets knew of the existence of the satellites of Mars.

## THE TERRIBLE ONES

Venus had a tail, considerably shortened since the time it was a comet, but still long enough to give the impression of a hanging flame, or smoke, or attached hair. When Mars clashed with Venus, asteroids,[4] meteorites, and gases were torn from this trailing part, and began a semi-independent existence, some following the orbit of Mars, some other paths.

These swarms of meteorites with their gaseous appendages were newborn comets; flying in bands and taking various shapes, they made an uncanny impression. Those which followed

[1] *Iliad*, xv. 119; *Georgics*, iii. 91. Horses were sacrificed to Mars (Plutarch, *Roman Questions*, xcvii) either because they are animals employed in war, or because of the trabants of Mars which looked like horses drawing a chariot.

[2] G. A. Atwater suggests that these might have been electrical effects.

[3] Asaph Hall, *The Satellites of Mars* (1878): "Of the various names that have been proposed for these satellites, I have chosen those suggested by Mr. Madan of Eton, England," Deimos and Phobos.

[4] Between Mars and Jupiter are over a thousand asteroids that have been thought to have once been a planet. G. A. Atwater queries whether they could have resulted from the encounter between Mars and Venus.

Mars closely looked like a troop following their leader. They also ran along different orbits, grew quickly from small to giant size, and terrorized the peoples of the earth. And when, soon after the impact of Venus and Mars, Mars began to threaten the earth, the new comets, running very close to the earth, added to the terror, continually recalling the hour of peril.

Ares of Homer, going into battle, is accompanied by never resting horrible creatures, Terror, Rout, and Discord. Terror and Rout yoke the gleaming horses of Ares, themselves dreadful beasts, also known by these names; Discord, "sister and comrade of man-slaying Ares, rageth incessantly; she at the first rears her crest but little, yet thereafter planteth her head in heaven, while her feet tread on earth."

Similarly, the Babylonians saw the planet Mars-Nergal in the company of demons, and wrote their hymns to Nergal: [1] "Great giants, raging demons, with awesome members, run at his right and at his left." These "raging demons" are pictured also in the Nergal-Eriskigal poem; [2] they bring pestilence and cause earthquakes.

It appears that the mythological figures of the Furies of the Latins or the Erinyes of the Greeks, with serpents winding about their heads and arms, flashing flame with their eyes, swinging torches around like wheels, grew out of the same prodigies which moved rapidly, changed their forms hourly, and acted violently. The Erinyes travelled in a group, like huntresses or like a "pack of savage hounds," [3] but sometimes they appeared to be split into two groups.[4]

To these comets, travelling in bands with Mars or Indra, are dedicated many Vedic hymns, indeed a great part of them. They are called Maruts "shining like snakes," "blazing in their strength," "brilliant like fires." [5]

> O Indra, O strong hero, grant thou glory to us
> with the Maruts, terrible with the terrible ones,
> strong and giver of victory.[6]

[1] Böllenrücher, *Gebete und Hymnen an Nergal*, p. 29.
[2] Fragments of this poem were found presumably at el-Amarnà. It is very likely that the Ethiopians, who subdued Egypt in the eighth century, occupied Akhet-Aten (Tell-el-Amarna), and that some parts of the archives may have been deposited by them.
[3] J. Geffcken, "Eumenides, Erinyes" in *Encyclopaedia of Religion and Ethics*, ed. J. Hastings, Vol. V.
[4] Euripides, *Iphigenia in Tauris*, l. 968; Aeschylus, *Eumenides*.
[5] *Vedic Hymns* (transl. F. Max Müller, 1891).
[6] *Ibid.*, Mandala I, Hymn 171.

And it is said that their "strength is like the vigour of their father."

> Your march, O Maruts, appears brilliant. . . .
> We invoke you, the great Maruts,
> the constant wanderers. . . .
> Like the dawn, they uncover the dark nights
> with red rays, the strong ones,
> with their brilliant light,
> as with a sea of milk. . . .
> Streaming down with rushing splendour,
> they have assumed their bright and brilliant colour.[1]

Stones were hurled by these comets.

> You the powerful, who shine with your spears,
> shaking even what is unshakable by strength . . .
> Hurling the stone in the flight! . . .
> All beings are afraid of the Maruts.[2]

> May your march be brilliant, O Maruts . . .
> Shining like snakes.
> May that straightforward shaft of yours, O Maruts,
> bounteous givers, be far from us,
> and far the stone which you hurl![3]

Meteorites, when entering the earth's atmosphere, make a frightful din. So did the Maruts:

> Even by day the Maruts create darkness. . . .
> Then from the shouting of the Maruts
> over the whole space of the Earth,
> men reeled forward.[4]

This darkness and this din were narrated in scriptural and rabbinical sources, in Roman traditions, and in hymns to Nergal. As the similarity of the description of the "terrible ones" in the Vedic hymns and in Joel is striking, but has not been noticed, a few more quotations should follow here.

The comets, just beginning to whirl, looked like revolving torches or writhing snakes; they assumed the form of spinning

[1] *Vedic Hymns* (transl. F. Max Müller, 1891), Hymn 172.
[2] *Ibid.*, Hymn 85.
[3] *Ibid.*, Hymn 172.
[4] *Ibid.*, Hymn 48.

wheels, and the celestial phantasmagoria appeared like swift chariots; changing their forms, the Maruts looked like horses racing along the sky, and then again like a host of warriors, leaping, climbing, irresistible.

The verses of the second chapter of Joel (2 : 2-11) are given in their order, interspersed with verses taken from a number of Vedic hymns dedicated to the Maruts.

JOEL 2 : 2 A day of darkness and of gloominess,
a day of clouds and of thick darkness,
as the morning spread upon the mountains:
a great people and a strong;
there hath not been ever the like,
neither shall be any more after it,
even to the years of many generations.

VEDIC HYMNS Even by day the Maruts create darkness.[1]
The terrible Marut-host
of ever-youthful heroes.[2]
All beings are afraid of the Maruts:
they are men terrible to behold, like kings.[3]

JOEL 2 : 3 A fire devoureth before them;
and behind them a flame burneth. . . .
Nothing shall escape them.

VEDIC HYMNS Like a blast of fire. . . .

Blazing in their strength,
brilliant like fires, and impetuous.[4]

JOEL 2 : 4 The appearance of them
is as the appearance of horses:
and as horsemen, so shall they run.

VEDIC HYMNS At their racings, the earth shakes,
as if broken,
when on the heavenly path
they harness for victory.

They wash their horses like racers in the courses,
they hasten with the points of the reed
on their quick steeds.[5]

[1] *Vedic Hymns* (transl. F. Max Müller, 1891). Mandala I, Hymn 38.
[2] *Ibid.*, Mandala V, Hymn 53.
[3] *Ibid.*, Mandala I, Hymn 85.
[4] *Ibid.*, Hymns 39, 172.
[5] *Ibid.*, Hymns 86, 172.

| | |
|---|---|
| JOEL 2 : 5 | Like the noise of chariots on the tops of mountains shall they leap,<br>like the noise of a flame of fire<br>that devoureth the stubble,<br>as a strong people set in battle array. |
| VEDIC HYMNS | They are like headlong charioteers<br>on their ways.<br><br>They who are brilliant, of terrible design,<br>powerful, and devourers of foes.<br><br>On your chariots charged with lightning . . .<br><br>Host of your chariots, terrible Marut host.[1] |
| JOEL 2 : 6 | Before their face the people shall be much pained:<br>all faces shall gather blackness. |
| VEDIC HYMNS | At your approach the son of man holds himself down. . . .<br>You have caused men to tremble,<br>you have caused mountains to tremble.[2] |
| JOEL 2 : 7 | They shall run like mighty men;<br>they shall climb the wall like men of war;<br>and they shall march every one on his ways,<br>and they shall not break their ranks. |
| VEDIC HYMNS | Your conquest is violent, splendid, terrible, full and crushing. . . .<br>The terrible train of untiring Maruts. . . .<br>Full of terrible designs, like giants.[3] |

Joel describes how these warriors, coming with fire and clouds, will run upon the wall, enter in at the windows, run to and fro in the city, and the sword can do them no harm. In similar terms the Vedic hymns describe the conquest by this terrible host.

If there is any doubt as to the nature of the "terrible ones," the following words would dissipate it:

| | |
|---|---|
| JOEL 2 : 10 | The earth shall quake before them;<br>the heavens shall tremble; |

[1] *Vedic Hymns* (transl. F. Max Müller, 1891). Mandala I, Hymns 172, 19, 36; Mandala V, Hymn 53.
[2] *Ibid.*, Mandala I, Hymn 37.
[3] *Ibid.*, Hymns 168, 64.

the sun and the moon shall be dark,
and the stars shall withdraw their shining.

Maruts are often called "shakers of heaven and earth."

VEDIC HYMNS You shake the sky.

The terrible ones . . . even what is firm and unshakable
is being shaken.

When they whose march is terrible have caused
the rocks to tremble,
or when the manly Maruts have shaken the back
of heaven.

Hide the hideous darkness,
make the light which we long for![1]

The earth groaned, the meteorites—the host of the Lord—filled the sky with a battle cry "over the whole space of the Earth," and "men reeled forward."

These were, in Joel's words, the "wonders in the heavens and in the earth, blood, and fire, and pillars of smoke," when the "sun is turned into darkness, and the moon into blood."

The clouds, the fire, the terrifying din, the darkness in the middle of the day; the fantastic figures on the sky of speeding chariots, running horses, marching warriors; the trembling of the earth, the reeling of the firmament, were visualized, felt, and feared on the shores of both the Mediterranean Sea and the Indian Ocean, for they were not local disturbances, but displays of cosmic forces in cosmic dimensions. Joel did not copy from the *Vedas* nor the *Vedas* from Joel. In more than this one instance it is possible to show that peoples, separated even by broad oceans, have described some spectacle in similar terms. These were pageants, projected against the celestial screen, that, a few hours after they were seen in India, appeared over Nineveh, Jerusalem, and Athens, shortly thereafter over Rome and Scandinavia, and a few hours later over the lands of the Mayas and Incas.

The spectators saw in the celestial prodigies either demons, as the Erinyes of the Greeks or the Furies of the Latins, or gods

[1] *Vedic Hymns* (transl. F. Max Müller, 1891). Mandala I, Hymns 168, 167, 106, 38, 86.

whom they invoked in prayers, as in the *Vedas* of the Hindus, or the executors of the Lord's wrath, as in Joel and Isaiah.

In the Section "Isaiah" we maintained that the army of the Lord was not the Assyrian host, but a celestial host. Isaiah called the army of the Most High "the terrible ones."

> And he will lift up an ensign to the nations from far,
> and will hiss unto them from the end of the earth:
> and, behold, they shall come with speed swiftly:
> None shall be weary nor stumble among them;
> none shall slumber nor sleep;
> neither shall the girdle of their loins be loosed,
> nor the latchet of their shoes be broken:
> Whose arrows are sharp, and all their bows bent,
> their horses' hoofs shall be counted like flint,
> and their wheels like a whirlwind.
> Their roaring shall be like a lion . . .
> they shall roar like young lions . . .
> like the roaring of the sea:
> and if one look unto the land,
> behold darkness and sorrow;
> and the light is darkened in the heavens thereof.[1]

The mighty roaring, the wheels revolving like a whirlwind, the horses with hoofs of flint, the light darkened in heaven are once more common features.

VEDIC HYMNS

> These strong, manly, strong armed Maruts
> do not strive among themselves;
> firm are the horns, the weapons on your chariot,
> and on your faces are splendours.[2]
>
> They who by their own might
> seem to have risen above heaven and earth . . .
> they are glorious like brilliant heroes,
> they shine forth like foe-destroying youths.[3]
>
> They who are roaring and hasting like winds,
> brilliant like the tongues of fire,
> powerful like mailed soldiers . . .
> who hold together like the spokes of chariot-wheels,
> who glance forward like victorious heroes,
> who are swift, like the best of horses.[4]

[1] Isaiah 5 : 26 ff.
[2] Mandala VIII, Hymn 20.
[3] Mandala X, Hymn 77.
[4] *Ibid.*, Hymn 78.

The dreadful figures scattered a hail of meteorites that bombarded walls with hot gravel and flew into windows; simultaneously cities were turned into heaps by the leaping ground.

"The multitude of the terrible ones" is "like small dust," their invasion "shall be at an instant suddenly," says Isaiah.[1] The Lord shall send his host "with thunder, and with earthquake, and great noise, with storm and tempest, and the flame of devouring fire."

> These Maruts are men brilliant with lightning,
> they shoot with thunderbolts,
> they blaze with the wind,
> they shake the mountains.[2]

Isaiah (25 : 4) says that "the blast of the terrible ones is as a storm against the wall."

> Thou [the Lord] shalt bring down the noises of the strangers . . . the branch of the terrible ones shall be brought low.[3]

The Maruts are often called "the terrible ones," the same term Isaiah used. "The terrible ones" of the *Vedas* were not common storm clouds, nor were the "terrible ones" of Joel and Isaiah human beings. Certainly only by chance did the similarity of names and pictures in the *Vedas* and the Prophets escape the attention of students of religion.

The Maruts are understood here as comets which in great numbers started to whirl in the sky on short orbits, after the impact of Mars and Venus. They followed and preceded the planet Mars. The name Mars (genitive, Martis) would be of the same origin as Marut. It is therefore gratifying to read that the philological relation has already been established.[4] It is even more satisfactory that this philological equation was made without knowledge of the actual relation between the planet Mars and "the terrible ones."

[1] Isaiah 29 : 5.
[2] *Vedic Hymns*, Mandala V, Hymn 54.
[3] Isaiah 25 : 5.
[4] "Why should we object to Mars, Martis as a parallel form of Maruts? I do not say the two words are identical, I only maintain that the root is the same. . . . If there could be any doubt as to the original identity of Marut and Mars, it is dispelled by the Umbrian name *cerfo Martio*, which, as Grassmann (Kuhn's *Zeitschrift*, XVI, 190, etc.) has shown, corresponds exactly to the expression

By comparing Hebrew historical, Chinese astronomical, and Latin ecclesiastical material, we have established that it was the planet Mars which caused a series of catastrophes in the eighth and seventh centuries before this era. The Greek epos explained how it happened that Venus ceased and Mars began to be a threat to the earth. In heavenly battles, Ares or Nergal, both known as the planet Mars, had an entourage of demoniac figures. The name Mars is derived from the Indian Marut; Maruts, "the terrible ones," are "the terrible ones" of Isaiah and Joel.

The origin of the Greek name Ares was debated by philologists,[1] and reasons against a common root with the identical Mars were admitted. It seems to me that just as Mars is derived from Marut, "the terrible ones" of the *Vedas*, so Ares was formed from the "terrible one" of the Hebrew, which, as used by Joel and Isaiah, is *ariz*.

In a no longer extant passage of Pliny there was something said about comets being produced by planets.[2] Also the Soochow Chart refers to occasions in the past when comets were born from planets, from Mars, Venus, and others.

## SAMPLES FROM THE PLANETS

In the Vedic hymns the Maruts are implored to "be far from us and far the stone which you hurl." When comets pass close to the earth, stones occasionally fall; the classic case is that of the meteorite that fell at Aegospotami when a comet shone in the sky.[3] The Hindu book of *Varahasanhita* sees in the meteorites portents of devastation by fire and earthquake.[4]

Since the planets were gods, stones hurled by them or by the comets created in their encounters, were feared as divine

---

*sardha-s maruta-s,* the host of the Maruts. Such minute coincidences can hardly be accidental." F. Max Müller, *Vedic Hymns* (1891), I, xxv.

[1] *Ibid.*, p. xxvi.

[2] Cf. Pauly-Wissowa, *Real-Encyclopädie,* Vol. XI, Col. 1156.

[3] Aristotle, *Meteorologica*, i. 7.

[4] Frazer, *Aftermath* (supplement to *The Golden Bough*) (1936), p. 312.

Two Greek cities, Bura and Helice, were destroyed by earthquake and tidal wave and swallowed by the earth and sea in the year -373, when a comet shone in the sky.

missiles,[1] and when they fell and were found, they were worshipped.

The stone of Cronus at Delphi,[2] the image of Diana at Ephesus, which, according to Acts (19 : 35), was the image which fell down from Jupiter, the stones of Amon and Seth at Thebes,[3] were meteorites. Also the image of Venus on Cyprus was a stone which fell from the sky.[4] The Palladium of Troy was a stone that fell on the earth "from Pallas Athene"[5] (the planet Venus). The sacred stone of Tyre, too, was a meteorite related to Astarte, the planet Venus. "Travelling about the world, she [Astarte] found a star falling from air, or sky, which she taking up, consecrated on the holy island [Tyre]."[6] At Aphaca in Syria a meteorite fell which "was thought to be Astarte herself," and a temple to Astarte was built there; festivals "were regularly timed to coincide with the appearance of Venus as the Morning or Evening Star."[7]

The stone on which the Temple of Solomon was built—Eben Shetiya, or fire stone—is a bolide that fell in the beginning of the tenth century, in the time of David, when a comet, which bore the appearance of a man with a sword, was seen in the sky.[8] The sacred shield of Numa at Rome, the ancile of Roman Mars, was a bolide; it fell from the sky[9] in the beginning of the seventh century and its origin was connected with Mars.

In the years when the planet Mars had long been pacified, its position was still watched when meteorites fell. Thus the Chinese wrote in −211: "The planet Mars being in the neighbourhood of Antares, a star fell at Toung-Kiun, and arriving to the ground, it changed to a stone."[10] The people of the place cut a prophecy of evil for the emperor on the stone, and the emperor had it destroyed. Carving messages to peoples or kings on fallen stones was known before and has been practised since.

[1] According to Mohammed, stones that fell on the sinful tribes were inscribed with the names of those whom they were destined to kill.

[2] G. A. Wainwright, "The Coming of Iron," *Antiquity*, X (1936), 6.

[3] Wainwright, *Journal of Egyptian Archaeology*, XIX (1933), 49-52.

[4] Olivier, *Meteors*, p. 3.

[5] Cf. Bancroft, *The Native Races*, III, 302.

[6] R. Cumberland, *Sanchoniatho's Phoenician History* (1720), p. 36. Lucian says that Astarte was the fallen star of Sanchoniathon. *Ibid.*, p. 321. See also F. Movers, *Die Phönizier*, I, 639.

[7] Frazer, *The Golden Bough*, V, 258 ff. Cf. the Section "Worship of the Morning Star," note 18.

[8] I Chronicles 21; II Samuel 24. See Tractate Yoma 5, 2; cf. Tractate Sota 48b; also Ginzberg, *Legends*, V, 15.

[9] Olivier, *Meteors*, p. 3.

[10] Abel-Rémusat, *Catalogue des bolides et des aérolithes observés à la Chine*, p. 7.

One of the stones that fell from the sky is still worshipped today—it is the black stone of Kaaba in Mecca. Now its surface is black from being touched and kissed innumerable times, but under its cover of dirt it retains its original reddish colour. It is the holiest thing in Mecca, built into the wall of Kaaba, and pilgrims travel thousands of miles to kiss it.

Kaaba is older than Mohammedanism. Mohammed, in the early part of his career, worshipped Venus (al-Uzza) and other planetary gods which even today enjoy great veneration among the Moslems as the "daughters of the god."[1]

The black stone of Kaaba, according to Moslem tradition, fell from the planet Venus;[2] but another legend says that it was brought down by the Archangel Gabriel.[3] Granted that this legend may conceal some information about the origin of the stone, we ought to ask ourselves: Who is the Archangel Gabriel?

## THE ARCHANGELS

In the Scriptures the destruction of the army of Sennacherib is said to have been caused by a "blast," and a few verses later it is said to have been the act of an angel of God.[4] The Talmudic and Midrashic sources, which relate that the army of Sennacherib was destroyed by a blast and scourge accompanied by a terrible din on the night following the day when the shadow of the sun returned ten degrees, are more specific: the scourge was inflicted by the Archangel Gabriel "in the guise of a column of fire."[5] In the present research it has been established that it was the work of Mars.

Are archangels planets? "An old tradition, dating back to Gaonic times, had it that there are seven archangels, each of whom is associated with a planet." [6] "The seven archangels were believed to play an important part in the universal order through their association with the planets and the constellations. There is some variation, in the different versions, in the angels assigned to the planets."[7] "In some medieval writings Gabriel is associated with the moon, but in one or two with

[1] Wellhausen, *Reste arabischen Heidentums*, p. 34.
[2] F. Lenormant, *Lettres assyriologiques* (1871-1872), II, 140.
[3] *Ibid.*
[4] II Kings 19 : 7 and 35; Isaiah 37 : 7; 37 : 36.
[5] Babylonian Talmud, Tractate Sanhedrin 95b; Tosefta Targum Isaiah 10 : 32; Aggadat Shir 5, 39 and 8; 45; Jerome on Isaiah 30 : 2.
[6] J. Trachtenberg, *Jewish Magic and Superstition* (1939), p. 98.
[7] *Ibid.*, p. 250.

Mars.[1] The following, however, makes the identification of Gabriel possible: Gabriel is connected with the foundation of Rome. The Jewish legend says that when Solomon took the daughter of Pharaoh to wife, "the Archangel Gabriel descended from heaven and inserted a reed in the sea. About this reed more and more earth was gradually deposited, and, on the day on which Jeroboam erected the golden calves, a little hut was built on the island. This was the first dwelling-place of Rome."[2] Here Gabriel is cast in the role the Romans ascribed to Mars, that of the founder of Rome.[3] Our assumption that it was the planet Mars which caused the destruction of the army of Sennacherib in the spring of −687 is implied also by rabbinical sources: Since the Archangel Gabriel is another name for the planet Mars, the ancient Jews knew the origin of the "blast" and the identity of "the angel of the Lord" who destroyed the Assyrian army.

Gabriel is the angel appointed over fire; he is also, according to Origen,[4] the angel of war. Thus we again recognize in him Mars-Nergal. The rabbinical tradition says that the Assyrians of the host of Sennacherib, before they died, were permitted by Gabriel to hear "the song of the celestials," which can be interpreted as the sound caused by a close approach of the planet. The words of Isaiah (33 : 3), "at the noise of the tumult [*hamon*] the people fled," should, according to the Jewish tradition as related by Jerome, refer to Gabriel, *Hamon* being another of his names.[5]

The planet Mars is red, and *Maadim* (the red or the one who reddens) is the name for Mars in the Hebrew astronomical texts. One text says: "The Holy One created Mars—Maadim —that he should throw them [the nations] down into hell." [6]

A few rabbinical sources attribute the destruction of Sennacherib's army to the action of the Archangel Michael; some ascribe it to both archangels.[7] Who, then, is the Archangel Michael?

[1] J. Trachtenberg, *Jewish Magic and Superstition* (1939), p. 251.

[2] Ginzberg, *Legends*, VI, 128 and 280, based on Tractate Shabbat 56b and other sources; also M. Grünbaum, *Gesammelte Aufsätze zur Sprach- und Sagenkunde* (1901), pp. 169 ff.

[3] Livy, *History of Rome*, i. Preface; Macrobius, *Saturnalia*, xii.

[4] Origen, *De principiis*, i. 8. "A particular office is assigned to a particular angel . . . to Gabriel the conduct of wars." Cf. Tractate Shabbat 24.

[5] Jerome on Isaiah 10 : 3; Aggadat Shir 5, 39; Ginzberg, *Legends*, VI, 363. Cf. V. Vikentiev, "Le Dieu 'Hemen,'" *Recueil de Travaux* (1930), Faculté des Lettres, Université Egyptienne, Cairo.

[6] Pesikta Raba 20, 38b.

[7] Midrash Shemot Raba (ed. Vilna, 1887) 18 : 5; Tosefta Targum II Kings 19 : 35,

The entire story of Exodus is connected with the Archangel Michael. In Exodus 14 : 19 the pillar of fire and of cloud is called Angel of God. According to the Midrash,[1] it was the Archangel Michael who made himself "a wall of fire" between the Israelites and the Egyptians. Michael is said to be made of fire. The Haggadah states: "Michael was appointed High Priest of the celestial sanctuary at the same time that Aaron was made high priest of Israel," that is, in the time of the Exodus. Michael was also the angel who appeared to Joshua, son of Nun.

The celestial struggle at the Sea of Passage is depicted in the familiar image of the Archangel Michael slaying the dragon. Michael produces fire by touching the earth, and it was the emanation of this archangel that was seen in the burning bush. He has his abode in heaven and is the forerunner of Shehina or God's presence, but as Lucifer, Michael falls from heaven and his hands are bound by God. All these attributes and acts of the Archangel Michael[2] lead us to recognize which planet he represents: it is Venus.

The Archangel Michael, or the planet Venus, and the Archangel Gabriel, or the Planet Mars, saved the people of Israel on two dramatic occasions. At the Sea of Passage, when the hosts of Egypt, pursuing the fleeing slaves, could be seen in the distance ("the children of Israel lifted up their eyes, and, behold, the Egyptians marched after them; and they were sore afraid" [3]), the sea was torn apart, and the slaves walked on the bottom of the sea and reached the other shore. Their enemies were thrown high by the released tides, which fell down when a spark passed between Venus and the earth.

Eight hundred years passed after the Exodus. The Assyrian hosts, which a generation earlier had removed the Ten Tribes of Israel to an exile from whence they never returned, invaded Judea with the express purpose of crushing rebellious Judah and removing him from his homeland and from the scene of history. A blast from the planet Mars fell upon the camp of the Assyrians and annihilated it. Those rabbinical sources which ascribed this act to both archangels were not wrong. Venus pushed Mars towards the earth, and thus both were instrumental in the destruction.

[1] Pirkei Rabbi Elieser 42.
[2] An extensive literature on the Archangel Michael can be found in Ginzberg, *Legends*, Index Volume, under "Michael."
[3] Exodus 14 : 10.

The author of the apocryphal book of the Ascension [Assumption] of Moses knew that "Venus and Mars are each as large as the whole Earth." [1]

Because of their intervention at moments when the national existence of Israel was at stake, Michael and Gabriel were looked upon as "guardian angels" of the eternal people.

Gabriel is the Hebrew Hercules (Heracles). Actually the classic authors made it clear that Hercules is another name for the planet Mars.[2] In the Gospel of Luke (1 : 26) Gabriel is the angel of Annunciation to the Virgin.

In the Roman Catholic Church Michael is the conqueror of Satan, "head of the host of heaven and first of the saints after Mary."

## PLANET WORSHIP IN JUDEA IN THE SEVENTH CENTURY

In the Northern Kingdom the process of disassociating the deity from the celestial object had not yet been completed when the Kingdom was destroyed (−723 or −722), and its population was led away into captivity, from which they did not return. "And they [the tribes of the Northern Kingdom] left all the commandments of the Lord their God, and made them molten images, even two calves, and made a grove, and worshipped all the host of heaven and served Baal" (II Kings 17 : 16).

Only a few years after the deliverance of Judea from the hand of Sennacherib, Manasseh, son of Hezekiah, "built altars for all the host of heaven in the two courts of the house of the Lord" (II Kings 21 : 5). "For he [Manasseh] built again high places which Hezekiah his father had broken down, and he reared up altars for Baalim, and he made groves, and worshipped all the host of heaven, and served them" (II Chronicles 33 : 3).

It was in the time of Josiah, grandson of Manasseh, and shortly before the exile of Judah to Babylon, that a pure monotheism emerged as an outcome of the progress the Jewish people had made during its long struggle for national existence,

[1] Ginzberg, *Legends*, II, 307.

[2] See note 1, the Section, "The Worship of Mars." Plutarch wrote in *Of the Fortune of Romans*, Chap. XII: "It is asserted that Hercules was conceived in a long night, *the day having been rolled back and retarded against the order of nature and the sun arrested.*"

on the one hand, and for purification of its concept of God, on the other. "And the king [Josiah] commanded Hilkiah the High Priest . . . to bring forth out of the Temple of the Lord all the vessels that were made for Baal and for the grove, and for all the host of heaven: and he burned them without Jerusalem in the fields of Kidron, and carried the ashes of them into Bethel. And he put down the idolatrous priests, whom the kings of Judah had ordained to burn incense in the high places in the cities of Judah, and in the places round about Jerusalem; them also that burned incense unto Baal, to the sun, and to the moon, and to the planets, and to all the host of heaven" (II Kings 23 : 4-5).

The Scriptures do not hide the fact that in Judea, as well as in Israel, the planetary cult was the official cult with the priests and with kings, with many prophets and with the people. Thus Jeremiah, contemporary of King Josiah, says: "At that time, saith the Lord, they shall bring out the bones of the kings of Judah, and the bones of his princes, and the bones of the priests, and the bones of the prophets, and the bones of the inhabitants of Jerusalem, out of their graves: and they shall spread them before the sun, and the moon, and all the host of heaven, whom they have loved, and whom they have served, and after whom they have walked, and whom they have sought, and whom they have worshipped" (Jeremiah 8 : 1-2). And again he says: "And the house of Jerusalem, and the houses of the kings of Judah, shall be defiled as the place of Tophet, because of all the houses upon whose roofs they have burned incense unto all the host of heaven" (Jeremiah 19 : 13).

In the days of Jeremiah and King Josiah, a scroll was found in a chamber of the Temple (II Kings 22). It is generally thought that it was the book of Deuteronomy, the last book of the Pentateuch. The text of the scroll made a strong impression on the king.

"And lest thou lift up thine eyes unto heaven, and when thou seest the sun, and the moon, and the stars, even all the host of heaven, shouldest be driven to worship them, and serve them, which the Lord thy God hath divided unto all nations under the whole heaven" (Deuteronomy 4 : 19).

"Thou shalt not make thee any graven image, or any likeness of any thing that is in heaven above, or that is in the earth beneath . . ." (5 : 8), which is a passage of the Decalogue (Exodus 20 : 4) verbatim.

"If there be found among you . . . man or woman, that hath wrought wickedness . . . and hath gone and served other gods, and worshipped them, either the sun, or moon, or any of the host of heaven, which I have not commanded . . . then shalt thou bring forth that man or that woman . . . and shall stone them with stones, till they die" (17 : 2-5).

Thus we see the centuries-long struggle for the Jewish God, Creator and not unanimated planet, itself a creation, being carried on in the closing decades before the exile to Babylon with the help of the book whose authorship was ascribed to Moses.

When the people of Jerusalem were exiled to Babylon, and groups of refugees succeeded in escaping to Egypt, taking with them Jeremiah, they said to him: "But we will certainly . . . burn our incense unto the queen of heaven, and to pour out drink offerings unto her, as we have done, we, and our fathers, our kings, and our princes, in the cities of Judah, and in the streets of Jerusalem: for then had we plenty of victuals and were well, and saw no evil. But since we left off to burn incense to the queen of heaven, and to pour out drink offerings unto her, we have wanted all things, and have been consumed by the sword and by the famine" (Jeremiah 44 : 17-18).

It is apparent from this passage that the population of Jerusalem that sought refuge in Egypt thought the national catastrophe fell upon their people, not because they had left the Lord God, but because in the days of Josiah and his sons they ceased to worship the planetary gods of Manasseh and especially the Queen of Heaven, the planet Venus.

Of this remnant of the people that went to Egypt in the beginning of the sixth century a military colony was established in Ebb (Elephantine) in southern Egypt. Documents (papyri) of this colony were unearthed in the beginning of this century. The Jewish colony in Elephantine faithfully worshipped Yahu (Yahwe), the Lord of the sky, as the theophoric names of many members of the colony testify. Scholars were puzzled, however, to find on one of the papyri the name Anat-Yahu; they were uncertain whether it belonged to a goddess or a place or a person. "Anat is the familiar name of the Canaanite goddess identified with Athene in a Cyprian inscription." [1] The historical facts revealed in the present research make the understanding

[1] E. Sachau, *Aramäische Papyrus and Ostraka aus einer jüdischen Militärkolonie zu Elephantine* (1911), p. xxv.

of such cult easier. The dark tradition that it was the planet Venus that played such an important role in the days when the forebears of these refugees in Egypt left that land and passed through cataclysms of fire and water, sea and desert, was responsible for this syncretism of names.

The Jewish people did not obtain all of its "supremacy"[1] in that one day at the Mountain of Lawgiving; this people did not receive the message of monotheism as a gift. It struggled for it; and step by step, from the smoke rising from the overturned valley of Sodom and Gomorrah, from the furnace of affliction of Egypt, from the deliverance at the Red Sea amid the sky-high tides, from the wandering in the cloud-enshrouded desert burning with naphtha, from the internal struggle, from the search for God and for justice between man and man, from the desperate and heroic struggle for national existence on its narrow strip of land against the overwhelming empires of Assyria and Egypt, it became a nation chosen to bring a message of the brotherhood of man to all the peoples of the world.

[1] S. A. B. Mercer, *The Supremacy of Israel* (1945).

CHAPTER VI

# A COLLECTIVE AMNESIA

> At any rate they seem to have been strangely forgetful of the catastrophe.
> —PLATO, *Laws* iii (transl. R. Bury)

IT IS AN established fact in the learning about the human mind that the most terrifying events of childhood (in some cases even of manhood) are often forgotten, their memory blotted out from the consciousness and displaced into the unconscious strata of the mind, where they continue to live and to express themselves in bizarre forms of fear. Occasionally they may be converted into symptoms of compulsion neuroses and even contribute to the splitting of the personality.

One of the most terrifying events in the past of mankind was the conflagration of the world, accompanied by awful apparitions in the sky, quaking of the earth, vomiting of lava by thousands of volcanoes, melting of the ground, boiling of the sea, submersion of continents, a primeval chaos bombarded by flying hot stones, the roaring of the cleft earth, and the loud hissing of tornadoes of cinders.

There occurred more than one world conflagration; the most horrible one was in the days of the Exodus. In hundreds of passages in their Bible, the Hebrews described what happened. Returning from the Babylonian exile in the sixth and fifth centuries before this era, the Hebrews did not cease to learn and repeat the traditions, but they lost sight of the fearful reality of what they learned. Apparently, the post-Exile generations looked upon all these descriptions as the poetical utterances of religious literature.

The talmudists in the beginning of this era disputed whether a deluge of fire, prophesied in old traditions, would take place or not; those who denied that it might come, based their argument on the divine promise found in the Book of Genesis, that the Deluge would not be repeated; those who argued to the contrary, reasoning that though the deluge of water would not recur, there might come a deluge of fire, were attacked for construing too narrowly the promise of the Lord.[1] Both sides overlooked the most prominent part of their traditions: the history of the Exodus and all the passages about the cosmic catastrophe, endlessly repeated in Exodus, Numbers, and the Prophets, and in the rest of the Scriptures.

The Egyptians in the sixth pre-Christian century knew about the catastrophes that overwhelmed other countries. Plato narrates the story which Solon heard in Egypt about the world destroyed in deluges and conflagrations: "You remember but one deluge, though many catastrophes had occurred previously." The Egyptian priests who said this and who maintained that their land was spared on these occasions, forgot what happened to Egypt. When, in the Ptolemaic age, the priest Manetho starts his story of the invasion of the Hyksos by acknowledging his ignorance of the cause and nature of the blast of heavenly displeasure that befell his land, it becomes apparent that the knowledge which was possibly alive in Egypt in the days when Solon and Pythagoras visited there, had already sunk into oblivion in the Ptolemaic age. Only some hazy tradition about a conflagration of the world was repeated, without knowing when or how it occurred.

The Egyptian priest, described by Plato as conversing with Solon, supposed that the memory of the catastrophes of the fire and flood had been lost because literate men perished in them, together with all the achievements of their culture, and these upheavals "escaped your notice because for many generations the survivors died with no power to express themselves in writing."[2] A similar argument is found in Philo the Alexandrian, who wrote in the first century of this era: "By reason of the constant and repeated destructions of water and fire, the later generations did not receive from the former the memory of the order and sequence of events."[3]

Although Philo knew about the repeated destructions of the world by water and fire, it did not occur to him that a catastrophe of conflagration was described in the Book of Exodus. Nor did he think that anything of this sort took place in the days of Joshua or even of Isaiah. He thought that the Book of

[1] Cf. Ginzberg, "Mabul shel esh" in *Ha-goren*, VIII, 35-51.
[2] Plato, *Timaeus*, 23 C.
[3] Philo, *Moses*, ii.

Genesis comprised the story of "how fire and water wrought great destruction of what is on the earth," and that the destruction by fire, about which he knew from the teachings of the Greek philosophers, was identical with the destruction of Sodom and Gomorrah.

The memory of the cataclysms was erased, not because of lack of written traditions, but because of some characteristic process that later caused entire nations, together with their literate men, to read into these traditions allegories or metaphors where actually cosmic disturbances were clearly described.

It is a psychological phenomenon in the life of individuals as well as whole nations that the most terrifying events of the past may be forgotten or displaced into the subconscious mind. As if obliterated are impressions that should be unforgettable. To uncover their vestiges and their distorted equivalents in the physical life of peoples is a task not unlike that of overcoming amnesia in a single person.

## FOLKLORE

> Day unto day uttereth speech, and night unto night showeth knowledge. There is no speech nor language, where their voice is not heard.
>
> —Psalms 19: 2-3

The scholars who dedicate their efforts to gathering and investigating the folklore of peoples are constantly aware that folk tales require interpretation, for, in their opinion, these tales are not innocent and unambiguous products of the imagination, but veil some inner and more significant meaning.

The legends of classic peoples, first among them the Greeks, also belong to folklore. As early as pre-Christian times these legends were subjected to interpretation, many interpreters recognizing the symbolic character of mythology.

With Macrobius in the fourth Christian century, there begins a tendency to see in many gods of Egyptian and Greek antiquity the personification of the sun. Macrobius compared Osiris to the sun, and Isis to the moon, disregarding the opinion of earlier authors. He also interpreted Jupiter as the sun.

As the role the planets played in the history of the world retreated ever further into oblivion, the interpretation of nature myths as referring to the sun or the moon became more and more widespread. In the nineteenth century it was the vogue to explain the old myths as inspired by the movement of the sun and the moon, during the day, night, month, and year. Not only Ra, Amon, Marduk, Phaëthon, and even Zeus,[1] but also king-heroes, like Oedipus, became solar symbols.[2]

This exclusive role of sun and moon in mythology is a reflection of their significance in nature. However, in former times the planets played a decidedly more important role in the imagination of peoples, to which fact their religions give testimony. True, sun and moon (Shamash and Sin, Helios, Apollo and Selene) were also numbered among the planet-gods, but usually they were not the most important ones. Their enumeration among the seven planets sometimes startles the modern scholar, because these two luminaries are so much more conspicuous than the other planets; the dominance of Saturn, Jupiter, Venus, and Mars must startle us even more as long as we do not know what was displayed on the celestial scene a few thousand years ago.

Modern folklorists occupy themselves mainly with the folklore of primitive peoples, material unspoiled by generations of copyists and interpreters. Being received at its source, it is supposed to shed light not only on the mentality of these primitive peoples, but also on many problems of sociology and psychology in general.

The sociological method explores mythology for evidence of social usages. Folklorists like James Frazer expended their efforts on this aspect. Freud, the psychologist, centred his attention on the motif of father-murder (patricide), presenting it as though it had been a regular institution in ancient times. He makes it appear a general practice in the past and a subconscious urge in present-day man.

However, regular institutions and practices in the life of the family would not give rise to myths. A writer on this subject

[1] In the Phaëthon story, Ovid makes it clear that Sun and Zeus are two separate deities.

[2] In a separate work I intend to trace the historical prototype of the legend of Oedipus Rex.

has correctly pointed out this fact: "What is quite normal in nature and society rarely excites the myth-making imagination which is more likely to be kindled by the abnormal, some startling catastrophe, some terrible violation of the social code."[1]

Even less than daily tribal life do the daily occurrences in nature give rise to legends. The sun rises every morning, it travels from east to west; the moon enters a new phase four times a month; the year has four seasons—such regular changes do not stir the imagination of peoples, because they contain nothing unexpected in themselves. Daily things do not evoke astonishment and influence but little a people's creative faculty. Sunrise and sunset, morning dew and evening mist, are common experiences, and if a single spectacle impresses itself upon us in the course of life, the many sunrises and the many sunsets in our memory pale and each looks like the other. Seasonal snowstorms or thunderstorms do not leave indelible memories. Only striking, perturbing experiences of a social or physical order are designed to stir the imagination of peoples. Seneca says: "It is for this very reason that the assembly of stars that lends beauty to the immense firmament does not compel the attention of the masses; but when a change occurs in the order of the universe, all looks are fixed on the sky." [2]

Even local catastrophes, regarded as very violent, do not serve for the creation of cosmic myths. First in power to impress the races of the earth are the cataclysms of the past, and on this we have dwelt at length. Comets, because of their casual relation to world catastrophes, and also because of their terrifying appearance, were the kind of phenomenon to kindle the imagination of peoples. But for some reason, the impression they must have made on the peoples of antiquity is not considered in explanation of myths and legends.

Since the invention of the printing press, the great agitation and mass hysteria caused by the more brilliant comets can be traced in contemporary books and pamphlets. Were the ancients immune to these feelings? If not, then why are the exegetes of the Bible and the commentators on the epic compositions of antiquity so remiss as not to think of phenomena that could not but impress the ancients? Or did no comets appear in the sky

[1] L. R. Farnell, "The value and the methods of mythological study," *Proceeding of the British Academy*, 1919-1920, p. 47.
[2] *Naturales quaestiones* vii.

during ancient times? This, of course, is only a rhetorical question.

Keeping this in mind, we shall be able to answer the question about the striking similarity of certain concepts among peoples of different cultures, sometimes separated by oceans.

## OF "PRE-EXISTING IDEAS" IN THE SOULS OF PEOPLES

The similarity of motifs in the folklore of various peoples on the five continents and on the islands of the oceans posed a difficult problem for the ethnologists and anthropologists. The migration of ideas may follow the migration of peoples, but how could unusual motifs of folklore reach isolated islands where the aborigines do not have any means of crossing the sea? And why did not technical civilization travel together with spiritual? Peoples still living in the stone age, possess the same, often strange, motifs as the cultured nations. The particular character of some of the contents of folklore makes it impossible to assume that it was only by mere chance that the same motifs were created in all corners of the world. The problem is so perplexing to the scientists that, for lack of a better proposition, an explanation was offered according to which the motifs of folklore are a pre-existing possession in the soul of peoples; peoples are born with these ideas just as an animal is born with an urge to propagate its kind, to nurse its offspring, to build a lair or a nest, and to travel in herds or migrate in flocks to far-away countries. But it is not so simple to explain in these terms why, for instance, the aborigines of America imagined a witch as a woman riding on a broom across the sky, exactly as the European peoples imagined her. "The Mexican witch, like her European sister, carried a broom on which she rode through the air, and was associated with the screech owl. Indeed, the queen of witches, Tlagoltiotl, is depicted as riding on a broom and as wearing the witch's peaked hat."[1] As with the witch on her broom, so also with hundreds of other odd fantasies and beliefs.

The answer to the problem of the similarity of the motifs in the folklore of various peoples is, in my view, as follows: A great many ideas reflect real historical content. There is a

[1] Lewis Spence, *The History of Atlantis* (1930), p. 224.

legend, found all over the world, that a deluge swept over the earth and covered hills and even mountains. We have a poor opinion of the mental abilities of our ancestors if we think that merely an extraordinary overflow of the Euphrates so impressed the nomads of the desert that they thought the entire world was flooded, and that the legend so born wandered from people to people. At the same time, geological problems of the origin and distribution of till, or diluvial deposit, are awaiting explanation.

The peoples of ancient times, who, like the primitive peoples of the present, lacked modern protection against the elements of nature, and who live in the insecurity of tropical storms and tornadoes or frost and snowstorms, must have been more accustomed to seasonal disturbances than we are, and would not have been impressed by the overflow of a river to such a degree as to carry their experience to all parts of the world as a story of cosmic upheaval.

Traditions about upheavals and catastrophes, found among all peoples, are generally discredited because of the shortsighted belief that no forces could have shaped the world in the past that are not at work also at the present time, a belief that is the very foundation of modern geology and of the theory of evolution. "Present continuity implies the improbability of past catastrophism and violence of change, either in the lifeless or in the living world; moreover, we seek to interpret the changes and laws of past time through those which we observe at the present time. This was Darwin's secret, learned from Lyell."[1] It has been shown in this book, however, that forces which at present do not act on the earth, did so act in historical times, and that these forces are of a purely physical character. Scientific principles do not warrant maintaining that a force which does not act now, could not have acted previously. Or must we be in permanent collision with the planets and comets in order to believe in such catastrophes?

## THE PAGEANTS OF THE SKY

Cosmic perturbations took place, catastrophes swept the globe, but did witches fly through the air on brooms? The reader would agree that cosmic catastrophes, if they occurred,

[1] H. F. Osborn, *The Origin and Evolution of Life* (1918), p. 24.

could leave, and must have left, similar memories all around the world; but there are fantastic images that do not appear to represent realities. We shall follow this rule: if there exists a fantastic image that is projected against the sky and that repeats itself all around the world, it is most probably an image that was seen on the screen of the sky by many peoples at the same time. On one occasion a comet took the striking form of a woman riding on a broom, and the celestial picture was so clearly defined that the same impression was imposed on all the peoples of the world. It is well known how, in modern times, the forms of comets impress people. One comet was said to look like "*un crucifix tout sanglant*," another like a sword; actually every comet has its peculiar shape which may also change during the visibility of the comet.

To illustrate what is said here by another example, it may be asked: What induced the Mayas to call by the name of Scorpion the constellation known to us and to the ancients by the same name? [1] The outlines of this constellation do not resemble the shape of this insect. It is "one of the most remarkable coincidences in nomenclature." [2]

The constellation, which is not at all like a scorpion, probably was called by this name because a comet that looked like a scorpion appeared in it. Actually, we read on one of the Babylonian astronomical tablets that "a star flared up and its light radiated bright as day, and as it blazed, it lashed its tail like an angry scorpion." [3] If it was not this particular appearance of a comet that caused the constellation to be called Scorpion, there must have been a similar occurrence on another date.

Another example is the dragon. All around the world this image is prominent in literature and art and also in the religion of peoples. There is probably no nation that does not use this symbol or this creature as an important motif, yet it does not exist. Several scholars thought that possibly it represented some extinct menace that impressed mankind to a much greater degree than any other creature since it appears on the Chinese flag, and in pictures showing Archangel Michael or

[1] Sahagun, in the fourth chapter of the seventh book of his historical work, says that the people of Mexico called the constellation Scorpion (Scorpio) by this very same name.

[2] Seler, *Ges. Abhand. zur amer. Sprach- und Alterthumskunde*, II (1903), 622. His surmise, disagreeing with the assertion of De Sahagun, was that Scorpion of the ancients was more to the south. However, with the displacement of the poles, the stars acquired new positions.

[3] Kugler, *Babylonische Zeitordnung*, p. 89.

St. George in battle with it, in Egyptian mythology, in Mexican hieroglyphics and bas-reliefs, and in Assyrian bas-reliefs. However, bones of this presumably extinct reptile have not been found.

From the description of the comet Typhon that spread like an animal over the sky with its many heads and winged body, with fire flaming from its mouths, as described in a previous chapter by quotations from Apollodorus and others, we recognize the origin of this widespread motif.

## THE SUBJECTIVE INTERPRETATION OF THE EVENTS AND THEIR AUTHENTICITY

What helped to discredit the traditions of the peoples about the catastrophes was their subjective and magical interpretation of the events. The sea was torn apart. The people attributed this act to the intervention of their leader; he lifted his staff over the waters and they divided. Of course, there is no person who can do this, and no staff with which it can be done. Likewise in the case of Joshua who commanded the sun and the moon to halt in their movements. Because the scientific mind cannot believe that a man can make the sun and the moon to stand still, it disbelieves also the alleged event. What contributes to this is the fact that least of all do we place faith in books that demand belief, religious books, though we swear on these.

The peoples of the past were prepared to see miracles in unusual occurrences; for this reason modern man, who does not believe in miracles, rejects the event together with the interpretation. But as we find the same event in the traditions of many peoples, and as each people has differently comprehended it, its historicity can be checked, and this in addition to the control offered by natural science. For example, if the geographical poles changed their location, or the axis its inclination, the ancient solar clock would not show the correct time; or, if the magnetic poles became reversed at some time in the past, the lava of earlier volcanic activity must show reversed magnetic orientation.

But there is also a check by folklore. Isaiah foretold to King Hezekiah, probably a few hours before the event, that the shadow of the sundial would return ten degrees. (As we know

now, the planet Mars was at that moment very close to the earth, and Isaiah could make an estimate based on experiences during previous perturbations of the earth by Mars.) The Chinese explained this phenomenon as having occurred to help their princes in their strategy, or to settle a quarrel among them. The Greek people thought the phenomenon was an expression of heavenly wrath at the crime of the Argive tyrants. The Latins thought the phenomenon was an omen associated with Romulus, son of Mars. In the Icelandic epos the same event has a different purpose, in the Finnish epos another, and yet others in Japan and Mexico and Polynesia. The American Indians say that the sun went backwards several degrees for fear of a boy who tried to snare it or because of some animal that terrified it. Precisely because there are great differences in the subjective evaluation of the causes or purposes of the phenomenon, we can assume that the folklore of different peoples deals with one and the same factual event, and only the magical explanations of the miracle are subjective inventions. Many accompanying details are preserved in the variants of different peoples, which could not have been invented without an adequate knowledge of the laws of motion and thermodynamics. It is inconceivable that the ancients or the primitive races would, for instance, by sheer chance invent the tale that a huge conflagration enveloped the American prairies and forests as soon as the sun, frightened off by the snarer, returned a little on its way.

If a phenomenon had been similarly described by many peoples, we might suspect that a tale, originating with one people, had spread around the world, and consequently there is no proof of the authenticity of the event related. But just because one and the same event is embodied in traditions that are very different indeed, its authenticity becomes highly probable, especially if the records of history, ancient charts, sundials, and the physical evidence of natural history testify to the same effect.

In the Section "Venus in the Folklore of the Indians" a few illustrations were offered to illuminate this thesis. In order to illustrate it with additional examples, we choose the nature-folkloristic motif of the sun being arrested in its movement across the firmament in the tales of the Polynesians, Hawaiians, and North American Indians.

The best known legend cycle on the Pacific islands is that

which has for its hero the semigod Maui.[1] This cycle comprises a trilogy: "Of the many exploits of Maui three seem to be most widely spread: they are fishing up of the land, snaring the sun and the quest of fire." [2] There are two versions of this cycle, one in New Zealand and one in Hawaii, but both are variants of a common tradition.

The Hawaiian version of the snaring of the sun runs thus: "Maui's mother was much troubled by the shortness of the day, occasioned by the rapid movement of the sun; and since it was impossible to dry properly the sheets of tapa used for clothing, the hero resolved to cut off the legs of the sun, so that he could not travel fast.

"Maui now went off eastward to where the sun climbed daily out of the underworld, and as the luminary came up, the hero noosed his legs, one after the other, and tied the ropes strongly to great trees. Fairly caught, the sun could not get away, and Maui gave him a tremendous beating with his magic weapon. To save his life, the sun begged for mercy, and on promising to go more slowly ever after, was released from his bonds."

The "fishing up of islands" or the appearance of new islands took place at the same time; the causal relation to the cosmic change in the sky is evident. In one of the versions told in Polynesia about the fishing up of the islands, it is said that a star was used as bait.

The following is a tale told by the Menomini Indians, an Algonquin tribe.[3] "The little boy made a noose and stretched it across the path, and when the Sun came to that point the noose caught him around the neck and began to choke him until he almost lost his breath. It became dark, and the Sun called out to the ma'nidos, 'Help me, my brothers, and cut this string before it kills me.' [4] The ma'nidos came, but the thread had so cut into the flesh of the Sun's neck that they could not sever it. When all but one had given up, the Sun called to the Mouse to try to cut the string. The Mouse came

[1] "Of all the myths from the Polynesian area, probably none have been more frequently quoted than those which recount the deeds and adventures of the semi-god Maui. The Maui cycle is one of the most important for the study of this whole area." Dixon, *Oceanic Mythology*, p. 41.

[2] *Ibid.*, p. 42.

[3] Hoffman, *Report of the Bureau of American Ethnology*, XIV, 181, reproduced by S. Thompson, *Tales of the North American Indians* (1929).

[4] Ma'nido is "a spirit or spiritual being; any person or subject endowed with spiritual power."

up and gnawed at the string, but it was difficult work, because the string was hot and deeply embedded in the Sun's neck. After working at the string a good while, however, the Mouse succeeded in cutting it, when the Sun breathed again and the darkness disappeared. If the Mouse had not succeeded, the Sun would have died."

The story about snaring the sun associates itself in our mind with one of the occasions when the sun was disrupted in its movement across the sky. The story contains an important detail and enables us to understand a natural phenomenon.

In a previous section we discussed the various versions of the annihilation of Sennacherib's army and the physical phenomena which caused it. According to the Scriptures, in the days of Isaiah the sun was interrupted in its course, turning back ten degrees on the sundial. That night the army of Sennacherib was destroyed by a blast. In Egypt this victory over the common enemy of the Jews and the Egyptians was observed in a festival at Letopolis, "the city of the thunderbolt"; the holy animal of the city was a mouse, and bronze mice inscribed with the prayers of pilgrims are found in its soil. Herodotus saw there a statue of a god with a mouse in his hand, commemorating the annihilation of the army of Sennacherib. The story he heard gave as the cause of the event an invasion of mice that gnawed the strings of the bows. He also told the story of the changed movements of the sun directly following the record of the destruction of the Assyrian army. We recognized that the image of the mouse must have had some relation to the cosmic drama. The best we could do was to interpret the mouse as a symbol of a simultaneous plague, exemplified by the illness of King Hezekiah.

The tale of the Indians that combines the snaring of the sun with the deed of the mouse explains the relation of these two elements to each other. Apparently the atmosphere of the celestial body that appeared in the darkness and was illuminated took on the elongated form of a mouse. This explains why the blast that destroyed the army of Sennacherib was commemorated by the emblem of a mouse. The Indian tale grew from the picture on the celestial screen where a great mouse freed the snared sun.

Thus we see how a folk story of the primitives can solve an unsettled problem between Isaiah and Herodotus.

A four-legged animal in the sky approaching the sun was visualized as a mouse by the Egyptians and the Menomini Indians. In the tale of the southern Ute Indians, the cottontail is the animal that is connected with the disruption of the movement of the sun.[1] He went to the east with the intention of breaking the sun in pieces. There he waited for the sun to rise. "The sun began to rise, but seeing the cottontail, it went down again. Then it rose slowly again and did not notice the animal. He struck the sun with his club, breaking off a piece, which touched the ground and set fire to the world.

"The fire pursued Cottontail, who began to flee. He ran to a log and asked if it would save him if he got inside. 'No, I burn up entirely.' So he ran again and asked a rock with a cleft in it. 'No, I cannot save you, when I am heated I burst . . .' At last he got to a river. The river said, 'No, I cannot save you; I'll boil and you will get boiled.' "

On the plain, Cottontail ran through the weed, but the fire came very close, the weed burned and fell on his neck, "where cottontails are yellow now."

" From everywhere he saw smoke rising. He walked a little way on the hot ground and one of his legs was burned up to the knee; before that he had been long-legged. He walked on two legs, and one of them burned off. He jumped on one till that also burned off."

In this version of the attack on the sun, two points worthy of mention are the world fire following the disruption in the movement of the sun, and the change in the world of animals accompanied by strong mutations. In the section, "Phaëthon," we wondered how the Roman poet Ovid could have known of the relation between the interrupted movement of the sun and a world fire unless such a catastrophe had really occurred. The same reasoning applies to the Indians. The story of snaring the sun or attacking the sun is told in many variants, but the world fire is a consistent result. Forests and fields burn, mountains smoke and vomit lava, rivers boil, caves in the mountains collapse, and rocks burst when the sun peeps above the horizon and then disappears and again comes over the horizon.

There is one instance more in the Indian story of the sun being impeded on its path and the ensuing world conflagration. Before the catastrophe, "the sun used to go round close to the

[1] R. H. Lowie, "Shoshonean Tales," *Journal of American Folk-lore*, XXXVII (1924), 61 ff.

ground." The purpose of the attack on the sun was to make " the sun shine a little longer: The days are too short." After the catastrophe "the days became longer."

The ancestors of the Shoshonean Indians, a tribe of Utah, Colorado, and Nevada, appear to have lived in the days of Sennacherib and Hezekiah at such a longitude that the sun was just on the eastern horizon when it changed its direction and went back and then came up again.

## CHAPTER VII

# POLES UPROOTED

WHAT CHANGES IN the motion of earth, moon, and Mars resulted from the contacts in the eighth and seventh centuries?

The moon, being smaller than Mars, would have been greatly influenced by Mars if it came close enough to that planet. It could have been drawn nearer to the earth or pulled away to a more remote orbit. It is therefore of interest to investigate whether, in the time shortly after -687, reforms of the lunar calendar were undertaken.

Also, the earth could have been "removed from her place," which would have meant a change in the orbital circumference and thus in the length of the year, or in the inclination of the terrestrial axis to the plane of the ecliptic and thus in the seasons, in the position of the poles on the terrestrial globe, in the velocity of axial rotation, and in the length of the day, and so on. Some of these changes could be traced if a chart of the sky, drawn in a period prior to -687, could be examined. Such a chart does exist; it is painted on the ceiling of the tomb of Senmut, the Egyptian vizier. As explained previously,[1] the tomb dates from a time following the Exodus but before the days of Amos and Isaiah.

The charts of Senmut show the sky over Egypt at two different epochs; one of them depicts the sky of Egypt before the poles were interchanged probably in the catastrophe that terminated the Middle Kingdom; the other represents the sky of Egypt in the lifetime of Senmut. The first chart startled the investigators because in it west and east are reversed. Their judgment of the other chart, in which west and east are not reversed, is as follows:

"It is not surprising to find that the celestial charts which have been preserved until our time did not correspond to direct observations, nor to the calculations made at the moment of erection of the monument on which these charts are pictured." [2]

[1] See the Section "East and West."

[2] A. Pogo, "Astronomie égyptienne du tombeau de Senmout," *Chronique d'Égypte*, 1931.

Modern astronomy does not admit, or even consider, the possibility that at some historical time east and west as well as south and north were reversed. Consequently, the first chart could not have been interpreted at all. The other chart, with its displaced constellations, suggested to the author of the above quotation that it depicted some more ancient tradition. The only change, according to modern astronomy, comes from the precession of the equinoxes or the slow movement of the polar axis which describes a circle in the course of about twenty-six thousand years. The computation of the precession is insufficient by far to explain the position of the constellations on the chart if we rely on the conventional chronology (and even more so if we follow the revised chronology, which brings the age of Senmut and Queen Hatshepsut closer to modern times).

The changes in the geographical position and cosmic direction of the poles caused by the catastrophes of the eighth and seventh centuries, as well as those brought about by the catastrophes of the fifteenth century, can be studied with the help of the astronomical charts of Senmut.

According to Seneca the Great Bear had been the polar constellation. After a cosmic upheaval shifted the sky, a star of the Little Bear became the polar star.

Hindu astronomical tablets composed by the Brahmans in the first half of the first millennium before the present era show a uniform deviation from the expected position of the stars at the time the observations were made (the precession of the equinoxes being taken into consideration).[1] Modern scholars wondered at this, in their opinion inexplicable, error. In view of the geometrical methods employed by Hindu astronomy and its detailed method of calculation, a mistake in observation equal to even a fraction of a degree would be difficult to account for.

In *Jaiminiya-Upanisad-Brahmana* it is written that the centre of the sky, or the point around which the firmament revolves, is in the Great Bear.[2] This is the same statement we found in *Thyestes* of Seneca.

In Egypt, too, "the Great Bear played the part of the Pole Star."[3] "The Great Bear never set."[4] Could it be that the

[1] J. Bentley, *A Historical View of the Hindu Astronomy* (1825), p. 76.
[2] Thibaut, "Astronomie, Astrologie und Mathematik," p. 6.
[3] G. A. Wainwright, "Orion and the Great Star," *Journal of Egyptian Archaeology*, XXII (1936).
[4] Wainwright, "Letopolis," *Journ. Egypt. Archaeol.*, XVIII (1932).

precession of equinoxes shifted the direction of the axis so that, three or four thousand years ago, the polar star was among the stars of the Great Bear?[1] No. If the earth moved all the time as it moves now, four thousand years ago the star nearest the North Pole must have been *a*-Draconis.[2] The change was sudden; the Great Bear "came bowing down."[3] In the Hindu sources it is said that the earth receded from its wonted place by 100 yojanas,[4] a yojana being five to nine miles. Thus the displacement was estimated at from 500 to 900 miles.

The origin of the polar star is told in many traditions all over the world. The Hindus of the *Vedas* worshipped the polar star, Dhrura, "the fixed" or "immovable." In the *Puranas* it is narrated how Dhrura became the polar star. The Lapps venerate the polar star and believe that if it should leave its place, the earth would be destroyed in a great conflagration.[5] The same belief is found among the North American Indians.[6]

The day on which the shortest shadow is cast at noon is the day of the summer solstice; the longest shadow at noon is cast on the day of the winter solstice. This method of determining the seasons by measuring the length of the shadows was applied in ancient China, as well as in other countries.

We possess the Chinese records of the longest and shortest shadows at noontime. These records are attributed to -1100. "But the shortest and the longest shadows recorded do not really represent the true lengths at present."[7] The old Chinese charts record the longest day with a duration which "does not represent the various geographical latitudes of their observatories," and therefore the figures are supposed to have been those of Babylonia, borrowed by ancient Chinese, a rather unusual conjecture.[8]

The length of the longest day in a year depends on the latitude, or the distance from the pole, and is different at

[1] Wainwright in the *Studies* presented to F. L. Griffith, pp. 379-380.
[2] Cf. H. Jeffreys, "Earth," *Encyclopaedia Britannica* (14th ed.).
[3] Wainwright, *Journ. Egypt. Archaeol.*, XVIII, p. 164.
[4] J. Hertel, *Die Himmelstore im Veda und im Awesta* (1924), p. 28.
[5] Kunike, "Sternmythologie," *Welt und Mensch*, IX-X; A. B. Keith, *Indian Mythology* (1917), p. 165.
[6] *The Pawnee Mythology* (collected by G. A. Dorsey; 1906) Pt. I, p. 135.
[7] J. N. Lockyer, *The Dawn of Astronomy* (1894), p. 62; cf. M. Cantor, *Vorlesungen über Geschichte der Mathematik* (2nd ed., 1894), p. 91. Laplace made efforts to find an explanation for these figures.
[8] Kugler, *Sternkunde und Sterndienst in Babel*, I, 226-227.

different places. Gnomons or sundials can be built with great precision.[1]

The Babylonian astronomical tablets of the eighth century provide exact data, according to which the longest day at Babylon was equal to 14 hours 24 minutes, whereas the modern determination is 14 hours 10 minutes, and 54 seconds.

"The difference between the two figures is too great to be attributable to refraction, which makes the sun still visible over the horizon after it has set. Thus, the greatest length of the day corresponds to latitude 34° 57′, and points to a place 2½° farther to the north; we stand therefore before a strange riddle [*vor einem merkwürdigen Rätsel*]. One tries to decide: either the tablets of System II do not originate from Babylon [though referring to Babylon], or this city actually was situated far [farther] to the north, about 35° away from the equator."[2]

Since the computations of the astronomical tablets did refer to Babylon, there is a possible solution that Babylon was situated at a latitude of 35° from the equator, much farther to the north than the ruins of this city.

Claudius Ptolemy, who, in his *Almagest*, made computation for contemporaneous and ancient Babylon, arrived at two different estimates of the longest day at that city, and consequently the latitude at which it was located,[3] one of his estimates being practically of the present-day value, the other coinciding with the figure of the ancient Babylonian tables, 14 hours 24 minutes.

The Arabian medieval scholar Arzachel computed from ancient codices that in more ancient times Babylon was situated at a latitude of 35° 0′ from the equator, while in later times it was situated more to the south. Johannes Kepler drew attention to this calculation of Arzachel and to the fact that between ancient and modern Babylon there was a difference in latitude.[4]

Thus Ptolemy, and likewise Arzachel, computed that in historical times Babylon was situated at latitude 35°. Modern

[1] A gnomon (277 feet high), built by Toscanelli in 1468, during the Renaissance, for the cathedral in Florence, shows midday to within half a second. R. Wolf, *Handbuch der Astronomie* (1890-1893), n. 164.

[2] Kugler, *Die babylonische Mondrechnung: Zwei Systeme der Chaldäer über den Lauf des Mondes und der Sonne* (1900), p. 80.

[3] Ptolemy, *Almagest*, Bk. 13 (ed. Halma); Bk. 4, Chap. 10; also *idem*, *Geography*, Bk. 8, Chap. 20. Cf. Kugler, *Die babylonische Mondrechnung*, p. 81; also Cantor, *Vorlesungen über Geschichte der Mathematik*, pp. 82 ff.

[4] J. Kepler, *Astronomi opera omnia* (ed. C. Frisch), VI (1866), 557; "Et quai altitudinem poli veteri Babyl. assignat 35° 0′, novae 30° 31′."

scholars arrived at identical results on the basis of ancient Babylonian computations. "This much, therefore, is certain: our tables [System II, and I also], and the astronomers mentioned as well, point to a place about 35° north latitude. Is it possible that they were mistaken by 2° to 2½°? This is scarcely believable."[1]

As there was but one Babylon, its location, at some historical time, at 35° north latitude signifies that at the longitude of Babylon the earth since then has turned toward the south, and the direction of the polar axis, or its geographical location, or both, have undergone displacement.

Some of the classic authors knew that the earth had changed its position and had turned toward the south; not all of them, however, were aware of the real cause of this perturbation. Diogenes Laërtius repeated the teaching of Leucippus: "The earth was bent or inclined toward the south because the northern regions grew rigid and inflexible by the snowy and cold weather which ensued thereon."[2] The same idea is found in Plutarch, who quoted the teaching of Democritus: "The northern regions were ill temperate, but the southern were well; whereby the latter becoming fruitful, waxed greater, and, by an overweight, preponderated and inclined the whole that way."[3] Empedocles, quoted by Plutarch, taught that the north was bent from its former position, whereupon the northern regions were elevated and the southern depressed. Anaxagoras taught that the pole received a turn and that the world became inclined toward the south.

As we have seen, Seneca in *Thyestes* correctly ascribed the displacement of the pole to a cosmic catastrophe.

## TEMPLES AND OBELISKS

In classic authors references can be found to the fact that the temples of the ancient world were built facing the rising sun.[4] Orientation toward the sun is, at the same time, orientation

[1] Kugler, *Die babylonische Mondrechnung*, p. 81.
[2] This is a translation by Whiston in his *New Theory of the Earth*. The modern version of L. D. Hicks differs greatly.
[3] Plutarch, "What Is the Cause of the World's Inclination?" in Vol. III of *Morals* (transl. revised by W. Goodwin).
[4] Plutarch, *Lives*, "Lives of Numa": "Temples face the east and the sun."

toward the visible planets, as all of them travel through the signs of the zodiac or in the ecliptic. The sun changes the point of its rising and setting from one day to another, and the ecliptic makes a corresponding slow swing from one solstice to another. Therefore, for the purposes of accurate observation of whether the terrestrial pole shifted in a sudden way, it was necessary to build the temple observatories, not simply facing the east and the west, but with a device that would permit checking the position of the sun on the days of the vernal and autumnal equinoxes, when the sun rises exactly in the east and sets exactly in the west.

The Tractate Erubin of the Jerusalem Talmud[1] records "the surprising fact"[2] that the Temple of Jerusalem was so built that on the two equinoctial days the first rays of the rising sun shone directly through the eastern gate; the eastern gate was kept closed during the year, but was opened on these two days for this very purpose. The first ray of the equinoctial sun shone through the eastern gate and into the very heart of the Temple.[3] There was no sun worship in this arrangement; it was dictated by the events of the past, when the position of the earth, in relation to the rising and setting points of the sun, was moved in world catastrophes. The fall equinox was observed as New Year's day. This ceremony with the equinoctial sun was old. The Babylonian temples, also, had "the gate of the rising sun" and "the gate of the setting sun."[4] With the growing belief that there would be no more changes in the world system, a belief expressed also by Deutero-Isaiah (66 : 22), the eastern gate of the Jerusalem Temple was closed forever: it will be opened in Messianic times.

Although unaware of these ancient practices and literary references to the orientation of the temples, a writer of the end of the nineteenth century came to the conclusion that the temples of the ancient world faced the sunrise.[5] He found considerable evidence in the position of temples, but he wondered also that there were deliberate changes in the orientation of the foundations of some older temples. "The many changes in direction of the foundations at Eleusis revealed

[1] Jerusalem Talmud, Tractate Erubin, V, 22c.
[2] J. Morgenstern, "The Book of the Covenant," *Hebrew Union College Annual* V, 1927, p. 45.
[3] Morgenstern, "The Gates of Righteousness," *Hebrew Union College Annual*, VII, 1929.
[4] Winckler, *Keilinschriftliche Bibliothek*, III, Part 2 (1890), p. 73.
[5] Lockyer, *The Dawn of Astronomy*.

by the French excavations were so very striking and suggestive" that the author asked "whether there was possible astronomical origin for the direction of the temple and the various changes in direction."[1]

Further investigation by other authors revealed the fact that generally only the temples of a later time faced the east, and that earlier temples, built before the seventh century, had their foundations purposely directed—the same orientation can be traced in a number of archaic foundations—away from the present east.[2]

Knowing by now that the earth repeatedly shifted the direction of the sunrise and sunset, we understand the changes in the orientation of the foundations as the result of changes in nature. Thus, we have in the foundations of the temples, like that of Eleusis, a record of the changing direction of the terrestrial axis and the position of the pole; the temple was destroyed by catastrophes and rebuilt each time with a different orientation.

Besides the temples and their gates, the obelisks also served the purpose of fixing the direction of east and west, or of sunrise and sunset on equinoctial days. As this purpose was not perceived, the object for which the obelisks were built seemed enigmatic: "The origin and religious significance of the obelisks are somewhat obscure."[3]

Two pillars were erected before the Temple of Solomon,[4] but their purpose is not revealed in the Scriptures.

In America, obelisk-pillars were built too. Sometimes a ring was set on the vertex of the pillar for the sun's rays to pass through. "The solstices and equinoxes were carefully observed. Stone pillars were erected eight on the east and eight on the west side of Cuzco, to observe the solstices. . . . At the heads of the pillars there were discs for the sun's rays to enter. Marks were made on the ground, which had been levelled and paved. Lines were drawn to mark the movement of the sun. . . .

"To ascertain the time of the equinoxes there was a stone column in the open space before the temple of the sun, in the centre of a large circle. . . . The instrument was called

[1] Lockyer, *The Dawn of Astronomy*, p. viii.

[2] H. Nissen, *Orientation, Studien zur Geschichte der Religion* (1906); E. Pfeiffer, *Gestirne und Wetter im griechischen Volksglauben* (1914), p. 7. See also F. G. Penrose, *Philosophical Transactions of the Royal Society of London*, CLXXXIV, 1893, 805-834, and CXC, 1897, 43-65.

[3] R. Engelbach, *The Problem of the Obelisks* (1923), p. 18.

[4] I Kings 7 : 15.

*inti-huatana*, which means the place where the sun is tied up or encircled. There are *inti-huatanas* on the height of Ollantay-tampu, at Pissac, at Hatuncolla, and in other places."[1]

The Egyptian obelisk could serve as a gnomon, or shadow clock. The length of the shadow and its direction would indicate the hour of the day. Obelisks placed in pairs served as a calendar. On the vernal and autumnal equinoxes their shadows would be continuous for the length of the day, the sun rising exactly in the east and setting exactly in the west.

That the purpose for which the obelisks were erected was to check on the shadow of the sun (and the position of the earth) can be plainly seen from this passage of Pliny:

"The obelisk [of Sesothis, brought from Egypt] that has been erected in the campus Martius [in Rome] has been applied to a singular purpose by the late Emperor Augustus: that of marking the shadow projected by the sun, and so measuring the length of the days and nights." There then follows this remark: "For nearly the last thirty years, however, the observations derived from this dial have been found not to agree: whether it is that the sun itself has changed its course in consequence of some derangement of the heavenly system; or whether that the whole earth has been in some degree displaced from the centre, a thing that, I have heard say, has been remarked in other places as well; or whether that some earthquake, confined to this city only, has wrenched the dial from its original position; or whether it is that in consequence of the inundations of the Tiber, the foundations of the mass have subsided."[2]

The passage indicates that Pliny envisaged every possible cause, not excluding the one known to have occurred in earlier times when, in the language of Plutarch, "the Pole received a turn or inclination," or in the words of Ovid, "Earth sank a little lower than her wonted place."

## THE SHADOW CLOCK

The poles changed their locations; all latitudes were displaced, the axis changed its direction; the number of days in the year increased from 360 to 365¼, a fact demonstrated in a

[1] Markham, *The Incas of Peru*, pp. 115, 116.
[2] Pliny, *Natural History*, xxxvi. 15 (transl. Bostock and Riley).

following section; the length of the day probably also altered. Of course, a sundial or shadow clock from before −687 can no longer serve the purpose for which it was devised, but it might well be of use in proving our assumption.

Such a clock, originating from the period between circa −850 and −720, was found in Faijum in Egypt at latitude 27°. A horizontal slab with hour marks has at one end a shadow-casting vertical hob.[1] This shadow clock cannot show correctly the change of time in Faijum or elsewhere in Egypt. A scholar who investigated its working came to the conclusion that it must have been kept with its head to the east in the forenoon and to the west in the afternoon, and several scholars agreed that this was the way to use the clock. But this arrangement by itself did not make it possible to read the time. "Since all actual hour shadows lie substantially closer to the hob than the corresponding marks of the instrument, the shadow-casting edge múst have been higher over the shadow-receiving plane than we find it to be. The upper edge cannot be the shadow-caster of the instrument; it must have been on a parallel line above this edge."[2] "The marks were also not made on the basis of actual observations, but must have been taken from some theory or other."[3] But, as a critic remarked, "this theory implies that at no season of the year did the clock denote the hours correctly, without an hourly alteration of the height of that part of the instrument which cast the shadow."[4]

As the clock has no device to adjust the height of the head, it is improbable that this hourly manipulation took place. Besides, in order to change the height of the head every hour, in itself an impractical method, it would have been necessary to have another clock to show the hours without any manipulation, thus indicating the exact moment when the first clock had to be adjusted. But if there was a clock that could show the hours correctly without adjustment, what purpose did the shadow clock serve?

Another explanation has therefore been offered for the manner in which the Egyptian sundial was used. The author of the new idea supposes that at some early date (the precession

[1] The Egyptian day was divided into hours that represented equal portions of time between sunrise and sunset, independently of the length of the day.

[2] L. Borchardt, "Altägyptische Sonnenuhren," *Zeitschrift für ägyptische Sprache und Altertumskunde*, XLVIII (1911), 14.

[3] *Ibid.*, p. 15.

[4] J. MacNaughton, "The Use of the Shadow Clock of Seti I," *Journal of the British Astronomical Association*, LIV, No. 7 (Sept. 1944).

of the equinoxes being taken into consideration) the shadow clock was used at some latitude in Egypt on the day of the summer solstice. He admits: "Account has, however, not been taken of change in the declination of the sun between sunrise and sunset. . . . For other seasons of the year it would be necessary at each hour or each clock reading, either to alter the height of the hob, or tilt the st't [clock] or both. Indeed, when the sun had south declination, and even when it had slight north declination, it would always be necessary to do both. The inference is, therefore, that the clock was originally used at or near the time of the summer solstice."[1] The problem of adjustment for each reading once more crops up in this explanation, again requiring some better means of knowing the exact time. The conclusion at which the author of this explanation arrives—that originally the clock was built for a single day—is rather odd and defies the very purpose for which clocks are constructed. And even if a clock were to be read only once a year, the author of this theory could not make the specimen found in Faijum work, but only a similar clock that had been found broken in pieces; and this he could do only by having recourse to the precession of the equinoxes and by referring the clock to a period many hundreds of years earlier than chronologists assume.

The shadow clock found at Faijum, built under the Libyan Dynasty, between about –850 and –720 before the present era, may help us to learn the length of the day, the inclination of the pole to the ecliptic, and the latitudes of Egypt in that historical period. A change in any of these three factors would have made the clock obsolete as an instrument for time reading, and probably all three factors did change.

We do not possess the sundial of King Ahaz, but we do have the shadow clock used in Egypt in the period before the last catastrophe of –687 and possibly before the catastrophe of –747.

## THE WATER CLOCK

Besides the gnomon or sundial, the Egyptians used the water clock, which had the advantage over the former of showing time during the night as well as during the day.

[1] J. MacNaughton, "The Use of the Shadow Clock of Seti I," *Journal of the British Astronomical Association*, LIV, No. 7 (Sept. 1944).

A complete example was found in the Amon Temple of Karnak (Thebes), 25.5° north of the equator. This water clock dates from the time of Amenhotep III of the Eighteenth Dynasty, father of Ikhnaton. The jar has an opening through which water flows out; marks are incised on the inner surface of the jar to indicate the time. Since the Egyptian day was divided into hours, which changed in length with the length of the day, the jar has different sets of markings for the various seasons of the year. Four time points are prominently important: the autumnal equinox, the winter solstice, the vernal equinox, and the summer solstice. The equinoxes have equal days and nights in all latitudes. But on the solstices, when either the day or the night is the longest of the year, the length of the daylight varies with the latitude: the farther from the equator, the greater is the difference between the day and the night on the day of the solstice. This difference also depends on the inclination of the equator to the plane of the orbit or ecliptic which is at present 23½°. Should this inclination change, or in other words, should the polar axis change its astronomical position (direction), or should the polar axis change its geographical position with each pole shifting to another point, the length of the day and night (on any day except the equinoxes) would change, too.

The water clock of Amenhotep III presented its investigator with a very strange time scale.[1] Calculating the length of the day of the winter solstice, he found that the clock was constructed for a day of 11 hours 18 minutes, whereas the day of the solstice at 25° north latitude is 10 hours 26 minutes, a difference of fifty-two minutes. Similarly, the builder of the clock reckoned the night of the winter solstice to be 12 hours 42 minutes, whereas it is 13 hours 34 minutes—fifty-two minutes too short.

On the summer solstice, the longest day, the clock anticipated a day of 12 hours 48 minutes, whereas it is 13 hours and 41 minutes, and a night of 11 hours 12 minutes, whereas it is 10 hours 19 minutes.

On the vernal and autumnal equinoxes the day is 11 hours and 56 minutes long, and the clock actually shows 11 hours and 56 minutes; the night is 12 hours 4 minutes long, and the clock shows exactly 12 hours 4 minutes.

The difference between the present values and the values

[1] L. Borchardt, *Die altägyptische Zeitrechnung* (1920), pp. 6-25.

of the day for which the clock is adjusted is very consistent: on the winter solstice the day of the clock is fifty-two minutes longer than the present day of the winter solstice in Karnak, and the night is fifty-two minutes shorter; on the summer solstice the day is fifty-three minutes shorter on the clock and the night fifty-three minutes longer.

The figures on the clock show a smaller difference between the length of daylight on the solstices or between the longest and the shortest days of the year than is observed at Karnak at the present time. Thus the water clock of Amenhotep III, if it was correctly built and correctly interpreted, indicates that either Thebes was closer to the equator or that the inclination of the equator toward the ecliptic was less than the present angle of $23\frac{1}{2}°$. In either case the climate of the latitudes of Egypt could not have been the same as it is in our age.

As we find from the present research, the clock of Amenhotep III became obsolete in the middle of the eighth century; and the clock that might have replaced it at that time would have been made obsolete in the catastrophes of the end of the eighth and the beginning of the seventh centuries, when once more the axis changed its direction in the sky and its position on the globe as well.

## A HEMISPHERE TRAVELS SOUTHWARD

> "Behold the world bowing with its massive dome—earth and expanse of sea and heaven's depth!"
>
> —VIRGIL *Eclogues* iv. 50

The change in the position of the poles carried the polar ice outside the new polar circle, while other regions were brought into the polar circle. There is nothing imperative in the present position of the pole or in the direction of the polar axis. No known astronomical or geological law requires the present direction of the axis and the present position of the pole. I find a similar thought in the writings of Schiaparelli: "The permanence of the geographical poles in the very same regions of the Earth cannot yet be considered as incontestably established by astronomical or mechanical arguments. Such permanence may be a fact today, but it remains a matter still to be proven for the preceding ages of the history of the globe." "Our problem, so important from the astronomical and mathematical standpoint, touches the foundations of geology and paleontology; its solution is tied to the [problem of the] most grandiose events in the history of the Earth."[1]

The present pole was not always the terrestrial pole, nor did the changes occur in a slow process. The glacial sheet was a polar cover; the ice ages terminated with catastrophic suddenness; regions of mild climate moved instantly into the polar circle; the ice sheet in America and Europe started to melt; great quantities of vapour rising from the surface of the oceans caused increased precipitation and the formation of a new ice cover. Gigantic waves that travelled across continents, more than the movement of the ice, were responsible for the drift, especially in the north, and for the boulders that were carried long distances and placed atop unrelated formations.

If we look at the distribution of the ice sheet in the Northern Hemisphere, we see that a circle, with its centre somewhere near the east shore of Greenland or in the strait between Greenland and Baffin Land near the present north magnetic pole, and a radius of about 3,600 kilometres, embraces the region of the ice sheet of the last glacial age. Northeastern Siberia is outside the circle; the valley of the Missouri down to 39° north latitude is within the circle. The eastern part of Alaska is included, but not its western part. Northwestern Europe is well within the circle; some distance behind the Ural Mountains, the line curves toward the north and crosses the present polar circle.

Now we reflect: Was not the North Pole at some time in the past 20° or more distant from the point it now occupies—and closer to America? In like manner, the old South Pole would have been roughly the same 20° from the present pole.[2]

The Brahman charts of the sky show a large difference from what modern astronomers would expect to find. Calcutta being removed 180° longitude from Baffin Land, the Brahman charts would rather correspond to a position of the earth in which the axis would pierce the globe at Baffin Land, close to the present magnetic pole. The change in latitude of other regions to the west and to the east of India would have been smaller.

[1] G. V. Schiaparelli, *De la rotation de la terre sous l'influence des actions géologiques* (St. Petersburg, 1889), p. 31.
[2] In the direction of Queen Mary Land of the Antarctic continent.

It is probable that twenty-seven centuries ago, or perhaps thirty-five, the present North Pole was at Baffin Land or close to the Boothia Felix Peninsula of the American mainland.

The sudden extermination of mammoths was caused by a catastrophe and probably resulted from asphyxiation or electrocution. The immediately subsequent movement of the Siberian continent into the polar region is probably responsible for the preservation of the corpses.[1]

It appears that the mammoths, along with other animals, were killed by a tempest of gases accompanied by a spontaneous lack of oxygen caused by fires raging high in the atmosphere. A few instants later their dying or dead bodies were moving into the polar circle. In a few hours northeastern America moved from the frigid zone of the polar circle into a moderate zone; northeastern Siberia moved in the opposite direction from the moderate zone to the polar circle. The present cold climate of northern Siberia started when the glacial age in Europe and America came to a sudden end.

It is assumed here that in historical times neither northeastern Siberia nor western Alaska were in the polar regions, but that as a result of the catastrophes of the eighth and seventh centuries this area moved into that region. This assumption implies that these lands, to the extent that they were not covered by the sea, were most probably places of human habitation. Archaeological work should be undertaken in northeastern Siberia with the purpose of establishing whether these now uninhabited tundras were sites of culture twenty-seven centuries ago.

In 1939 and 1940 "one of the most startling and important finds of the century" (E. Stefansson) was made at Point Hope in Alaska, on the shores of Bering Strait: an ancient city of about eight hundred houses, whose population had been larger than that of the modern city of Fairbanks, was discovered there, north of 68°, about 130 miles within the Arctic Circle.[2]

"Ipiutak, as the location of this ancient city is called by the present Eskimos, must have been built before the Christian era; two thousand years is thought a conservative estimate of

[1] Greek authors referred to the mummifying quality of ambrosia; they described the process of pouring the fluid ambrosia into the noses of the dead; this was the process used by the Egyptians also in applying their drugs for mummification; the Babylonians used honey for that purpose.

[2] By F. G. Rainey and his colleagues under the sponsorship of the American Museum of Natural History in New York; the results of their expedition were published in the anthropological papers of the museum.

its age. The excavations have yielded beautiful ivory carvings unlike any known Eskimo or other American Indian culture of the northern regions. Fashioned of logs, the strange tombs gave up skeletons which stared up at the excavators with artificial eyeballs carved of ivory and inlaid with jet. . . . Numerous delicately made and engraved implements, also found in the graves, resembled some of those produced in North China two or three thousand years ago; others resemble carvings of the Ainu peoples in northern Japan and the natives of the Amur River in Siberia. The material culture of these people was not a simple one, of the kind usually found in the Arctic, but elaborate and that of a sophisticated people, in this sense more advanced than any known Eskimos, and clearly derived from eastern Asia."[1]

In Central Alaska, where the ground has been frozen for many centuries, animals with flesh still attached to their bones have been excavated. "Bones of extinct as well as living species of mammals have been found in most of the regions. . . . They remain not as fossilized bones but in a frozen state, and in some cases, ligaments, skin, and flesh adhere to the bones."[2] During the season of 1938, "almost the entire skin of a super-bison, the hair remaining," was found in the Fairbanks area.

"Some of the artifacts found after the stripping at depths of 18 to 20 metres below the original surface may have been on or near the surface originally, but the position of others tends to associate them with extinct animal bones at great depths. The recognizable artifacts are implements of chipped stone, bone and ivory."[3]

In 1936–1937, in a small area designated as Ester, several implements were found, as well as numerous burned stones, in association with mammoth, mastodon, bison, and horse bones, at the bottom of the muck deposits in Ester Creek, some twenty metres below the original surface.[4] In 1938 similar finds were made at Engineer Creek at the bottom of the muck, forty metres below the original surface of the soil.[5]

These vestiges of life and culture far beneath the surface of the ground are, for the most part, remnants buried in catastrophes prior to that described in the present chapter; among

[1] Description by Evelyn Stefansson in her book, *Here Is Alaska* (1943), pp. 138 ff.

[2] F. G. Rainey, "Archaeology in Central Alaska," *Anthropological Papers of the Museum of Natural History*, XXXV, Pt. IV (1939), 391 ff.

[3] *Ibid.*, p. 393.

[4] By P. Maas.

[5] By J. L. Giddings.

them are also remains of culture and life engulfed in the cataclysms of the eighth and seventh centuries. When the earth's rotation was disturbed, waves of translation moved eastward, because of inertia, and poleward, because of the recession of the waters from the equatorial bulge where they are held by the rotation of the earth. Thus Alaska must have been swept by waters from the Pacific.

Towns similar to those unearthed in Alaska, and possibly larger ones, will most likely be found in Kamchatka, or farther to the north on the Koluma or Lena rivers flowing into the Arctic Ocean. The conditions that preserved mammoths with flesh and skin on their bones must have had the same effect on human beings, and it is not excluded that human bodies encased in ice will be found, too.

A problem the archaeologists will have to solve is that of clarifying whether the extermination of life in these regions of northwest America and northeast Asia, resulting in the death of mammoths, took place in the eighth and seventh or fifteenth century before the present era (or earlier)—in other words, whether the herds of mammoths were annihilated in the days of Isaiah or in the days of the Exodus.

CHAPTER VIII

# THE YEAR OF 360 DAYS

PRIOR TO THE last series of cataclysms, when, as we assume, the globe spun on an axis pointed in a different direction in space, with its poles at a different location, on a different orbit, the year could not have been the same as it has been since.

Numerous evidences are preserved which prove that prior to the year of 365¼ days, the year was only 360 days long. Nor was that year of 360 days primordial; it was a transitional form between a year of still fewer days and the present year.

In the period of time between the last of the series of catastrophes of the fifteenth century and the first in the series of catastrophes of the eighth century, the duration of a seasonal revolution appears to have been 360 days.[1]

In order to substantiate my statement, I invite the reader on a world-wide journey. We start in India.

The texts of the *Veda* period know a year of only 360 days. "All Veda texts speak uniformly and exclusively of a year of 360 days. Passages in which this length of the year is directly stated are found in all the Brahmanas."[2] "It is striking that the Vedas nowhere mention an intercalary period, and while repeatedly stating that the year consists of 360 days, nowhere refer to the five or six days that actually are a part of the solar year."[3]

This Hindu year of 360 days is divided into twelve months of thirty days each.[4] The texts describe the moon as crescent for fifteen days and waning for another fifteen days; they also say that the sun moved for six months or 180 days to the north and for the same number of days to the south.

The perplexity of scholars at such data in the Brahmanic literature is expressed in the following sentence: "That these

[1] W. Whiston, in *New Theory of the Earth* (1696), expressed his belief that before the Deluge the year was composed of 360 days. He found references in classic authors to a year of 360 days, and as he recognized only one major catastrophe, the Deluge, he related these references to the antediluvian era.

[2] Thibaut, "Astronomie, Astrologie und Mathematik," *Grundriss der indo-arischen Philologie und Alterthumskunde* (1899), III, 7.

[3] *Ibid.*

[4] *Ibid.*

are not conventional inexact data, but definitely wrong notions, is shown by the passage in *Nidana-Sutra*, which says that the sun remains 13½ days in each of the 27 *Naksatras*, and thus the actual solar year is calculated as 360 days long. "Fifteen days are assigned to each half-moon period; that this is too much is nowhere admitted." [1]

In their astronomical works, the Brahmans used very ingenious geometric methods, and their failure to discern that the year of 360 days was 5¼ days too short seemed baffling. In ten years such a mistake accumulates to fifty-two days. The author whom I quoted last was forced to conclude that the Brahmans had a "wholly confused notion of the true length of the year." Only in a later period, he said, were the Hindus able to deal with such obvious facts. To the same effect wrote another German author: "The fact that a long period of time was necessary to arrive at the formulation of the 365-day year is proved by the existence of the old Hindu 360-day Savana-year and of other forms which appear in the Veda literature." [2]

Here is a passage from the *Aryabhatiya*, an old Indian work on mathematics and astronomy: "A year consists of twelve months. A month consists of 30 days. A day consists of 60 nadis. A nadi consists of 60 vinadikas." [3]

A month of thirty days and a year of 360 days formed the basis of early Hindu chronology used in historical computations.

The Brahmans were aware that the length of the year, of the month, and of the day changed with every new world age. The following is a passage from *Surya-siddhanta*, a classic of Hindu astronomy. After an introduction, it proceeds: "Only by reason of the revolution of the ages, there is here a difference of times." [4] The translator of this ancient manual supplied an annotation to these words: "According to the commentary, the meaning of these last verses is that in successive Great Ages . . . there were slight differences in the motion of the heavenly bodies." Explaining the term *bija*, which means a correction of time in every new age, the book of *Surya* says "time is the destroyer of the worlds."

[1] Thibaut, "Astronomie, Astrologie und Mathematic," *Grundriss der indo-arischen Philologie und Alterthumskunde* (1899), III, 7.

[2] F. K. Ginzel, "Chronologie," *Encyklopädie der mathematischen Wissenschaften* (1904-1935), Vol. VI.

[3] *The Aryabhatiya of Aryabhata*, an ancient Indian work on mathematics and astronomy (transl. W. E. Clark, 1930), Chap. 3, "Kalakriya or the Reckoning of Time," p. 51.

[4] *Surya-siddhanta: A Text Book of Hindu Astronomy* (transl. Ebenezer Burgess, 1860).

The sacerdotal year, like the secular year of the calendar, consisted of 360 days composing twelve lunar months of thirty days each. From approximately the seventh pre-Christian century on, the year of the Hindus became 365¼ days long, but for temple purposes the old year of 360 days was also observed, and this year is called *savana*.

When the Hindu calendar acquired a year of 365¼ days and a lunar month of twenty-nine and a half days, the older system was not discarded. "The natural month, containing about twenty-nine and a half days mean solar time, is then divided into thirty lunar days (*tithi*), and this division, although of so unnatural and arbitrary a character, the lunar days beginning and ending at any moment of the natural day and night, is, to the Hindu, of the most prominent practical importance, since by it are regulated the performances of many religious ceremonies, and upon it depend the chief considerations of propitious and unpropitious times, and the like."[1]

The double system was the imposition of a new time measure upon the old.

The ancient Persian year was composed of 360 days or twelve months of thirty days each. In the seventh century five *Gatha* days were added to the calendar.[2]

In the *Bundahis*, a sacred book of the Persians, the 180 successive appearances of the sun from the winter solstice to the summer solstice and from the summer solstice to the next winter solstice are described in these words: "There are a hundred and eighty apertures [*rogin*] in the east, and a hundred and eighty in the west . . . and the sun, every day, comes in through an aperture, and goes out through an aperture. . . . It comes back to Varak, in three hundred and sixty days and five Gatha days."[3]

Gatha days are "five supplementary days added to the last of the twelve months of thirty days each, to complete the year; for these days no additional apertures are provided . . . . This arrangement seems to indicate that the idea of the apertures is older than the rectification of the calendar

[1] *Surya-siddhanta: A Text Book of Hindu Astronomy* (transl. Ebenezer Burgess, 1860), comment by Burgess in note to p. 7.
[2] "Twelve months . . . of thirty days each . . . and the five Gatha-days at the end of the year." "The Book of Denkart," in H. S. Nyberg, *Texte zum mazdayasnischen Kalender* (Uppsala, 1934), p. 9.
[3] *Bundahis* (transl. West), Chap. V.

which added the five Gatha days to an original year of 360 days."[1]

The old Babylonian year was composed of 360 days.[2] The astronomical tablets from the period antedating the Neo-Babylonian Empire compute the year at so many days, without mention of additional days. That the ancient Babylonian year had only 360 days was known before the cuneiform script was deciphered: Ctesias wrote that the walls of Babylon were 360 furlongs in compass, "as many as there had been days in the year."[3]

The zodiac of the Babylonians was divided into thirty-six decans, a decan being the space the sun covered in relation to fixed stars during a ten-day period. "However, the 36 decans with their decades require a year of only 360 days."[4] To explain this apparently arbitrary length of the zodiacal path, the following conjecture was made: "At first the astronomers of Babylon recognized a year of 360 days, and the division of a circle into 360 degrees must have indicated the path traversed by the sun each day in its assumed circling of the earth."[5] This left over five degrees of the zodiac unaccounted for.

The old Babylonian year consisted of twelve months of thirty days each, the months being computed from the time of the appearance of the new moon. As the period between one new moon and another is about twenty-nine and a half days, students of the Babylonian calendar face the perplexity with which we are already familiar in other countries. "Months of thirty days began with the light of the new moon. How agreement with astronomical reality was effected, we do not know. The practice of an intercalary period is not yet known."[6] It appears that in the seventh century five days were added to the Babylonian calendar; they were regarded as unpropitious, and people had a superstitious awe of them.

The Assyrian year consisted of 360 days; a decade was called a *sarus*; a *sarus* consisted of 3,600 days.[7]

[1] Note by West on p. 24 of his translation of the *Bundahis*.
[2] A. Jeremias, *Das Alter der babylonischen Astronomie* (2nd ed., 1909), pp. 58 ff.
[3] *The Fragments of the Persika of Ktesias* (Ctesiae Persica), ed. J. Gilmore (1888), p. 38; Diodorus ii. 7.
[4] W. Gundel, *Dekane und Dekansternbilder* (1936), p. 253.
[5] Cantor, *Vorlesungen über Geschichte der Mathematik*, I, 92.
[6] "Sin" in Roscher, *Lexikon der griech. und röm. Mythologie*, Col. 892.
[7] Georgius Syncellus, ed. Jacob Goar (Paris, 1652), pp. 17, 32.

"The Assyrians, like the Babylonians, had a year composed of lunar months, and it seems that the object of astrological reports which relate to the appearance of the moon and sun was to help to determine and foretell the length of the lunar month. If this be so, the year in common use throughout Assyria must have been lunar. The calendar assigns to each month thirty full days; the lunar month is, however, little more than twenty-nine and a half days."[1] "It would hardly be possible for the calendar month and the lunar month to correspond so exactly at the end of the year."[2]

Assyrian documents refer to months of thirty days only, and count such months *from crescent to crescent.*[3] Again, as in other countries, it is explicitly the *lunar month* that is computed by the Assyrian astronomers as equal to thirty days. How could the Assyrian astronomers have adjusted the length of the lunar months to the revolutions of the moon, modern scholars ask themselves, and how could the observations reported to the royal palace by the astronomers have been so consistently erroneous?

The month of the Israelites, from the fifteenth to the eighth century before the present era, was equal to thirty days, and twelve months comprised a year; there is no mention of months shorter than thirty days, nor of a year longer than twelve months. That the month was composed of thirty days is evidenced by Deuteronomy 34 : 8 and 21 : 13, and Numbers 20 : 29, where mourning for the dead is ordered for a "full month," and is carried on for thirty days. The story of the Flood, as given in Genesis, reckons in months of thirty days; it says that one hundred and fifty days passed between the seventeenth day of the second month and the seventeenth day of the seventh month.[4] The composition of this text apparently dates from the time between the Exodus and the upheaval of the days of Uzziah.[5]

The Hebrews observed lunar months. This is attested to by

[1] R. C. Thompson, *The Reports of the Magicians and Astrologers of Nineveh and Babylon in the British Museum*, II (1900), xix.
[2] *Ibid.*, p. xx.
[3] Langdon and Fotheringham, *The Venus Tablets of Ammizaduga*, pp. 45-46; C. H. W. Johns, *Assyrian Deeds and Documents*, IV (1923), 333; J. Kohler and A. Ungnad, *Assyrische Rechtsurkunden* (1913), 258, 3; 263, 5; 649, 5.
[4] Genesis 7 : 11 and 24; 8 : 4.
[5] The other variant of the story of the Flood (Genesis 7 : 17; 8 : 6) has the Deluge lasting 40 days instead of 150.

the fact that the new-moon festivals were of great importance in the days of Judges and Kings.[1] "The new moon festival anciently stood at least on a level with that of the Sabbath."[2] As these (lunar) months were thirty days long, with no months of twenty-nine days in between, and as the year was composed of twelve such months, with no additional days or intercalated months, the Bible exegetes could find no way of reconciling the three figures: 354 days, or twelve lunar months of twenty-nine and a half days each; 360 days, or a multiplex of twelve times thirty; and 365¼ days, the present length of the year.

The Egyptian year was composed of 360 days before it became 365 by the addition of five days. The calendar of the Ebers Papyrus, a document of the New Kingdom, has a year of twelve months of thirty days each.[3]

In the ninth year of King Ptolemy Euergetes, or -238, a reform party among the Egyptian priests met at Canopus and drew up a decree; in 1866 it was discovered at Tanis in the Delta, inscribed on a tablet. The purpose of the decree was to harmonize the calendar with the seasons "according to the present arrangement of the world," as the text states. One day was ordered to be added every four years to the "three hundred and sixty days, and to the five days which were afterwards ordered to be added."[4]

The authors of the decree did not specify the particular date on which the five days were added to the 360 days, but they do say clearly that such a reform was instituted on some date after the period when the year was only 360 days long.

On a previous page I referred to the fact that the calendar of 360 days was introduced in Egypt only after the close of the Middle Kingdom, in the days of the Hyksos. The five epigomena must have been added to the 360 days subsequent to the end of the Eighteenth Dynasty. We have no mention of "five days" in all the numerous inscriptions of the Eighteenth Dynasty; the epigomena or, as the Egyptians called them, "the five days which are above the year,"[5] are known from the

[1] I Samuel 20 : 5-6; II Kings 4 : 23; Amos 8 : 5; Isaiah 1 : 13; Hosea 2 : 11; Ezekiel 46 : 1, 3.

In the Bible the month is called *hodesh*, or "the new (moon)," which testifies to a lunation of thirty days.

[2] J. Wellhausen, *Prolegomena to the History of Israel* (1885), p. 113.

[3] Cf. G. Legge in *Recueil de travaux relatifs à la philologie et à l'archéologie égyptiennes et assyriennes* (La Mission française du Caire, 1909).

[4] S. Sharpe, *The Decree of Canopus* (1870).

[5] E. Meyer, "Ägyptische Chronologie," *Philos. und hist. Abhandlungen der Preuss. Akademie der Wissenschaften* (1904), p. 8.

documents of the seventh and following centuries. The pharaohs of the late dynasties used to write: "The year and the five days." The last day of the year was celebrated, not on the last of the epagomena, but on the thirtieth of Mesori, the twelfth month."[1]

In the fifth century Herodotus wrote: "The Egyptians, reckoning thirty days to each of the twelve months, add five days in every year over and above the number, and so the completed circle of seasons is made to agree with the calendar."[2]

The *Book of Sothis*, erroneously ascribed to the Egyptian priest Manetho,[3] and Georgius Syncellus, the Byzantine chronologist,[4] maintain that originally the additional five days did not follow the 360 days of the calendar, but were introduced at a later date,[5] which is corroborated by the text of the Canopus Decree.

That the introduction of epagomena was not the result of progress in astronomical knowledge, but was caused by an actual change in the planetary movements, is implied in the Canopus Decree, for it refers to "the amendment of the faults of the heaven." In his *Isis and Osiris*[6] Plutarch describes by means of an allegory the change in the length of the year: "Hermes playing at draughts with the moon, won from her the seventieth part of each of her periods of illumination, and from all the winnings he composed five days, and intercalated them as an addition to the 360 days." Plutarch informs us also that one of these epagomena days was regarded as inauspicious; no business was transacted on that day, and even kings "would not attend to their bodies until nightfall."

The new-moon festivals were very important in the days of the Eighteenth Dynasty. On all the numerous inscriptions of that period, wherever the months are mentioned, they are reckoned as thirty days long. The fact that the new-moon festivals were observed at thirty-day intervals implies that the lunar month was of that duration.

Recapitulating, we find concordant data. The Canopus Decree states that at some period in the past the Egyptian year was only 360 days long, and that the five days were added at

[1] E. Meyer, "Agyptische Chronologie," *Philos. und hist. Abhandlungen der Preuss. Akademie der Wissenschaften* (1904), p. 8.
[2] Herodotus, *History*, Bk. ii. 4 (transl. A. D. Godley).
[3] See volume of Manetho in Loeb Classical Library.
[4] *Georgii Monachi Chronographia* (ed. P. Jacobi Goar, 1652), p. 123.
[5] In the days of the Hyksos King Aseth. But see the Section "Changes in the Times and the Seasons."
[6] Translated by F. C. Babbit.

some later date; the Ebers Papyrus shows that under the Eighteenth Dynasty the calendar had a year of 360 days divided into twelve months of thirty days each; other documents of this period also testify that the lunar month had thirty days, and that a new moon was observed twelve times in a period of 360 days. The Sothis book says that this 360-day year was established under the Hyksos, who ruled after the end of the Middle Kingdom, preceding the Eighteenth Dynasty.

In the eighth or seventh century the five epagomena days were added to the year under conditions which caused them to be regarded as unpropitious.

Although the change in the number of days in the year was calculated soon after it occurred, nevertheless, for some time many nations retained a civil year of 360 days divided into twelve months of thirty days each.

Cleobulus, who was counted among the seven sages of ancient Greece, in his famous allegory represents the year as divided into twelve months of thirty days: the father is one, the sons are twelve, and each of them has thirty daughters.[1]

From the days of Thales, another of the seven sages, who could predict an eclipse, the Hellenes knew that the year consists of 365 days; Thales was regarded by them as the man who discovered the number of days in the year. As he was born in the seventh century, it is not impossible that he was one of the first among the Greeks to learn the new length of the year; it was in the beginning of that century that the year achieved its present length. A contemporary of Thales and also one of the seven sages, Solon was regarded as the first among the Greeks to find that a lunar month is less than thirty days.[2] Despite their knowledge of the correct measure of the year and the month, the Greeks, after Solon and Thales, continued to keep to the obsolete calendar, a fact for which we have the testimony of Hippocrates ("Seven years contain 360 weeks"), Xenophon, Aristotle, and Pliny.[3] The persistence of reckoning by 360 days is accounted for not only by a certain reverence for the earlier astronomical year, but also its convenience for every computation.

[1] See Diogenes Laërtius, *Lives of Eminent Philosophers*, "Life of Thales."

[2] Proclus, *The Commentaries on the Timaeus of Plato* (1820); Diogenes Laërtius, *Lives*, "Life of Solon"; Plutarch, *Lives*, "Life of Solon."

[3] Aristotle, *Historia animalium*, vi. 20; Pliny, *Natural History*, xxxiv. 12 (transl. Bostock and Riley).

The ancient Romans also reckoned 360 days to the year. Plutarch wrote in his "Life of Numa" that in the time of Romulus, in the eighth century, the Romans had a year of 360 days only.[1] Various Latin authors say that the ancient month was composed of thirty days.[2]

On the other side of the ocean, the Mayan year consisted of 360 days; later five days were added, and the year was then a *tun* (360-day period) and five days; every fourth year another day was added to the year. "They did reckon them apart, and called them the days of nothing: during the which the people did not anything," wrote J. de Acosta, an early writer on America.[3]

Friar Diego de Landa, in his *Yucatan before and after the Conquest*, wrote: "They had their perfect year like ours, of 365 days and six hours, which they divided into months in two ways. In the first the months were of 30 days and were called *U* which signifies the moon, and they counted from the rising of the new moon until it disappeared."[4] The other method of reckoning, by months of twenty days' duration (*uinal hunekeh*), reflects a much older system, to which I shall return when I examine more archaic systems than that of the 360-day year. De Landa also wrote that the five supplementary days were regarded as "sinister and unlucky." They were called "days without name."[5] Although the Mexicans at the time of the conquest called a thirty-day period "a moon," they knew that the synodical period is 29.5209 days,[6] which is more exact than the Gregorian calendar introduced in Europe ninety years after the discovery of America. Obviously, they adhered to an old tradition dating from the time when the year had twelve months of thirty days each, 360 days in all.[7]

In ancient South America also the year consisted of 360 days, divided into twelve months.

"The Peruvian year was divided into twelve Quilla, or

[1] Plutarch, *Lives*, "The Life of Numa," xviii.
[2] Cf. Geminus, *Elementa astronomiae*, viii; cf. also Cleomedes, *De motu circulari corporum celestium*, xi. 4.
[3] J. de Acosta, *The Natural and Moral Histories of the Indies*, 1880 (Historia natural y moral de las Indias, Seville, 1590).
[4] Diego de Landa, *Yucatan*, p. 59.
[5] D. G. Brinton, *The Maya Chronicles* (1882).
[6] Gates' note to De Landa, *Yucatan*, p. 59.
[7] R. C. E. Long, "Chronology–Maya," *Encyclopaedia Britannica* (14th ed.): "They [the Mayas] never used a year of 365 days in counting the distance of time from one date to another."

moons of thirty days. Five days were added at the end, called Allcacanquis."[1] Thereafter, a day was added every four years to keep the calendar correct.

We cross the Pacific Ocean and return to Asia. The calendar of the peoples of China had a year of 360 days divided into twelve months of thirty days each.[2]

A relic of the system of 360 days is the still persisting division of the sphere into 360 degrees; each degree represented the diurnal advance of the earth on its orbit, or that position of the zodiac which was passed over from one night to the next. After 360 changes the stellar sky returned to the same position for the observer on the earth.

When the year changed from 360 to 365¼ days, the Chinese added five and a quarter days to their year, calling this additional period *Khe-ying;* they also began to divide a sphere into 365¼ degrees, adopting the new year-length not only in the calendar, but also in celestial and terrestrial geometry.[3]

Ancient Chinese time reckoning was based on a coefficient of sixty; so also in India, Mexico, and Chaldea, sixty being the universal coefficient.

The division of the year into 360 days was honoured in many ways,[4] and, indeed, it became an incentive to progress in astronomy and geometry, so that people did not readily discard this method of reckoning when it became obsolete. They retained their "moons" of thirty days, though the lunar month

[1] Markham, *The Incas of Peru*, p. 117.

[2] Joseph Scaliger, *Opus de emendatione temporum*, p. 225; W. Hales, *New Analysis of Chronology* (1809-1812), I, 31; W. D. Medhurst, notes to pp. 405-406 of his translation of *The Shoo King* (Shanghai, 1846).

[3] H. Murray, J. Crawfurd, and others, *An Historical and Descriptive Account of China* (p. 235); *The Chinese Classics*, III, Pt. 2, ed. Legge (Shanghai, 1865), note to p. 21.

Cf. also Cantor, *Vorlesungen*, p. 92. "Zuerst wurde von den Astronomen Babylons das Jahr von 360 Tagen erkannt, und die Kreisteilung in 360 Grade sollte den Weg versinnlichen welchen die Sonne bei ihrem vermeintlichen Umlaufe um die Erde jeden Tag zurücklegte."

[4] C. F. Dupuis (*L'Origine de tous les cultes* [1835-1836], the English compendium being *The Origin of All Religious Worship* [1872], p. 41) gathered material on the number 360, "which is that of the days of the year without the epigomena." He refers to the 360 gods in the "theology of Orpheus," to the 360 eons of the gnostic genii, to the 360 idols before the palace of Dairi in Japan, to 360 statues "surrounding that of Hobal," worshipped by the ancient Arabs, to the 360 genii who take possession of the soul after death, "according to the doctrine of the Christians of St. John," to the 360 temples built on the mountain of Lowham in China, and to the wall of 360 stadia "with which Semiramis surrounded the city" of Babylon. This material did not convey to its collector the idea that an astronomical year of 360 days had been the reason for the sacredness of the number 36.

in fact became shorter, and they regarded the five days as not belonging to the year.

All over the world we find that there was at some time the same calendar of 360 days, and that at some later date, about the seventh century before the present era, five days were added at the end of the year, as "days over the year," or "days of nothing."

Scholars who investigated the calendars of the Incas of Peru and the Mayas of Yucatan wondered at the calendar of 360 days; so did the scholars who studied the calendars of the Egyptians, Persians, Hindus, Chaldeans, Assyrians, Hebrews, Chinese, Greeks, or Romans. Most of them, while debating the problem in their own field, did not suspect that the same problem turned up in the calendar of every nation of antiquity.

Two matters appeared perplexing: a mistake of five and a quarter days in a year could certainly be traced, not only by astronomers, but even by analphabetic farmers, for in the short span of forty years—a period that a person could readily observe—the seasons would become displaced by more than two hundred days. The second perplexity concerns the length of a month. "It seems to have been a prevailing opinion among the ancients that a lunation or synodical month lasted thirty days."[1] In many documents of various peoples, it is said that the month, or the "moon," is equal to thirty days, and that the beginning of such a month coincides with the new moon.

Such declarations by ancient astronomers make it clear that there was no such thing as a conventional calendar with an admitted error; as a matter of fact, the existence of an international calendar in those days is extremely unlikely. After centuries of open sea lanes and international exchange of ideas, no uniform calendar for the whole world has as yet been devised: the Moslems have a lunar year, based on the movements of the moon, which is systematically adjusted every few years to the solar year by intercalation; many other creeds and peoples have systems of their own containing many vestiges of ancient systems. The reckoning of months as equal to thirty and thirty-one days is also a relic of older systems; the five supplementary days were divided among the old lunar months. But at present the almanac does not ascribe an interval of thirty days between two lunations or a period of 360 days for twelve lunations.

[1] Medhurst, *The Shoo King*.

The reason for the universal identity of time reckoning between the fifteenth and the eighth centuries lay in the actual movement of the earth on its axis and along its orbit, and in the revolution of the moon, during that historical period. The length of a lunar revolution must have been almost exactly 30 days, and the length of the year apparently did not vary from 360 days by more than a few hours.

Then a series of catastrophes occurred that changed the axis and the orbit of the earth and the orbit of the moon, and the ancient year, after going through a period marked by disarranged seasons, settled into a "slow-moving year" (Seneca) of 365 days, 5 hours, 48 minutes, 46 seconds, a lunar month being equal to 29 days, 12 hours, 44 minutes, 2.7 seconds, mean synodical period.

## DISARRANGED MONTHS

As a result of repeated perturbations, the earth changed from an orbit of 360 days' duration to one of 365¼ days, the days probably not being exactly equal in both cases. The month changed from thirty to twenty-nine and a half days. These were the values at the beginning and at the end of the century of "the battle of the gods." As a result of the perturbations of this century, there were intermediary values of the year and the month. The length of the year probably ranged between 360 and 365¼ days, but the moon, being a smaller (or weaker) body than the earth, suffered greater perturbations from the contacting body, and the intermediate values of the month could have been subjected to greater changes.

Plutarch declares that in the time of Romulus the people were "irrational and irregular in their fixing of the months," and reckoned some months at thirty-five days and some at more, "trying to keep to a year of 360 days," and that Numa, Romulus' successor, corrected the irregularities of the calendar and also changed the order of the months. This statement suggests the question: Might it not have been that during the period between consecutive catastrophes the moon receded to an orbit of thirty-five or thirty-six days' duration?

If, in the period of confusion, the moon actually changed for a while to such an orbit, it must have been an ellipse or a circle of a radius larger than before. In the latter case, each of the four moon phases must have been of nine days' duration. It is of interest, therefore, to read that in many sagas dealing with the moon, the number nine is used in measures of time.[1]

A series of scholars found that nine days was for a while a time period of many ancient peoples: the Hindus, the Persians,[2] the Babylonians,[3] the Egyptians,[4] and the Chinese.[5] In religious traditions, literature, and astrological works, seven days and nine days compete as the measure of the month's quarter.

In the time of the Homeric epics, the nine-day week became prevalent in the Greek world. The seven-day week and the nine-day week are both found in Homer.[6] The Romans, too, retained the recollection of a time when the week had been of nine days' duration.[7]

The change from a seven-day phase to a nine-day phase is found in the traditions of the peoples of Rumania, Lithuania, and Sardinia, and among the Celts of Europe, the Mongols of Asia, and the tribes of West Africa.[8]

In order to explain this strange phenomenon in time reckoning, obviously connected with the moon, the suggestion was made that, in addition to the seven-day phase of the moon, a nine-day phase was also observed, which is a third of the month.[9] But this idea must be rejected, because a third part of a month of twenty-nine and a half days would more nearly be ten days and not nine.[10] Besides, the quarter-month phases

[1] "The number nine occurs conspicuously in so many sagas which, for other reasons, I recognized to be moon sagas, that I am convinced that the holiness of this number has its origin in its very ancient application in time division." The author of this passage (E. Siecke, *Die Liebesgeschichte des Himmels, Untersuchungen zur indogermanischen Sagenkunde* [1892] did not suppose a change in the nature of the lunar cycles, and also was not aware of the work of the scholar referred to in the following footnote, yet he was forced to believe that nine was connected with a time subdivision of a month.

[2] A. Kaegi, "Die Neunzahl bei den Ostarien," in the volume dedicated to H. Schweizer-Sidler (1891).

[3] Kugler, "Die Symbolik der Neunzahl," *Babylonische Zeitordnung*, p. 192.

[4] E. Naville, *Transactions of the Society of Biblical Archaeology*, IV (1875), 1-18.

[5] Roscher, *Die enneadischen und hebdomadischen Fristen und Wochen*, Vol. XXI, No. 4, of *Abhandlungen der philol.-histor. Klasse der Kgl. sächs. Ges. der Wissenschaften* (1903).

[6] Roscher, *Die Sieben- und Neunzahl im Kultus und Mythus der Griechen, ibid.*, Vol. XXIV, No. 1 (1904): "Die beiden Arten von Fristen schon bei Homer und ebenso auch im ältesten Kultus nebeneinander vorkommen" (p. 54). "In der Zeit des älteren Epos herrschend gewordene 9-tägige Woche" (p. 73).

[7] Cf. Ovid, *Metamorphoses*, vii. 23 ff.; xiii. 951; xiv. 57.

[8] Roscher, *Die Sieben- und Neunzahl.*

[9] Roscher, *Fristen und Wochen.*

[10] The sidereal month, or the period of time during which the moon completes a revolution in relation to the fixed stars is 27 days, 7 hours, 43 minutes. But the phases of the moon change according to the synodical month of 29 days, 12 hours, 44 minutes; after a synodical month the moon returns to the same position in relation to the sun as viewed from the earth.

are easily observable periods during which the moon increases from new moon to half moon, to full moon, and then decreases accordingly; but a nine-day period falls between these phases.

Therefore, and in view of the vast material from many peoples, we conclude that at one time during the century of perturbations, for a period between two catastrophes, the moon receded to an orbit of thirty-five to thirty-six days' duration. It remained on such an orbit for a few decades until, at the next upheaval, it was carried to an orbit of twenty-nine and a half days' duration, on which it has proceeded since then.

These "perturbed months" occurred in the second half of the eighth century, at the beginning of Roman history.[1] What is more, we have actual dates like "the 33rd day of the month," cited in the Babylonian tablets of that period.[2]

Thus the month which was equal to thirty days, changed to thirty-six and then to twenty-nine and a half days. The last change was simultaneous with the change of the terrestrial orbit to one of 365¼ days' duration.

## YEARS OF TEN MONTHS

When the month was about thirty-six days and the year between 360 and 365¼ days, the year must have been composed of only ten months. This was the case.

According to many classical authors, in the days of Romulus the year consisted of ten months, and in the time of Numa, his successor, two months were added: January and February. Ovid writes: "When the founder of the city [Rome] was setting the calendar in order, he ordained that there should be twice five months in his year. . . . He gave his laws to regulate the year. The month of Mars was the first, and that of Venus the second. . . . But Numa overlooked not Janus and the ancestral shades [February] and so to the ancient months he prefixed two."[3]

Geminus, a Greek astronomer of the first century before the present era, says similarly that it was Romulus who (in the eighth century) established the year of ten months.[4] Aulus

[1] It was probably these changes that caused the gods in *The Clouds* of Aristophanes to accuse the moon of having brought disorder in the calendar and in the cult. Aristophanes, *The Clouds*, ll. 615 ff.

[2] Kugler, *Babylonische Zeitordnung*, p. 191, note.

[3] Ovid, *Fasti*, i. 27 ff.

[4] Geminus, "Introduction aux phénomènes" in Petau, *Uranologion* (1630).

Gellius, a second century author, writes in his *Attic Nights*: "The year was composed not of twelve months, but of ten."[1] Plutarch remarks that in his day there was a belief that the Romans, in the time of Romulus, computed the year "not in twelve months, but in ten, by adding more than thirty days to some of the months."[2] At the beginning of Numa's reign the ten-month year was still the official one.[3] "March was considered the first month until the reign of Numa, the full year before that time containing ten months," wrote Procopius of Caesarea, who lived in the closing years of the Roman Empire.[4] The fact that, in Romulus' time, the first month was named in honour of Mars and the second in honour of Venus shows the importance of these two deities in that period of history. July was named Quintilis (the fifth). The difference of two months still survives in the names September, October, November, and December, which denote the seventh, eighth, ninth, and tenth months, but according to present-day reckoning they are the ninth, tenth, eleventh, and twelfth months, respectively.

Not only was the year divided into fewer than twelve months, but also the zodiac, or the path of the sun and the moon across the firmament, at present consisting of twelve signs, at one time had eleven and at another time ten signs. A zodiac of fewer than twelve signs was employed by the astrologers of Babylonia, ancient Greece, and other countries.[5] A Jewish song in the Aramaic language which is included in the Seder Service refers to eleven constellations of the Zodiac.

The calendars of the primitive peoples disclose their early origin by the fact that many of them are composed of ten months, and some of eleven months. If the time of the lunar revolution was thirty-five days and some hours, the year was something over ten months long.

The Yurak Samoyeds reckon eleven months to the year.[6] The natives of Formosa, too, have a year of eleven months.[7] The year of the Kamchadals is made up of ten months, "one of

[1] Aulus Gellius, *Noctes Atticae*, iii. 16.

[2] Plutarch, *The Roman Questions*, xix.

[3] Eutropius, *Brevarium rerum romanorum*, i. 3 says: "Numa Pompilius divided the year into ten months." This must refer to the beginning of Numa's reign, when the calendar of Romulus was still valid.

[4] Procopius of Caesarea, *History of the Wars*, Bk. V, "The Gothic War" (transl. H. B. Dewing, 1919), Sec. 31.

[5] Boll, *Sternglaube und Sterndeutung*, p. 92; A. del Mar, *The Worship of Augustus Caesar*, pp. 6, 11, with references to Ovid, Virgil, Pliny, Servius, and Hyginus.

[6] M. P. Nilsson, *Primitive Time-Reckoning* (1920), p. 89.

[7] A. Wirth, "The Aborigines of Formosa," *The American Anthropologist*, 1897.

which is said to be as long as three."[1] The inhabitants of the Kingsmill Islands in the Pacific, also called the Gilbert Islands, near the equator, use a ten-month period for their year.[2] In the Marquesas (in Polynesia south of the equator) ten months form a year (*tau* or *puni*), but the actual year of 365 days is also known.[3]

The Toradja of the Dutch East Indies compute time in moon-months. Each year, however, a period of two or three months is not brought into the computation at all, and is omitted in time reckoning.[4]

The Chams of Indo-China have a calendar of only ten months to the year.[5] The natives in some islands of the Indian Ocean also observe ten months to the year.[6]

The aborigines of New Zealand do not count two months in the year. "These two months are not in the calendar: they do not reckon them; nor are they in any way accounted for."[7]

"Among the Yoruba of South Nigeria the three months—February, March, April—are generally given no specific name."[8]

These calendars of primitive peoples are similar to the old Roman calendar. They were not invented in disregard of the solar year ("Years with less than twelve months are to us the strangest of phenomena"[9]); their fault is that they are more constant than the revolution of the earth on her orbit around the sun. The work of adapting the old systems to a new order is still evident in the systems of the aborigines of Kamchatka, South Nigeria, the Dutch East Indies, and New Zealand. Instead of introducing two additional months, as in the reform of Numa, one of the months is extended to triple its length, or a period equivalent to two months is not counted at all in the calendric system.

The abundance of proofs of the existence of a ten-month year is even embarrassing. Since the period when the year was composed of ten months of thirty-five to thirty-six days each

[1] A Schiefner, *Bulletin de l'Académie de St. Petersbourg, Hist.-phil. Cl.*, XIV (1857), 198, 201 f.
[2] H. Hale, *Ethnography and Philology: U.S. Exploring Expedition*, 1838-42, VI (1846), 106, 170.
[3] G. Mathias, *Lettres sur les Isles Marquises* (1843), 211.
[4] N. Adriani and A. C. Kruijt, *De Bare'e-sprekende Toradja's* (1912-1914), II, 264.
[5] Frazer, *Ovid's Fasti* (1931), p. 386. [6] *Ibid.*
[7] W. Yate (English missionary in the early part of the nineteenth century), quoted in Frazer, *Ovid's Fasti*, p. 386
[8] *Ibid.*
[9] Nilsson, *Primitive Time-Reckoning*, p. 89.

was short, how could this ten-month year leave so many vestiges in the calendar systems all over the world? The answer to this question will become simple when we shall find that this was the second time in the history of the world that the year was composed of ten months. In a much earlier age, when the year was of an entirely different length, one revolution of the earth was also equal to ten revolutions of the moon. We shall trace this period in history in a succeeding volume of this work.

## THE REFORMING OF THE CALENDAR

In the middle of the eighth century the calendar then in use became obsolete. From the year −747 until the last of the catastrophes on the twenty-third of March, −687, the solar and lunar movements changed repeatedly, necessitating adjustments of the calendar. Reforms undertaken during this time soon became obsolete in their turn, and were replaced by new ones; only after the last catastrophe of −687, when the present world order was established, did the calendar become permanent.

Some of the clay tablets of Nineveh found in the royal library of that city[1] contain astronomical observations made during the period before the present order in the planetary system was established. One tablet fixes the day of the vernal equinox as the sixth of Nisan: "On the sixth of the month Nisan, the day and night are equal." But another tablet places the equinox on the fifteenth of Nisan. "We cannot explain the difference," wrote a scholar.[2] Judging by the accurate methods employed and the precision achieved in their observations, the stargazers of Nineveh would not have erred by nine days.

In the astronomical tablets of Nineveh "three systems of planets" are extensively represented; single planets are followed in all their movements in three different schedules. For the movements of the moon there are two different systems.[3] Each of these systems is carried out down to the smallest detail, but only the last system of the planets and of the moon conforms to the present world order.

[1] The palace of Nineveh was the residence of Sargon II, Sennacherib, Esarhaddon and Assurbanipal.

[2] J. Menant, *La Bibliothèque du palais de Ninive* (1880), p. 100.

[3] Kugler, *Die babylonische Mondrechnung: Zwei Systeme der Chaldäer über den Lauf des Mondes und der Sonne*, pp. 207-209.

According to Tablet No. 93, the perihelion, or the point on the earth's orbit that is nearest the sun, is defined as the twentieth degree of the sign of the zodiac called the Archer; at aphelion, when the earth is farthest from the sun, the sun is said to be at the twentieth degree of Gemini. Accordingly, these points are designated as stations of the fastest and slowest solar motion. "But the real position of the apsides decidedly contradicts these statements."[1] Another tablet, No. 272, seventy years younger than the first, gives very different data for the perihelion and aphelion, and scholars wonder at this.

All the numerous data on solar movements in one of the systems lead to one and the same conclusion. "The solstitial and equinoctial points of the ecliptic lay 6° too far to the east."[2]

"The distances travelled by the moon on the Chaldean ecliptic from one new moon to the next are, according to Tablet No. 272, on the average 3° 14′ too great."[3] This means that during a lunar month the moon moved a greater distance in relation to the fixed stars than present observation shows.

In Tablet No. 32, the movement of the sun along the zodiac is precisely calculated in degrees, and the station of the sun at the beginning of each lunar month is determined exactly; but it is "a perplexing presentation of the ununiform movement of the sun. The question is insistent: Why is it that the Babylonians formulated the nonuniformity of the solar movement precisely in this way?"[4]

As the various systems recorded in the astronomical tablets of Nineveh show, the world order changed repeatedly in the course of a single century. Hence, the Chaldean astronomers had the task of repeatedly readjusting the calendar. "From certain passages in the astrological tablets it is easy to see that the calculation of times and seasons was one of the chief duties of the astrologers in Mesopotamia." [5] The scholars ask: How could these men, employed for that very purpose, have made the egregious mistakes recorded in the tablets, and carried these mistakes over into systems in which the movements of the sun, the moon, and the five planets were recorded with repetitions at regular intervals, these movements and intervals being

[1] Kugler, *Die babylonische Mondrechnung: Zwei Systeme der Chaldäer über den Lauf des Mondes und der Sonne*, p. 90.
[2] *Ibid.*, p. 72.
[3] *Ibid.*, p. 90.
[4] *Ibid.*, p. 67.
[5] R. C. Thompson, *The Reports of the Magicians and Astrologers of Nineveh and Babylon*. II. xviii.

consistently different from those of the present celestial order? How could the stargazers who composed the earlier tablets be so careless as to maintain that the year is 360 days long, a mistake that in six years accumulates to a full month of divergence; or how could the astronomers of the royal observatories announce to the king the movements of the moon and its phases on wrong dates, though a child can tell when the moon is new,[1] and then record all this in very scholarly tablets requiring advanced mathematical knowledge?[2] Hence scholars speak "of enigmatic mistakes."[3]

However, it appears to us that the tablets with their changing astronomical systems reflect the changing order of the world and consequent attempts to adjust the calendar to the changes.

When the cataclysm of the 23rd of March, –687, brought about another disturbance in the length of the year and the month, the new standards remained uncertain until they could be calculated anew in a series of investigations.

From the time of that catastrophe until about the year –669 or –667, no New Year festivals were observed at Babylon.[4] "Eight years under Sennacherib, twelve years under Esarhaddon: for twenty years . . . the New Year's festival was omitted," says an ancient chronicle on a clay tablet.[5] According to cuneiform inscriptions, in the days of Sargon II a new world age began, and in the days of his son Sennacherib another world age.[6] In the days of Assurbanipal, son of Esarhaddon, son of Sennacherib, the planetary movements, the precession of the equinoxes, and the periodic returns of the eclipses were recalculated, and these new tablets, together with the older ones or copies of the older ones, were stored in the palace library at Nineveh. The tablets from Nineveh provide the best possible opportunity to learn how the order of the world changed in the eighth and seventh centuries.

[1] "The class of magicians who calculated the length of the months and published information concerning them formed a very important section of the Babylonian and Assyrian priesthood." *Ibid.*, p. xxiii.

[2] C. Bezold, "Astronomie, Himmelschau und Astrallehre bei den Babyloniern," in *Sitzungsberichte der Heidelberger Akademie der Wissenschaften, philos.-histor. Klasse*, 1911, expresses the opinion that before the sixth century the Babylonians were unaware of the relative lengths of the solar year and 12 lunar months. See also Gundel, *Dekane und Dekansternbilder*, p. 379.

[3] Kugler, *Die Mondrechnung*, p. 90.

[4] S. Smith, *Babylonian Historical Texts*, p. 22.

[5] *Ibid.*, p. 25.

[6] A. Jeremias, *Der alte Orient und die ägyptische Religion* (1907), p. 17; Winckler, *Forschungen*, III, 300.

Repeated changes in the course of the sun across the firmament led the astronomers of Babylonia to distinguish three paths of the sun: the Anu path, the Enlil path, and the Ea path. These three paths created much difficulty for the writers on Babylonian astronomy, and many explanations were offered and as many rejected.[1] The Anu, Enlil, and Ea paths of the planets across the sky appear to denote the successive ecliptics in various world ages. Like the sun, the planets in different times moved along the Anu, Enlil, and Ea paths.

In the Talmud [2] a number of scattered passages deal with a calendric change made by Hezekiah. The Talmud was written about a thousand years after Hezekiah, and not all details of the reform are preserved; it states that Hezekiah doubled the month of Nisan.

In later times, in order to adjust the lunar year to the solar year, an intercalary month was added every few years by doubling the last month of the year, Adar. This system of an intercalary Adar is preserved in the Hebrew calendar to this day.

The rabbis wondered why Hezekiah added another Nisan (the first month). The story is told in the Scriptures that Hezekiah, instead of celebrating Passover in the first month, put off the feast to the second month.[3] The Talmud explains that it was not the second month, but an additional Nisan.

It must be noted that in Judea in the days of Hezekiah the months were not called by Babylonian names, and therefore the situation should be stated as follows: Hezekiah, after the death of Ahaz, and before the second invasion of Sennacherib, added a month and postponed the feast of Passover. According to the Talmud this was done to make the lunar year correspond more closely to the solar year. As we shall see, there appears to be some similarity between this action and that by Numa at about the same time.

What permanent changes Hezekiah introduced in the calendar is not stated, but it is apparent that at that time calendar reckoning became a complicated matter. As Moses in his day "could not understand how to compute the calendar until God showed him the movements of the moon plainly," so in the

[1] Bezold, *Zenit und Aequatorialgestirne am babylonischen Fixsternhimmel* (1913), p. 6; M. Jastrow, *The Civilization of Babylonia and Assyria* (1915), p. 261.
[2] Tractate Berakhot 10b; Pesahim 56a; other sources in Ginzberg, *Legends*, VI, 369.
[3] II Chronicles 30.

days of Hezekiah the determination of the month and of the year became a matter, not of calculation, but of direct observation, and could not be performed much in advance. Isaiah called the astrologers "the monthly prognosticators."[1]

As we have already said, there is in the Talmud[2] the information that the Temple of Solomon was built so that on the equinoctial days of the year the direction of the rays of the rising sun could be tested. A gold plate or disc was affixed to the eastern gate: through it the rays of the rising sun fell into the heart of the Temple. The Festival of the Tabernacle (Sukkoth) "was originally an equinoctial festival as Exodus 23 : 16 and 34 : 22 state explicitly, celebrated during the last seven days of the year, and immediately preceding the New Year's Day, the day of the fall equinox, upon the tenth of the seventh month."[3] In other words, New Year's Day, or the day of the autumnal equinox, was observed on the tenth day of the seventh month, the day when the sun rose exactly in the east and set exactly in the west, the Day of Atonement falling on the same day.[4] Thereafter, the day of the New Year was moved back to the first day of the seventh month. We may note that not only on the Jewish calendar, but also according to the Babylonian tablets, the equinoctial dates were displaced by nine days: one tablet says that in the spring day and night are equal on the fifteenth of the month Nisan; another tablet says that it takes place on the sixth of the same month. This indicates that the change in the calendar of the feasts observed in Jerusalem followed astronomical changes.

The eastern gate of the Temple of Jerusalem was no longer correctly oriented after the cardinal points had become displaced. On his accession to the throne following the death of Ahaz, Hezekiah "inaugurated a sweeping religious reformation."[5] II Chronicles 29 : 3 ff. says: "He in the first year of his reign, in the first month, opened the doors of the house of the Lord and repaired them." Apparently the natural changes in terrestrial rotation which took place in the days of Uzziah and again on the day of the burial of Ahaz, necessitated a reform. Hezekiah therefore gathered the priests "into the east

[1] Isaiah 47 : 13.
[2] Talmudic references may be found in the article cited in the following footnote.
[3] Morgenstern, "The Gates of Righteousness," *Hebrew Union College Annual*, VI (1929), p. 31.
[4] Morgenstern says: "Upon the tenth of the seventh month Israel celebrated originally, not the Day of Atonement, but the New Year's Day." *Ibid.*, p. 37.
[5] *Ibid.*, p. 33.

street" and spoke to them, saying that "our fathers have trespassed" and "have shut up the doors of the porch."

In the pre-Exilic period it was held "to be of imperative necessity that on two days of the year the sun shone directly through the eastern gate," and "through all the eastern gates of the Temple arranged in line, directly into the very heart of the Temple proper."[1] The eastern gate, also called "sun gate," served not only to check on the equinoxes, when the sun rises exactly in the east, but on the solstices as well: a device on the eastern gate was designed to reflect the first rays of the sun on the summer and winter solstices, when the sun rises in the southeast and the northeast, respectively. According to Talmudic authorities, the early prophets experienced much difficulty in making this arrangement work.[2]

From biblical times vestiges of three calendar system remain,[3] and this assumes a special interest in view of the fact we noted some pages back, namely, the tablets from Nineveh record three different systems of solar and planetary movements, each of which is complete in itself and differs from the others at every point.

It appears that the adjustment of the calendar, following the initiation of the new world order in the days of Hezekiah, was a long and tedious process. As late as one hundred years after Hezekiah, during the Babylonian exile, in the days of Solon and Thales, Jeremiah, Baruch, and Ezekiel drew up the calendar from year to year.[4]

When the Jews returned from the Babylonian exile, they brought with them their present calendar, in which the months are called by Assyro-Babylonian names.

"For as the new heavens and the new earth, which I will [do] make, shall remain before me, saith the Lord, so shall your seed and your name remain," reads the closing chapter of the Book of Isaiah. All flesh will come to worship the Lord "from one new moon to another, and from one sabbath to another." The "new heavens" means a sky with constellations or luminaries in new places. The prophet promises that the new sky will be everlasting and that the months will keep forever their established order.

[1] *Ibid.*, pp. 17, 31.
[2] The Jerusalem Talmud, Tractate Erubin 22c.
[3] Morgenstern, "The Three Calendars of Ancient Israel," *Hebrew Union College Annual*, I (1924), 13-78.
[4] The Jerusalem Talmud, Tractate Sanhedrin I, 19a.

Daniel, the Jewish sage at the court of Nebuchadnezzar, king of the Exile, when blessing the Lord, said to the king: "He changeth the times and the seasons."[1] This is a remarkable sentence which is also preserved in many Jewish prayers. By the change of seasons or "appointed dates" (*moadim*) is meant an alteration in the order of nature, with shifting of solstitial and equinoctial dates and the festivals connected with them. "The change of times" could refer not only to the last change, but to the previous ones also, and it was "the change of the times and the seasons" that was followed by calendar reforms.

The old Hindu astronomical observations offer a set of calculations different from those of the present day. "What is extraordinary are the durations assigned to the synodical revolutions. . . . To meet in Hindu astronomy with a set of numerical quantities widely differing from those generally accepted is indeed so startling that one at first feels strongly inclined to doubt the soundness of the text. . . . Moreover, each figure is given twice over."[2]

In the astronomical work of Varaha Mihira, the recorded synodical revolutions of the planets, which are easy to calculate against the background of the fixed stars, are about five days too short for Saturn, over five days too short for Jupiter, eleven days too short for Mars, eight or nine days too short for Venus, less than two days too short for Mercury. In a solar system in which the earth revolves around the sun in 360 days, the synodical periods of Jupiter and Saturn would be about five days shorter than they are at present, and that of Mercury less than two days shorter. But Mars and Venus of the synodical table of Varaha Mihira must have had orbits different from their present ones, even if the terrestrial year was only 360 days.

Calendric changes in India were effected in the seventh century: at that time, as in China also, the ten-month year was supplanted by a twelve-month year.[3]

In the eighth century a calendar reform was made in Egypt. We have already referred to a cataclysm during the reign of the Pharaoh Osorkon II of the Libyan Dynasty;

[1] Daniel 2 : 21.
[2] G. Thibaut, p. xlvii of his translation of the *Panchasiddhantika*, the astronomical work of Varaha Mihira (Benares, 1889).
[3] A. del Mar, *The Worship of Augustus Caesar*, p. 4.

another disturbance of a cosmic nature took place a few decades later, still in the time of the Libyan Dynasty.

In the fifteenth year of the reign of Sosenk III "there occurred a remarkable prodigy of uncertain nature, but in some way connected with the moon."[1] The contemporaneous document written by the royal son, the high priest Osorkon, reads: "In the year 15, fourth month of the third season, 25th day, under the majesty of his august father, the divine ruler of Thebes, before heaven devoured (*or:* not devoured) the moon, great wrath arose in this land."[2] Soon thereafter Osorkon "introduced a new calendar of offerings."[3] The mutilated condition of the inscription makes it impossible to determine the exact nature of the calendric reform.[4]

It appears that the same or a similar disturbance in the movement of the moon is the subject of an Assyrian inscription, which speaks of the moon being obstructed on its way. "Day and night it was handicapped. In its august station it did not stand." Because of the duration of the phenomenon, it is concluded that "it could not mean an eclipse of the moon." [5] The reference to the moon's unwonted position also precludes such an interpretation.

At the end of the eighth or the beginning of the seventh century before the present era, the people of Rome introduced a calendar reform. In the preceding section we referred to Ovid's statement in *Fasti* concerning the reform of Romulus, who divided the year into ten months, and the reform of Numa, who "prefixed" two months. Plutarch's "Life of Numa" contains the following passage, part of which has already been quoted: "He [Numa] applied himself, also, to the adjustment of the calendar, not with exactness, and yet not altogether without careful observation. For during the reign of Romulus, they had been irrational and irregular in their fixing of the months, reckoning some at less than twenty days, some at thirty-five, and some at more; they had no idea of the inequality in the annual motions of the sun and moon, but held to the principle only, that the year should consist of three hundred and sixty days."[6]

[1] Breasted, *Records of Egypt*, IV, Sec. 757.
[2] *Ibid.*, Sec. 764. See controversy in *Zeitschrift für ägyptische Sprache*, VI (1868).
[3] Breasted, *Records of Egypt*, IV, Sec. 756.
[4] A. Erman, *Zeitschrift für ägyptische Sprache*, XLV (1908), 1-7.
[5] P. Jensen, *Die Kosmologie der Babylonier*, p. 39.
[6] Plutarch, *Lives*, "The Life of Numa" (transl. B. Perrin).

Numa reformed the calendar, and the "correction of the inequality which he made was destined to require other and greater corrections in the future. He also changed the order of the months."[1]

Numa was a contemporary of Hezekiah.[2]

In the second half of the seventh century before the present era, the length of the new month and the new year was calculated by the Greeks.

Diogenes Laërtius regarded Thales the Milesian, one of the "seven sages of antiquity," as the man who discovered the number of days in the year and the length of the seasons. In his *Life of Thales* he wrote: "He was the first to determine the sun's course from solstice to solstice." And again: "He is said to have discovered the seasons of the year and to have divided it into 365 days."[3] He was "the first to predict eclipses of the sun and to fix the solstices."[4] Thales is said to have written two treatises, one "On the Solstice" and the other "On the Equinox," neither of which is extant.

If the natural year always was what it is now, it is very strange that this discovery should have been attributed to a sage who lived as late as the seventh century, when Egypt and Assyria were already very old kingdoms, and when the dynasty of David was in its last decades. The longest and shortest days of the year, and thus the length of the year, are easily determined by the length of the shadow. Thales is said to have been born in the first year of the thirty-fifth Olympiad or -640. The progress of culture would hardly leave to one and the same person the calculation of the days in a year, which is a simple matter, and the calculation of forthcoming eclipses, which is an advanced achievement. Similarly, the fact, as stated by Plutarch and Diogenes Laërtius, that Solon, another sage of the same period, adjusted the months to the motion of the moon after finding that the time from one new moon to another is half a day shorter than thirty days, must be understood as an adjustment of the calendar to the new order in nature. The span of time from one new moon to another is a natural time division, almost as easily observable as day and night; primitive

[1] Plutarch, *Lives*, "The Life of Numa" (transl. B. Perrin).
[2] Cf. Augustine, *The City of God*, Bk. XVIII, Chap. 27.
[3] Diogenes Laërtius, *Lives of Eminent Philosophers* (English transl. R. D. Hicks, 1925).
[4] *Ibid.*, see also Herodotus i. 74.

peoples, unable to read or write, know that the period is less than thirty days.

On the other side of the globe, the people of Peru reckoned time from the day of the last cataclysm, and this method of computation was in use when the Europeans reached that country from the beginning of the sixteenth century.[1]

After the last cataclysm, the times and the seasons were computed anew. King Inti-Capac-Yupanqui ordered astronomical observations and calculations to be made, the result of which was a calendar reform, and the year, previously of 360 days, "was changed to 365 days and 6 hours."[2]

"This Ynca appears to have been the first to order and settle ceremonies. . . . He it was who established the twelve months of the year, giving a name to each, and ordaining the ceremonies that were to be observed in each. For although his ancestors used months and years counted by the *quipus*, yet they were never previously regulated in such order until the time of this lord."[3]

"All Toltec histories mention an assembly of sages and astrologers that was convoked in the city of Huehue-Tlapallan for the purpose of working on the correction of the calendar, and the reforming of the computation of the year, which was recognized as erroneous and which had been employed until that time."[4]

Half a meridian away, across the Pacific Ocean, a calendar was introduced in Japan in –660, and the reckoning of years in that country starts from that year.

In China, the astronomer Y-hang in the year –721 announced to the Emperor Hiuen-tsong that the order of the sky and the movements of the planets had changed which made it impossible to predict eclipses; and he referred to other authorities who asserted that in the time of Tsin the planet Venus used to move 40 degrees to the south of the ecliptic and eclipse the star Sirius. Y-hang explained that the course of the planet Venus changed in the days of Tsin.[5]

[1] Brasseur, *Manuscrit Troano*, p. 25.
[2] F. Montesinos (fl. 1628-1639), *Memorias antiguas historiales del Perú*, II, Chap. 7.
[3] Christoval de Molina (fl. 1570 to 1584), *An Account of the Fables and Rites of the Yncas*, transl. and ed. C. R. Markham (1873), p. 10.
[4] Brasseur, *Histoire des nations civilisées du Mexique*, p. 122. Among his sources were Ixtlilxochitl, *Sumaria relación*, etc.; M. Veytia (1718-1779), *Historia antigua de México*, I (1944), Chap. 2.
[5] A. Gaubil, *Histoire de l'astronomie chinoise* (1732), pp. 73-86.

All around the globe the years following –687 saw activity directed toward reforming the calendar. Between –747 and –687 the calendar was in a chaotic state, the length of the year and of the month, and probably also of the day, repeatedly changing. Before the eighth century there was a comparatively long span of time when the year had 360 days and the lunar month consisted of almost exactly thirty days.

Neither the calendar, nor the celestial charts, nor the sundials, nor the water clocks of the time before –687 were adequate for their purpose after that year. Values subsequently established in different parts of the terrestrial globe have remained practically unchanged down to the present save for very small improvements resulting from the more precise calculations of modern times. This stability of the calendar is due to the fact that the celestial order has remained unaltered: no changes in the heavenly order were observed except for minor perturbations between the planets which have no visible effect on their motion. Thus we are lulled into the belief—which is wishful thinking—that we live in an orderly universe. In the language of a modern scientist: "Though the order of the succession of events in the heavens is often somewhat complex, it is nevertheless systematic and invariable. The running of no clock ever approached in precision the motions of the sun, the moon, and the stars. In fact, to this day clocks are regulated by comparing them with the apparent diurnal motions of the heavenly bodies. Since not merely a few but hundreds of celestial phenomena were long ago found to be perfectly orderly, it was gradually perceived that majestic order prevails universally in those regions in which, before the birth of science, capricious gods and goddesses were believed to hold domain."[1]

However, as we have learned from the records of ancient times, the order today is not the primeval order; it was established less than twenty-seven centuries ago

> when the moon was placed in orbit,
> when the silver sun was planted,
> when the Bear was firmly stationed.[2]

[1] F. R. Moulton, *The World and Man as Science Sees Them*, p. 2.
[2] *Kalevale*, Rune, 3.

CHAPTER IX

# THE MOON AND ITS CRATERS

THE MOON REVOLVES around the earth and, together with the earth, around the sun, showing one and the same face to the inhabitants of the earth. It can be seen in the telescope that the surface of the moon is covered with seas of dried lava and with great crater-like formations. Since it has no atmosphere, the contours of its surface are clearly visible, and a city or village, if it existed there, could be seen through the Palomar telescope. But it is a dead planet and very inhospitable. For a half-month any place on it is cold night and for the other half-month in hot sunshine. There is no water on the planet, no vegetation, and probably no life at all. The ancients were interested to know whether the moon had human settlements, but moderns are concerned with the problem of the origin of the lunar craters.

There are two theories: one sees in them great extinct volcanoes; the other, formations produced by the bombardment of great meteorites on the semi-liquid mass of the moon before it solidified. There are more than thirty thousand such craters, small and large. Some of these circular crests rise as high as 20,000 feet above the plain—their height is measured by the length of their shadows; some, like Clavius near the moon's south pole, are one hundred and fifty miles in diameter. This tremendous width surpasses anything comparable among volcanoes on earth. It is therefore questioned whether these circular mountain formations represent true volcanoes. The largest known crater produced by the impact of a body that fell on the earth is in Arizona; it is four-fifths of a mile in diameter and much smaller than the crater formations on the moon.

As is readily seen, both theories of lunar craters imply a great catastrophic occurrence. For such craters to have been formed, tremendous forces must have acted from inside or from without; if these formations were caused by impact on a viscous mass, great meteors must have come flying from many directions.

Bright streaks or "rays" up to ten miles wide radiate from some of the craters; their origin, too, is not known. There are also clefts, irregular in form, about half a mile wide and of unknown depth.

In the cosmic catastrophes described in this book the moon was repeatedly involved. Together with the terrestrial globe it passed through the fabric of the great comet of the time of the Exodus, and in the conflicts of the eighth century before the present era, the moon was more than once displaced from its orbit by Mars. During these catastrophes the moon's surface flowed with lava and bubbled into great circular formations, which rapidly cooled off in the long lunar night, unprotected by an atmosphere from the coolness of cosmic spaces. In these cosmic collision or near contacts the surface of the moon was also marked with clefts and rifts.

The "play" of Mars with the moon was regarded by the Greeks and the Romans as a love affair.[1] From the *Iliad* we learn that Aphrodite (the Greek goddess of the moon) was warned by Jupiter-Zeus not to battle Ares-Mars, but to leave this task to Hera-Earth and Pallas Athene, being herself predestined to the sweet work of love.

Interplanetary contacts in the celestial sphere are in some respects similar to congress and germination in the biological world. In these contacts the bodies of the planets overflow with lava—fertile ground for vegetation—and comets born of such contacts fly across the solar system and rain gases and stones and possibly also spores, germs, or larvae on planets. Thus the notion of the ancients that love affairs were being carried on among the planetary gods and goddesses is a tale for the common people and a philosophical metaphor for the instructed.

The great seas of dried lava and the great craters on the dead planet devoid of air and water bespeak the dreadful devastations, even death itself, that interplanetary contacts can leave in their wake. The great formations of craters, mountains, rifts, and plains of lava on the moon were formed not only in the upheavals described in this book, but also in those which took place in earlier times. The moon is a great unmarked cemetery flying around our earth, a reminder of what can happen to a planet.

[1] Mars had near contacts with the moon and with the planet Venus, and as a result of these two "romances" the goddess Venus (Aphrodite) became associated in mythology with the moon as well as with the planet of that name.

# THE PLANET MARS

The planet Mars, at the present time, completes one revolution around the sun in 687 terrestrial days. Its orbit is entirely outside the earth's orbit, and is an ellipse, like that of the earth, but more stretched out, so that the planet's distance from the sun varies considerably during a revolution.

When Mars and the earth are on different sides of the sun, the distance between them rises to over 200,000,000 miles and may reach 248,600,000 miles. From this moment on, as the distance between the two planets diminishes, Mars nightly grows more and more luminous, changing from an inconspicuous point of light to a most brilliant star, brighter than any fixed star. During a period of little more than a year, it grows fifty-five times brighter. Among the planets it exceeds then even Jupiter in brilliance.

The earth and Mars approach each other every 780 days, this being the synodical period of Mars. But because of the ellipticity of the two orbits and the difference in the direction in which their longer radii are turned, the closeness of Mars and the earth is not the same at every opposition. At each seventh approach, which occurs every fifteen years, when Mars passes through that part of its orbit which is closest to the sun, and the earth simultaneously passes the segment of its orbit which is farthest from the sun, the conjunction of the two planets is especially close and is "the favourable opposition." These occasions are eagerly awaited by astronomers, for no celestial body, with the exception of the moon, is more readily observable than Mars when at "favourable opposition."

The distance between Mars and the earth at the oppositions varies from 61,000,000 miles to 35,500,000 miles ("favourable opposition"); the distance at various times during the period of fifteen years varies greatly, from 248,600,000 to 35,500,000 miles.

The cosmic disturbances recorded by Hebrew tradition—one on the day when Hezekiah's father, Ahaz, was entombed; the other, when Sennacherib's army invaded Palestine—were separated by a period of fourteen or fifteen years, if the figure in II Kings 18 : 13 refers to the invasion which ended in the disaster. A seemingly arbitrary period of fifteen years of grace, mentioned in Isaiah 38 : 5 and in II Kings 20 : 6, may also have had some relation to the periodicity of the catastrophes. The years -776, -747, -717 or -702, and -687 apparently were years of favourable oppositions of Mars, when perturbations, a regular phenomenon in oppositions, reached catastrophic dimensions.

If, because of other reasons, contact between Mars and the earth in the past is admitted, the combined shape of the orbits, with points of nearest approach being reached at present every fifteen years, could be regarded as a vestige of a contact or series of contacts at similar intervals in the past between the two planets then revolving on curved orbits that were closer to each other.

Mars bears a striking resemblance to the earth in the inclination of its axis of rotation to the plane of its orbit and in the period of its diurnal rotation. Whereas the equator of the earth is inclined 23½ degrees to the plane of the ecliptic, the equator of Mars is inclined 24 degrees to the plane of its orbit, a similarity unequalled among other planets in the solar system. The mean time of axial rotation of the earth is 23 hours, 56 minutes, 4 seconds, that of Mars 24 hours, 37 minutes, 23 seconds. No other two planets are so alike in the duration of their day, conceding that no conclusive data are available for the length of the day on Venus.

Is it possible that the axis of rotation and the velocity of rotation of Mars, stabilized and supported in their present position and rate by certain forces, were influenced originally by the earth at the time of contact? Mars, being small as compared with the earth, influenced to a lesser degree the rotation of the earth and the position of its poles.

The surface of Mars is crisscrossed with a network of "canals." Their discoverer, Schiaparelli, assumed that geological forces were a factor in their formation; on the other hand, he was "very careful not to combat this supposition, which includes nothing impossible," of the presence of intelligent beings on Mars who could have built these canals.

Percival Lowell spent his life in a crusade to convince fellow scholars and other contemporaries that intelligent human beings live on Mars and that the canals are their work. From his observatory in Flagstaff, Arizona, he believed he discovered water on Mars. He interpreted the polar caps as ice masses;

because of the dearth of water, the intelligent beings dug the canals to bring water to desert areas.[1]

In the early years of the twentieth century, plans were devised to communicate by light signals with the hypothetical men on Mars; according to one plan a series of light-sending stations was to be built into a geometric figure on the plains of Siberia. The figure was to represent the Pythagorean theorem of the relation of the three sides of a right-angle triangle. If there are intelligent beings on Mars, some writers argued, they should be able to notice and interpret the signals; if they are not intelligent enough to notice the signals and understand their meaning, we should not be so eager to communicate with them. The experiment was not carried out.

The contacts of Mars with other planets larger than itself and more powerful make it highly improbable that any higher forms of life, if they previously existed there, survived on Mars. It is, rather, a dead planet; every higher form of life, of whatever kind it might have been, most probably had its Last Day. Their work could not survive either. The "canals" on Mars appear to be a result of the play of geological forces that answered with rifts and cracks the outer forces acting in collisions.

## THE ATMOSPHERE OF MARS

The atmosphere of Mars is invisible. If there are any living creatures on that planet, and if they are endowed with organs of sight, they see a black sky, not a blue one as we do.

The atmosphere of Mars was the object of many investigations which produced conflicting and apparently unsatisfactory results. This gaseous envelope is transparent, permitting clear observation on the contours of the planet. Mars' seasonal polar caps are products of distillation; a polar cap disappears when summer arrives in its hemisphere and reappears in winter. It is not known whether these caps are composed of carbon dioxide or of ice, whether they are clouds floating over polar regions or layers of coagulated masses.

The general question as to the presence of water vapour in the atmosphere of Mars was answered in the affirmative by one group of observers (Lowell Observatory), and in the negative by another group (Lick Observatory). At present it is regarded

[1] P. Lowell, *Mars* (3rd ed. 1897); *idem, Mars and Its Canals* (1906).

as almost certain that there is on Mars only a low absolute content of water vapour, about one-twentieth of that in the atmosphere of the earth. This is the view supported by results announced by astronomers at Mount Wilson Observatory.

The observations concerning oxygen in the atmosphere of Mars are somewhat inconclusive; it is generally supposed that oxygen on Mars, if there is any, is less than 0.1 per cent of the oxygen content in the atmosphere of the earth per unit of surface area.[1]

The difficulty of a spectral analysis of the atmosphere of the planets lies in the fact that their light is the reflected light of the sun, and consequently it has in it the spectral picture of the atmosphere of the sun (absorption lines of spectrum), and also in the fact that the atmosphere of the earth, through which this reflected light travels, impresses its own characteristic spectral lines (of absorption) on the light reflected from the planets. The conclusion drawn and communicated to the general public is that "Mars' spectrum is practically that of reflected sunlight only" (E. Doolittle). This would suggest that there is no atmosphere on Mars or that it is very tenuous. However, there is a change in the distribution of light through the spectrum as compared with the light that arrives directly from the sun. The presence of an atmosphere on Mars can be proved by another set of observations, which indicate that it extends to about sixty miles above the surface of the planet. Also, its supposed thinness is in contradiction to findings obtained by photographs made in violet and in red light. One series of clouds is seen in the photographs taken in violet light, but not in those taken in red light; a second series of clouds is seen in the red, but not in the violet, light.

In the present study of cosmic catastrophes the endeavour has been to establish the fact that in the eighth and seventh centuries before this era the earth was repeatedly approached by a celestial body; that this body was the planet Mars; that previously Mars had been displaced from its path by contact with Venus, which up to that time had crossed the orbit of the earth, and that Venus, the earth, and Mars, as a consequence, assumed new positions in the solar system. In all these contacts between Venus, the earth, and Mars there was an exchange of atmospheres, the earth acquiring the carbon

[1] W. S. Adams and T. Dunham, *Contributions from the Mount Wilson Observatory*, No. 488 (1934).

clouds of Venus and also some of the atmosphere of Mars. The white precipitated masses on Mars, which form the polar caps, are probably of the nature of carbon having been acquired from the trailing part of Venus, and only the difference in atmospheric conditions on Mars as compared with the earth, together with a difference in temperature, keeps this "manna" from being permanently dissolved under the rays of the sun.

The main ingredients of the atmosphere of Mars must be present in the atmosphere of the earth. Mars, "the god of war," must have left part of his property on his visits. As oxygen and water vapour are not the main ingredients of the atmosphere of Mars, some other elements of the terrestrial atmosphere must be the main components of its atmosphere. It could be nitrogen, but the presence of nitrogen on Mars—or its absence—has not yet been established.

Besides oxygen and nitrogen, the main components of the terrestrial atmosphere, argon and neon are present in detectable quantities in the air. These rare gases excite spectral lines only when in a hot state; consequently, they cannot be detected through lines of emission from a comparatively cool body such as Mars. The absorption lines of argon and neon have not yet been investigated. When a study of these lines will make possible a spectral search for these rare gases on planets, Mars should be submitted to the test. If analysis should reveal them in rich amounts, this would also answer the question: What contribution did Mars make to the earth when the two planets came into contact?

## THE THERMAL BALANCE OF MARS

The equatorial diameter of Mars is about 4,200 miles; when compared with that of the earth, the ratio in volume is 15 to 100; the ratio in mass is supposed to be 10.8 to 100. Mars is one-sixth the volume of Venus, and Venus is considered to be seven and a half times heavier than Mars.

Due to the eccentricity of Mars' orbit, the insolation at aphelion is much smaller than at perihelion (the ratio being about 5 : 6), and in the southern hemisphere the summer is much hotter but much shorter than in the northern hemisphere. Because of the greater mean distance of Mars from the sun, it is supposed to receive less than half the light and warmth per unit of area that the earth receives; and for this reason its temperature must be some 65° C. below that of the earth, and never above freezing. The mean temperature of a year on the equatorial latitudes of Mars must be similar to that of the polar regions of the earth.

The radiometric measurement of the temperature of Mars actually shows an excess of heat.[1] Mars emits more heat than it receives from the sun. Does this excess of heat come from the interior of the planet? Mars is a smaller body than the earth; it has more surface per unit of volume, and it must have cooled down quicker than the earth, especially if it was released from the nebulous sun by a centrifugal force before the earth was (Kant-Laplace), but also if they both originated as planets simultaneously millions of years ago (tidal theory). What, then, is the cause of the excess of heat in Mars?

The assumed contacts with the earth would have caused much greater changes in and on Mars than in and on the earth, because of the difference in mass. An interplanetary contact must have caused a conversion of motion into heat, and consequently resulted in an excess of thermal radiation over the quantity of heat brought to the planet by insolation.

The contacts of Mars with Venus, and in a lesser degree with the earth, less than three thousand years ago probably are responsible for the present temperature of Mars; interplanetary electric discharges could also initiate atomic fissions with ensuing radioactivity and emission of heat.

## THE GASES OF VENUS

A part of the gaseous trail of Venus remained attached to the earth, another part was torn away by Mars, but the main mass of gases followed the head of the comet. Of the part which remained with the earth, some became a deposit of petroleum; some, in the form of clouds, enveloped the earth for many years, slowly precipitating. The part retained by Venus burned or smoked for a long time, as long as the oxygen carried from the earth lasted; what remained forms today the envelope of carbon clouds of the Morning Star. To the depth penetrated by spectroscopic analysis, oxygen and water vapour are absent.

[1] W. W. Coblentz and C. O. Lampland at the Lowell Observatory, and E. Pettit and S. B. Nicholson at the Mount Wilson Observatory.

The planet is covered with clouds of dust. Carbon dioxide is an ingredient of Venus' atmosphere.[1]

The brilliant envelope of Venus is the remnant of its tail of the days when, three thousand years ago, it was a comet. The reflecting power (albedo) of Venus is greater than that of any other planet. It is 0.75 as compared with 0.22 for Mars, and 0.13 for the moon.[2] The reflecting capacity of Venus is not only much greater than that of desert sand, but is almost equal to that of newly fallen snow.

On the basis of this research, I assume that Venus must be rich in petroleum gases. If and as long as Venus is too hot for the liquefaction of petroleum, the hydrocarbons will circulate in gaseous form. The absorption lines of the petroleum spectrum lie far in the infra-red where usual photographs do not reach. When the technique of photography in the infra-red is perfected so that hydrocarbon bands can be differentiated, the spectrogram of Venus may disclose the presence of hydrocarbon gases in its atmosphere, if these gases lie in the upper part of the atmosphere where the rays of the sun penetrate.

If the petroleum that poured down on the earth on its contact with the comet Venus was formed by means of electrical discharges from hydrogen and gaseous carbon, Venus must still have petroleum because of the discharges that passed, as we assume, between the head and tail of the comet when it was intercepted by the earth and in other celestial contacts.

Some indirect conclusion can also be drawn concerning the presence of liquid petroleum on Jupiter. If, as is assumed here, Venus was thrown off from Jupiter in a violent expulsion, and if Venus has petroleum gases, then Jupiter must have petroleum. The fact that methane has been discovered in the atmosphere of Jupiter—the only known constituents of its atmosphere are the poisonous gases methane and ammonia—makes it rather probable that it has petroleum; the so-called "natural gas" found in and near oil fields consists largely of methane.

The modern theory of the origin of petroleum, based upon its polarizing quality, regards petroleum as originating from

[1] C. E. St. John and J. B. Nicholson, "The Spectrum of Venus," *Contributions from the Mount Wilson Observatory*, No. 249 (1922).

The supposition has been advanced that Venus is covered with formaldehyde (R. Wildt) although no spectral lines of this compound have been identified in the atmosphere of Venus.

[2] These figures are from Arrhenius, *Das Schicksal der Planeten* (1911), p. 6. E. A. Antoniadi (*La planète Mercure* [1939], p. 49) gives 0.63 for Venus, 0.17 for Mars, and 0.10 for the moon.

organic, not inorganic matter. Consequently, if I am not mistaken, Venus and Jupiter must possess an organic source of petroleum. On preceding pages it was shown that there are some historical indications that Venus—and therefore also Jupiter—is populated by vermin; this organic life can be the source of petroleum.

## THE THERMAL BALANCE OF VENUS

Radiometric observations at the Mount Wilson and Flagstaff observatories in 1922 have shown that "a considerable amount of heat" is emitted by the dark part of the disc of the planet Venus.

Venus, being nearer to the sun than the earth, turns in succession its illuminated and shaded parts toward the earth: it shows phases like the moon. The temperature of the day and night sides of Venus was measured by a radiometric method and it was found that there is "a nearly uniform temperature over the planet's surface both on the illuminated and dark hemispheres." "This sentence [of E. Pettit and S. B. Nicholson] is a terse statement of what is perhaps the most valuable single discovery ever made with respect to the planet Venus." [1] Similar results were also obtained independently and almost simultaneously by a second pair of researchers. [2]

What explanation can be given for the phenomenon of the nearly uniform temperature of the day and night hemispheres of Venus? The conclusion drawn was this: The daily rotation of the planet Venus is very rapid and during the short night the temperature cannot fall to any considerable extent. But this conclusion stands in complete contradiction to what was believed to be the established fact of the nonrotation of Venus (with respect to the sun, or of a rotation in relation to the fixed stars with a period equal to the time of one revolution on its planetary orbit or 225 terrestrial days). Due to the cover of clouds over Venus, it is impossible to have a direct impression as to whether Venus has a day-night rotation or not. The spectrographic data suggest that the planet revolves always with the same side to the sun, just as the moon revolves always

[1] F. E. Ross, "Photographs of Venus," *Contributions from the Mount Wilson Observatory*, No. 363 (1928).
[2] Coblentz and Lampland, *Journal of Franklin Institute*, Vol. 199 (1925), 804.

with the same side to the earth, or that, at most, it rotates very slowly.[1] In any case, a short period of rotation is excluded by the spectrographic data.

"If the period of rotation of Venus is 225 days, as many observers have been led to believe, it is difficult to see how the high temperature of the rotating layer of the night side can be maintained."[2]

Compromise does not satisfy either side. Neither the radiometric data, which suggest a short period of rotation, nor the precise spectroscopic data, which indicate a long period of rotation, may be ignored, and "they will undoubtedly furnish material for discussion and debate for many years."[3]

In reality there is no conflict between the two methods of physical observation. The night side of Venus radiates heat because Venus is hot. The reflecting, absorbing, insulating, and conducting properties of the cloud layer of Venus modify the heating effect of the sun upon the body of the planet; but at the bottom of the problem lies this fact: Venus gives off heat.

Venus experienced in quick succession its birth and expulsion under violent conditions; an existence as a comet on an ellipse which approached the sun closely; two encounters with the earth accompanied by discharges of potentials between these two bodies and with a thermal effect caused by conversion of momentum into heat; a number of contacts with Mars and probably also with Jupiter. Since all this happened between the third and the first millennia before the present era, the core of the planet Venus must still be hot. Moreover, if there is oxygen present on Venus, petroleum fires must be burning there.

These conclusions are drawn from the history of Venus as established in this research.

[1] E. St. John and S. B. Nicholson, "The Spectrum of Venus," *Astrophysical Journal*, Vol. LVI (1922).
[2] Ross, "Photographs of Venus," p. 14.
[3] *Ibid.*

# THE END

> This world will be destroyed; also the mighty ocean will dry up; and this broad earth will be burnt up. Therefore, sirs, cultivate friendliness; cultivate compassion.
>
> —"World Cycles" in *Visuddhi-Magga*

The solar system is not a structure that has remained unchanged for billions of years; displacement of members of the system occurred in historical times. Nor is there justification for the excuse that man cannot know or find out how this system came into being because he was not there when it was arranged in its present pattern.

Catastrophes have repeatedly reduced civilization on this earth to ruins. But our earth has fared well in comparison with Mars; and judged by the state of civilization at which mankind has arrived, conditions for life processes have been improved in some respects. But if events of this kind happened in the past, they may happen again in the future, with perhaps a different —fatal—result.

The earth has come in contact with other planets and comets. At present no planet has a course that endangers the earth, and only a few asteroids—mere rocks, a few kilometres in diameter—have orbits that cross the path of the earth. This was discovered, to the amazement of scholars, only recently. But in the solar system there exists a possibility that at some date in the future a collision between two planets will occur, not a mere encounter between a planet and an asteroid. The orbit of Pluto, the farthest of the planets from the sun, though much larger than Neptune's, crosses that of Neptune. True, the plane of the orbit of Pluto is inclined 17° to the ecliptic, and therefore the danger of a collision is not impending. However, since the long axis of Pluto's orbit changes its direction, future contact between the two planets is probable if no comet intervenes to disrupt the intersecting orbits of these bodies. Astronomers will see the planets stop or slow down in their rotation, cushioned in the magnetic fields about them; a spark will fly from one planet to another, and thus an actual crushing collision of the lithospheres will be avoided; then the planets will part and change their orbits. It may happen that Pluto will become a satellite of Neptune. There is also the possibility that Pluto may encounter, not Neptune, but Triton, Neptune's satellite and about one-third as large as Pluto. Whether Pluto will become another moon of Neptune or will be thrown into a position much closer to the sun, or whether it will free Triton from being a satellite are matters of conjecture.

Another case of intersection may be found among the moons of Jupiter. The orbit of the sixth satellite is interlocked with the orbit of the seventh, and the eighth satellite is highly erratic and crosses the path of the ninth. One should be able to calculate how long the sixth and seventh satellites have moved on their present paths; the figures will probably not be large.

Each collision between two planets in the past caused a series of subsequent collisions, in which other planets became involved. The collision between major planets, which is the theme of the sequel to *Worlds in Collision*, brought about the birth of comets. These comets moved across the orbits of other planets and collided with them. At least one of these comets in historical times became a planet (Venus), and this at the cost of great destruction on Mars and on the earth. Planets, thrown off their paths, collided repeatedly until they attained their present positions, where their orbits do not intersect. The only remaining cases of intersection are those of Neptune and Pluto, the satellites of Jupiter, and some planetoids (asteroids) that cross the orbits of Mars and the earth.

Moreover, comets may strike the earth, as Venus did when it was a comet; in that major catastrophe it was fortunate that Venus is a slightly smaller body than the earth. A large comet arriving from interstellar spaces may run into one of the planets and push it from its orbit; then chaos may start anew. Also, some dark star, like Jupiter or Saturn, may be in the path of the sun, and may be attracted to the system and cause havoc in it.

The scholarly world assumed that in some hundreds of millions of years the heat of the sun would be exhausted, and then, as Flammarion frightened his readers, the last pair of human beings would freeze to death in the ice of the equator. But this is far off in the future. In view of modern knowledge that heat is discharged in the process of breaking up atoms, scientists are now prepared to credit the sun with an immense reserve of heat. The fear, if any, is focused on the possibility that the sun may explode; a few minutes later the earth will become aware of this, and soon thereafter will no longer exist. But the one end, that of freezing, is very remote; the other end, that of explosion, is very improbable; and the world is thought to have billions of peaceful years ahead. It is believed that the world has gone through eons of undisturbed evolution, and equally long eons are before us. Man can go far in such a span of time, considering that his entire civilization has endured less than ten thousand years, and in view of the great technological progress he has made in the last century.

The average man is no longer afraid of the end of the world. Man clings to his earthly possessions, registers his landholdings and fences them in; peoples carry on wars to preserve and to enlarge their historical frontiers. Yet the last five or six thousand years have witnessed a series of major catastrophes, each of which displaced the borders of the seas, and some of which caused sea-beds and continents to interchange places, submerging kingdoms, and creating space for new ones.

Cosmic collisions are not divergent phenomena, or phenomena that, in the opinion of some modern philosophers, take place in defiance of what is supposed to be physical laws; they are more in the nature of occurrences implicit in the dynamics of the universe, or, in terms of that philosophy, convergent phenomena.

"Lest by chance restrained by religion"—and we may read 'science' instead of 'religion'—"you should think that earth and sun, and sky, sea, stars, and moon must needs abide for everlasting, because of their divine body," think of the catastrophes of the past; and then "look upon seas, and lands, and sky; their threefold nature . . . their three textures so vast, one single day shall hurl to ruin; and the massive form and fabric of the world held up for many years, shall fall headlong."[1]

"And the whole firmament shall fall on the divine earth and on the sea, and then shall flow a ceaseless cataract of raging fire, and shall burn land and sea, and the firmament of heaven and the stars and creation itself it shall cast into one molten mass and clean dissolve. Then no more shall there be the luminaries' twinkling orbs, no night, no dawn, no constant

[1] Lucretius, *De rerum natura*, v (transl. C. Bailey, 1924).

days of care, no spring, no summer, no winter, no autumn."[1]

"A single day will see the burial of all mankind. All that the long forbearance of fortune has produced, all that has been reared to eminence, all that is famous and all that is beautiful, great thrones, great nations—all will descend into one abyss, will be overthrown in one hour."[2]

> The vehemence of flames will burst asunder
> the framework of the earth's crust.[3]

[1] *The Sibylline Oracles*, transl. Lanchester.
[2] Seneca, *Naturales quaestiones*, III, xxx (transl. J. Clarke).
[3] Seneca, *Epistolae morales*, Epistle xcl (transl. R. M. Gummere).

# EPILOGUE

EPILOGUE

# FACING MANY PROBLEMS

IN THIS BOOK, containing the first part of an historical cosmology, I have endeavoured to show that two series of cosmic catastrophes took place in historical times, thirty-four and twenty-six centuries ago, and thus only a short time ago not peace but war reigned in the solar system.

All cosmological theories assumed that the planets have revolved in their places for billions of years; we claim that they have been travelling along their present orbits for only a few thousand years. We maintain also that one planet—Venus—was formerly a comet and that it joined the family of planets within the memory of mankind, thus offering an explanation of how one of the planets originated. We conjectured that the comet Venus originated in the planet Jupiter; then we found that smaller comets were born in contacts between Venus and Mars, thus offering an explanation of the principle of the origin of the comets of the solar system. That these comets are only a few thousand years old explains why, despite dissipation of the material of their tails in space, they have not yet disintegrated entirely. From the fact that Venus was once a comet we learned that comets are not nearly immaterial bodies or "rien visible," as was thought because stars are usually seen through their tails, and on the passage of one or two of them in front of the sun, their heads were not perceptible.

We claim that the earth's orbit changed more than once and with it the length of the year; that the geographical position of the terrestrial axis and its astronomical direction changed repeatedly, and that at a recent date the polar star was in the constellation of the Great Bear. The length of the day altered; the polar regions shifted, the polar ice became displaced into moderate latitudes, and other regions moved into the polar circles.

We arrived at the conclusion that electrical discharges took place between Venus, Mars, and the earth when, in very close contacts, their atmospheres touched each other; that the magnetic poles of the earth became reversed only a few thousand years ago; and that with the change in the moon's orbit, the length of the month changed too, and repeatedly so. In the period of seven hundred years between the middle of the second millennium before the present era and the eighth century the year consisted of 360 days and the month of almost exactly thirty days, but earlier the day, month, and year were of different lengths.

We offered an explanation of the fact that the nocturnal side of Venus emits as much heat as the sunlit side; and we explained the origin of the canals of Mars and the craters and seas of lava on the moon as brought about in stress and near collisions.

We believe we came close to solving the problem of mountain building and the irruption of the sea; the exchange of place between sea and land; the rise of new islands and volcanic activity; sudden changes in climate and the destruction of quadrupeds in northern Siberia and the annihilation of entire species; and the cause of earthquakes.

Furthermore, we found that excessive evaporation of water from the surface of the oceans and seas, a phenomenon that was postulated to explain the excessive precipitation and formation of ice covers, was caused by extraterrestrial agents. Though in such occurrences we see the origin of the Fimbulwinter, we are inclined to regard the erratic boulders and till, or gravel, clay, and sand on the substratum of rock as having been carried, not by ice, but by onrushing gigantic tides caused by change in the rotation of the terrestrial globe; thus we accounted for moraines that migrated from the equator toward higher latitudes and altitudes (Himalayas) or from the equator across Africa toward the South Pole.

We recognized that the religions of the peoples of the world have a common astral origin. The narrative of the Hebrew Bible concerning the plagues and other wonders at the time of the Exodus is historically true and the prodigies recorded have a natural explanation. We learned that there was a world conflagration and that naphtha poured from the sky; that only a small proportion of people and animals survived; that the passage of the sea and the theophany at Mount Sinai are not inventions; that the shadow of death or twilight of the gods (*Götterdämmerung*) refers to the time of the wandering in the desert; that manna or ambrosia really fell from the sky, from the clouds of Venus.

We found also that Joshua's miracle with the sun and the moon is not a tale for the credulous. We learned why there are common ideas in the folklore of peoples separated by oceans, and we recognized the importance of world upheavals in the content of legends and why the planets were deified and which planet was represented by Pallas Athene, and what is the celestial plot of the *Iliad* and in what period this epic was created, and why the Roman people made Mars their national god and progenitor of the founders of Rome. We came to understand the real meaning of the messages of the Hebrew prophets Amos, Isaiah, Joel, Micah, and others. We were able also to ascertain the year, month, and day of the last cosmic catastrophe and to establish the nature of the agent that destroyed Sennacherib's army. We discerned the cause of the great wanderings of peoples in the fifteenth and eighth centuries. We learned the origin of the belief in the chosenness of the Jewish people; we traced the original meaning of the archangels, and the source of eschatological beliefs in doomsday.

In giving this enumeration of the claims made and problems dealt with in this book, we are aware that more problems have arisen than have been solved.

The question before historical cosmogony is this: If it is true that cosmic catastrophes occurred such a short time ago, how about the more remote past? What can we find out concerning the Deluge, at present thought to have been a local flooding of the Euphrates that impressed the Bedouins coming from the desert? In general, what can be brought to light concerning the world's more distant past and earlier celestial battles?

As explained in the Preface, the story of the catastrophes as they can be reconstructed from the records of man and of nature is not completed in this volume. Here are presented only two chapters—two world ages—Venus and Mars. I intend to go further back into the past and piece together the story of some earlier cosmic upheavals. This will be the subject of another volume. There I hope to be able to tell a little more of the circumstances preceding the birth of Venus from the body of Jupiter and narrate at length why Jupiter, a planet which only a few persons out of a crowd know how to find in the sky, was the main deity of the peoples of antiquity. In that book an attempt will be made to answer some more of the questions raised in the first pages of the Prologue of this volume.

Historical cosmogony offers a chance to employ the fact that there were catastrophes of global extent in establishing a synchronized history of the ancient world. Previous efforts to build chronological tables on the basis of astronomical calculations—new moon, eclipses, heliacal rising or culmination of certain stars—cannot be correct because the order of nature has changed since ancient times. But great upheavals of cosmic character may serve as points of departure for writing a revised history of the nations.

Such a synchronization of the histories of the ancient world is attempted in *Ages in Chaos*. Its starting point is the simultaneity of physical catastrophes in the countries of the ancient East and the comparison of records referring to such catastrophes among the peoples of antiquity. For the rest, I have proceeded by collating political records of archeological material of the ancient East covering a period of over a thousand years, from the end of the Middle Kingdom in Egypt to the time of Alexander of Macedonia: going step by step from century to century, the research arrives at an entirely revised sequence of events in ancient history and discloses a discrepancy of a number of centuries in the conventional chronology.

The development of religion, including the religion of Israel, comes under a new light. The facts established here may help in tracing the origin and the growth of planetary worship, animal worship, human sacrifices—also the source of astrological beliefs. The author feels an obligation to expand the scope of his work in order to include the problem of the birth of religion and of monotheism in particular. Investigation should be made into why and how the Jewish people, who had the same experiences as other peoples and who started with an astral religion like the rest of the nations, early cast off astral deities and forbade the worship of images.

The Scriptures invite a new approach to Bible criticism, one that will make it possible to see the process of transition from an astral religion to monotheism with its idea of a single Creator, not a star, not an animal, and not a human being.

An intriguing problem presents itself in psychology. Freud

searched for primordial urges in modern man. According to him, in the primitive society of the stone age, when the sons grew up, they looked for a chance to dispose of the father, once all-powerful and now ageing, and to work their will on their mother; and this urge is part of the heritage that modern man carries over from his prehistoric ancestors. According to the theory of another psychologist, Carl Jung, there exists a collective unconscious mind, a receptacle and carrier of ideas deposited there in primeval times, which plays an important role in our concepts and actions. In the light of these theories, we may well wonder to what extent the terrifying experiences of world catastrophes have become part of the human soul and how much, if any, of it can be traced in our beliefs, emotions, and behaviour as directed from the unconscious or subconscious strata of the mind.[1]

In the present volume geological and paleontological material was discussed only occasionally—when we dealt with rocks being carried considerable distances and placed on top of foreign formations; with mammoths being killed in a catastrophe; with the changes of climate, the geographical contours of the polar ice in the past, moraines in Africa, and remains of human culture in the north of Alaska; with the source of a substantial part of oil deposits, the origin of volcanoes, the cause of earthquakes. However, geological, paleontological, and anthropological material related to the problems of cosmic catastrophes is vast and may give a complete picture of past events no less than historical material.

What can we establish concerning the disappearance of species and even of genera, the theory of evolution versus the theory of catastrophic mutations, and the development of animal and plant life in general, or the time when giants lived or when brontosauri populated the earth?

The submersion and emersion of land, the origin of the salt in the sea, the origin of deserts, of gravel, of coal deposits in Antarctica, and the palm growth in the arctic regions; the building of sedimentary rocks; the intrusion of igneous rock above levels containing bones of marine and land animals and of iron in the superficial layers of the earth's crust, the times of geological epochs and the age of man on the earth—all these

[1] In connection with my idea of collective amnesia, G. A. Atwater suggests a search for the vestiges of terrifying experiences of the past in the present behaviour of man.

ask for treatment in the light of the theory of cosmic catastrophism.

Then there are physical problems. The accounts given in this book about planets changing their orbits and the velocities of their rotation, about a comet that became a planet, about interplanetary contacts and discharges, indicate a need for a new approach to celestial mechanics.

The theory of cosmic catastrophism can, if required to do so, conform with the celestial mechanics of Newton. Comets and planets pushing one another could change their orbits, although it is singular how, for instance, Venus could achieve a circular orbit, or how the moon, also forced from its place, could hold to an almost circular orbit. Nevertheless, there are precedents for such a concept. The planetesimal theory postulates innumerable collisions between small planetesimals—that flew out of the sun, gradually rounded their orbits, and formed planets and satellites; the tidal theory also regards the planets as derivatives of the sun swept by a passing star into a direction and with a force that, together with the gravitational attraction of the sun, created nearly circular orbits, the same having occurred to the moons in relation to their parent planets.[1] Another precedent for circular orbits formed under extraordinary circumstances can be found in the theory that regards the retrograde satellites as captured asteroids which succeeded, after being captured, in achieving approximately circular orbits.

If such effects from contacts between two stars or from capture of a smaller body by a larger body are not incompatible with celestial mechanics, then the orbits resulting from worlds in collision should be regarded as in harmony with it, too.

The physical effects of retardation or reversal of the earth in its diurnal rotation are differently evaluated by various scientists. Some express the opinion that a total destruction of the earth and volatilization of its entire mass would follow such slowing down or stasis. They concede, however, that destruction of such dimensions would not occur if the earth continued to rotate and only its axis were tilted out of its position. This could be caused by the earth's passing through a strong magnetic field at an angle to the earth's magnetic

[1] One of the authors of the tidal theory, Harold Jeffreys, writes that first among the "several striking facts" which "still remain unexplained" by the tidal theory is "the smallness of the eccentricities of the orbits of the planets and satellites" (*The Earth*, 2nd ed. [1929], p. 48).

axis. A rotating steel top, when tilted by a magnet, continues to rotate. Theoretically, the terrestrial axis could be tilted for a certain length of time, and at any angle, and also in such a fashion that it would lie in the plane of the ecliptic. In that case, one of the two hemispheres—the northern or the southern—would remain in prolonged day, the other, in prolonged night.

The tilting of the axis could produce the visual effect of a retrogressing or arrested sun; a greater tilting, a multiple day or night; and in the case of still greater tilting, a reversal with east and west exchanging places; all this without a substantial disruption in the mechanical momentum of the rotation or revolution of the earth.

Other scientists maintain that a theoretical slowing down or even stoppage of the earth in its diurnal rotation would not by itself cause the destruction of the earth. All parts of the earth rotate with the same angular velocity, and if the theoretical stoppage or slowing down did not upset the equality of the angular velocity of the various parts of the solid globe, the earth would survive the slowing down, or stasis, or even a reversal of rotation. However, the fluid parts—the air and the water of the oceans—would certainly have their angular velocity disrupted, and hurricanes and tidal waves would sweep the earth. Civilizations would be destroyed, but not the globe.

According to this explanation, the actual results of such a slowing down of the angular velocity of rotation would depend on the manner in which it occurred. If the application of an external medium, say a thick cloud of dust, acted equally on all parts of the surface of the globe, the globe would change its speed of rotation or might even cease rotating, and the energy of its rotation would be transferred to the cloud of dust; heat would develop as the result of the bombardment by the particles of dust striking the atmosphere and the ground. The earth would be buried under such a thick layer of dust that its mass would noticeably increase.

The cessation of the diurnal rotation could also be caused—and most efficiently—by the earth's passing through a strong magnetic field; eddy currents would be generated in the surface of the earth,[1] which in turn would give rise to magnetic fields,

[1] In this connection see the description of a sudden calamity in Numbers 16 : 45-49, in which thousands of Israelites roaming in the desert were "consumed as in a moment."

and these, interacting with the external field, would slow down the earth or bring it to a rotational stasis.

It is possible to calculate the mass of a cloud of particles and also the strength of the magnetic field that would cause the earth to stop rotating or to slow down, say, to half its original velocity. A rough calculation shows that if the mass of this cloud were equal to the mass of the earth and consisted of iron particles magnetized close to saturation, it would create a magnetic field strong enough to stop the rotation of the earth; if the magnetic field were half as strong it would slow the rotation of the earth to half its original velocity. However, if the cloud were electrically charged, the strength of its magnetic field would depend on its charge.

If the interaction with the magnetic field caused the earth to renew its spinning, it would almost certainly not be renewed at the same speed. If the magma inside the globe continued to rotate at a different angular velocity than the shell, it would tend to set the earth rotating slowly. In the tidal theory the origin of the earth's rotation is ascribed to the action of meteorites.

If the angular velocity of the various strata or segments of the globe were disrupted by some stress, these strata or segments would shift, and heat would be created as the result of the friction. Cracks and rifts would appear, seas would erupt, land would submerge or rise in mountain ridges, with "the midmost of the earth trembling with terror and the upper layers of the earth falling away." [1]

The stresses between the various strata that would result in all this might also convert some of the energy of rotation, not into heat, but into other forms of energy, including electrical. A discharge of great magnitude between the earth and the outer body (or cloud) could take place in this way.

Thus celestial mechanics does not conflict with cosmic catastrophism. I must admit, however, that in searching for the causes of the great upheavals of the past and in considering their effects, I became sceptical of the great theories concerning the celestial motions that were formulated when the historical facts described here were not known to science. The subject deserves to be discussed in detail and quantitatively. All that I would venture to say at this time and in this place is the following: The accepted celestial mechanics, notwithstanding

[1] See p. 83.

the many calculations that have been carried out to many decimal places, or verified by celestial motions, stands only *if* the sun, the source of light, warmth, and other radiation produced by fusion and fission of atoms, *is as a whole an electrically neutral body*, and also if the planets, in their usual orbits, are neutral bodies.

Fundamental principles in celestial mechanics including the law of gravitation, must come into question if the sun possesses a charge sufficient to influence the planets in their orbits or the comets in theirs. In the Newtonian celestial mechanics, based on the theory of gravitation, electricity and magnetism play no role.

When physicists came upon the idea that the atom is built like a solar system, the atoms of various chemical elements differing in the mass of their suns (nuclei) and the number of their planets (electrons), the notion was looked upon with much favour. But it was stressed that "an atom differs from the solar system by the fact that it is not gravitation that makes the electrons go round the nucleus, but electricity" (H. N. Russell).

Besides this, another difference was found: an electron in an atom, on absorbing the energy of a photon (light), jumps to another orbit, and again to another when it emits light and releases the energy of a photon. Because of this phenomenon, comparison with the solar system no longer seemed valid. "We do not read in the morning newspapers that Mars leaped to the orbit of Saturn, or Saturn to the orbit of Mars," wrote a critic. True, we do not read it in the morning papers; but in ancient records we have found similar events described in detail, and we have tried to reconstruct the facts by comparing many ancient records. The solar system is actually built like an atom; only, in keeping with the smallness of the atom, the jumping of electrons from one orbit to another, when hit by the energy of a photon, takes place many times a second, whereas in accord with the vastness of the solar system, a similar phenomenon occurs there once in hundreds or thousands of years. In the middle of the second millennium before the present era, the terrestrial globe experienced two displacements; and in the eighth or seventh century before the present era, it experienced three or four more. In the period between, Mars and Venus, and the moon also, shifted.

Contacts between celestial bodies are not limited to the

domain of the solar system. From time to time a nova is seen in the sky, a blazing fixed star which until then had been small or invisible. It burns for weeks or months and then loses its light. It is thought that this may be the result of a collision between two stars (a phenomenon that, according to the tidal theory, occurred to the sun or to its theoretical companion). Comets arriving from other solar systems may have been born in such collisions.

If the activity of an atom constitutes a rule for the macrocosm, then the events described in this book were not merely accidents of celestial traffic, but normal phenomena like birth and death. The discharges between the planets, or the great photons emitted in these contacts, caused metamorphoses in inorganic and organic nature. Of these things I intend to write in another volume, where problems of geology and paleontology and the theory of evolution will be discussed.

Having discovered some historical facts and having solved a few problems, we are faced with more problems in almost all fields of science; we are not free to stop and rest on the road on which we started when we wondered whether Joshua's miracle of stopping the sun was a natural phenomenon. Barriers between sciences serve to create the belief in a scientist in any particular field that other scientific fields are free from problems, and he trusts himself to borrow from them without questioning. It can be seen here that problems in one area carry over into other scientific areas, thought to have no contact with each other.

We realize the limitations which a single scholar must be aware of on facing such an ambitious programme of inquiry into the architectonics of the world and its history. In earlier centuries philosophers not infrequently attempted a synthesis of knowledge in its various branches. Today, with knowledge becoming more and more specialized, whoever tries to cope with such a task should ask in all humility the question put at the beginning of this volume: *Quota pars operis tanti nobis committitur*—Which part of this work is committed to us?

# INDEX

# INDEX